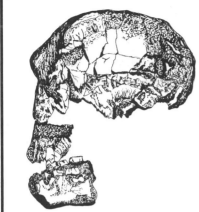

Human Evolution

An Introduction to Man's Adaptations

Third Edition

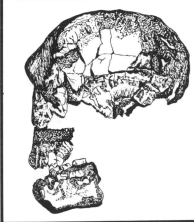

Human Evolution

**An Introduction to
Man's Adaptations**

Third Edition

Bernard Campbell

ALDINE
Publishing Company

New York

About the Author

Bernard Campbell is Adjunct Professor of Anthropology at the University of California, Los Angeles. He has been a visiting lecturer at Harvard and Cambridge Universities, and has taught and conducted research in East and South Africa. Dr. Campbell has achieved a position of eminence in the field of anthropology and is the author/coauthor of eight books including *Human Evolution, Sexual Selection and the Descent of Man,* and the definitive three-volume work *Catalogue of Fossil Hominids.*

Aldine Publishing Company
200 Saw Mill River Road
Hawthorne, New York 10532

Library of Congress Cataloging in Publication Data

Campbell, Bernard Grant.
  Human evolution.

  Bibliography: p.
  Includes index.
  1. Human evolution. I. Title.
GN281.C35 1985   573.2   85-1267
ISBN 0-202-02023-1
ISBN 0-202-02024-X (pbk.)

Printed in the United States of America
10 9 8 7 6 5 4 3 2 1

To Susan Ann, who makes all things possible.

Contents

3   THE PRIMATE RADIATION

4   THE FOSSIL EVIDENCE: THE HOMINIDAE

5   BODY STRUCTURE AND POSTURE

6   LOCOMOTION AND THE HINDLIMB

Contents   *xi*

Preface

This book reviews our present knowledge of the evolution of humankind. During the last twenty years, since the first edition was prepared,
the amount of evidence has grown remarkably, so that which was once
little more than the study of a few fossils, has grown into a complex
research undertaking involving hundreds of workers from many countries and a very wide range of disciplines. The anatomical basis, which
comprises such studies, still fills a central position, however, because
the fossils of human ancestors and related forms still form the only *direct*
evidence for our evolution. I have spread my net, however, and have
included much other evidence which is relevant to this story, and which
helps to make sense of such a mysterious biological process. Biochemical, geological, archaeological, as well as anatomical data, need consideration in such an attempt. The story leads us out of biology, into
history and philosophy, and moves from science into ethics. It concerns
the ultimately unreachable mysteries of human nature and the human
mind, and the ultimately unknowable questions of prehistory. Human
evolution is the most complex and mysterious phenomenon in all of
biology.

Any such brief account as this must be quite inadequate in its attempt
to explain the transformation of animal into human. Such an extraordinary chapter in the story of evolution can only be treated superficially in a single volume of this size. The book is no more than an
attempt to introduce a subject of awesome significance, in a simple,
step-by-step manner.

The topic is also of extraordinary interest and importance, central as it
is to all human affairs; for if we are ever to make decisions to secure our

future, they can only be made on the basis of an understanding of our past and our deepest nature, which was formed in those distant times. This book is intended as a small contribution toward such understanding.

The taxonomic framework of a book of this kind poses some key problems. Some introductory comments are therefore appropriate. It is over twenty years since Morris Goodman and others showed that on biochemical grounds the African apes were very closely related to humans (as Darwin saw in 1871) and therefore should be placed in the same taxonomic family as humans—the Hominidae (Goodman, 1963). This move has been resisted by most authorities, although the biochemical evidence has mounted. The reason, perhaps, is that the traditional, more restricted, use of the word Hominidae for the human lineage alone, has been convenient and was blessed by a number of distinguished authors (e.g., Le Gros Clark, 1964; Simpson, 1963). Today, however, not only is the biochemical and chromosomal/genetic evidence of the close relationship of humans and African apes very impressive, but the evidence of the comparative anatomy of living and fossil forms seems to point increasingly to such a relationship. In a recent publication on primate evolution, taxonomists from the American Museum of Natural History (Szalay and Delson, 1979) have placed not only the African but the Asian apes in the Hominidae. According to their scheme, the Hominidae include three subfamilies, the Homininae, Ponginae, and Hylobatinae, thus combining in the Ponginae, the orangutan and the African apes. This seems an unnatural arrangement, for their morphological and adaptive differences are very striking and deep seated, and they are not as closely related as the African apes and humans.

Andrews and Cronin (1982) use the family term Hominidae to include the three recent African branches of the Hominoidea only: those that gave rise to the chimpanzees (*Pan troglodytes* and *P. paniscus*), the gorilla (*Pan gorilla*), and humans (*Homo sapiens*). In this scheme it follows that the human line, since its separation from that leading to the African Great Apes, should fall into the subfamily Homininae.

There are, however, two main reasons for retaining the traditional meaning for the family name Hominidae and restricting it to the human lineage only. The first is the value of maintaining constancy in nomenclature and the importance of tradition in successful communication about zoological classification. The second, and perhaps more important, is that the human lineage, since its split from the African apes, has undergone very profound changes in both anatomy and behavior, through its evolution of bipedalism and a material and symbolic culture. This means that the group has diverged very markedly from its African relatives and indeed all other Hominoidea, and such divergence may be signified by a rank of at least family level.

I have therefore concluded that it is best to retain the family name Hominidae for the human lineage since its split from the African apes. Following Mayr (1950, 1963), Simpson (1963), and Szalay and Delson (1979), I place the gorilla in the genus *Pan* with the chimpanzee, a classification which was adopted in the 1974 edition of this book. The African apes now fall into the family Gorillidae; the Asian apes remain the Pongidae and Hylobatidae.

The taxonomy of the Hominidae themselves has been greatly simplified by the discovery of further remains of *Ramapithecus*, which has turned out (Cronin and Andrews, 1982; Pilbeam, 1982), to be much closer to an ancestral orangutan (*Pongo*) than to an ancestral human. It is included here within *Sivapithecus*. This means that we can look not to Asia, as we once proposed (Campbell and Bernor, 1976), but to Africa for the rootstock of the hominid line. Indeed, the discoveries from Hadar in northern Ethiopia bring the origin of that line closer than ever before to an ancestor shared with the chimpanzee. Today even the most prejudiced paleontologist, who had little time for a consideration of the biochemical evidence for this relationship, must now agree that the common ancestor of gorilla, chimpanzee, and human may not be very much older than 6 million years BP.

I am grateful to my readers for the chance to rewrite this book. The demand for the previous editions has been continuous over 20 years. This new edition has been a difficult task, because of its wide-ranging nature: no one individual can now possibly be expert in all the different fields covered in this volume. Nevertheless, it constitutes a richer mixture than other texts, in that I have tried to present a more rounded account of human evolution than any other that I know of. I have benefited enormously from reader's letters in the preparation of this volume, and I hope that readers of this edition will not hesitate to send me further comments and suggestions.

Bernard Campbell

Acknowledgments

In preparing this book, I have required help from many people. Ten years of discussions and correspondence since the preparation of the second edition have been vital to my understanding of the present state-of-the-art. My colleagues have been most helpful in sending reprints and answering my questions and without their cooperation this book could not have been put together. I extend my thanks to them all.

I wish to extend special thanks to Bernard Wood who has kindly read and commented upon the central, anatomical chapters of the book. I also wish to thank two people who have read the entire typescript and have given me an enormous amount of help and advice. Russel Tuttle went to immense trouble to comment on all parts of the book, especially the anatomical chapters and glossary. I am deeply grateful to him for the time he spent and the care with which he worked on my text. Leslie Aiello read the entire typescript and sent me very detailed comments and suggestions. Her help was especially valuable as she nudged me into making some fundamental changes and brought to my attention some important publications of which I was not aware. Without her aid I would still be struggling, and I am very grateful to her for all she did to help me with this new edition. I should add, however, that I am entirely responsible for the final text and its errors and failings are, of course, my own.

I wish to thank Anne Armitage, who has not only typed the entire book, but made a host of very constructive suggestions, and has often improved my style enormously. I am deeply grateful for her patience, diligence, and all the help she has provided me.

I also wish to thank Sheila Johnston and Kyle Wallace, of Aldine Publishing Company, who showed great tolerance and patience during the preparation of this book, and to Brett and Amanda Wallace for their cover design.

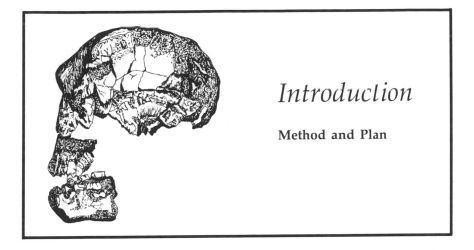

Introduction

Method and Plan

Science does not claim to discover the final truth but only to put forward hypotheses based on the evidence that is available at the time of their presentation. Well-corroborated hypotheses are often treated as facts, and such a fact is that of organic evolution. If a hypothesis is fairly general in its presentation, and rests on limited observations, it is difficult to test, but a hypothesis like that of organic evolution, with a vast array of detailed evidence, is readily susceptible to disproof.

The theory of evolution has now been developed over more than a century as a result of an enormous amount of painstaking research. The evidence that living organisms have evolved over many millions of years is today very strong and convincing. Science builds up such hypotheses or theories on the basis of a vast range of accumulated *evidence* derived from experiment and observation. Each new piece of evidence has corroborated the central theory. No evidence presently known either falsifies or undermines the theory of organic evolution.

Creationism (mis-named "creation-science"), which posits the separate creation of every species, is based on *belief*—a system of belief developed without a scientific assessment of evidence. It is a modern version of traditional beliefs which are based on the book of Genesis. Only by the selection of a very limited range of observational evidence can any sort of pseudoscientific case be made for it. It is therefore not a scientific theory but a statement of religious belief, which for support draws on the Biblical texts and the work of a few biologists, where such work can be manipulated to clothe the belief in a pseudoscientific light.

The theory of evolution and a belief in special creation are not rival explanations of organic life that have comparable status as scientific hypotheses; they are quite distinct approaches to the problem of the origin of species.

Although it was seen in the last century as a devastating threat to fundamentalist religious belief, the theory of evolution does not in any way negate the existence of God. It merely describes the mode in which the creation of living species occurred. Because we are beginning to understand some of the mechanics of this process of creation, it is no less miraculous, no less full of wonder. As Charles Darwin wrote on the last page of *The Origin of Species*:

> There is a grandeur in this view of life, with its several powers, having been originally breathed into a few forms or into one; and that . . . from so simple a beginning endless forms most beautiful and most wonderful have been, and are being, evolved.

As part of organic evolution, the phenomenon of human evolution (though it has often been questioned by laypeople) also amounts to a fact, but as yet its detailed path is not known with certainty. We shall not aim merely at showing that human evolution has occurred, for this has already been demonstrated (see, for example, Le Gros Clark, 1978; Campbell, 1982). My intention here is to examine the evidence we have available for the detailed path of human evolution, in an attempt to discover the mode of origin of humankind. Such a detailed hypothesis as I will present is likely to prove a fallible achievement, but its fallibility will not detract from its value. Not only does the presentation of such a hypothesis have a high heuristic value, but it is by the erection and testing of hypotheses that science progresses. They may indeed be tested and found wanting; science demands only that they should be consistent with all the available evidence and at the same time self-consistent.

With the heuristic as well as the scientific value of the exercise in mind, I have attempted to synthesize into a coherent account the evidence now available for the course of human evolution. When we examine the evidence carefully, however, we realize, of course, that directly pertinent facts are limited, but inferences from all available data (not a selected part of them) must be made as to what happened in time past. The evidence is often indirect; the whole detailed truth about past events can never be known, but that does not negate the value of building a hypothesis on the basis of all the evidence we *can* gather about human evolution.

Hypotheses about human evolution rest broadly on four kinds of evidence:

1. *Fossils.* The first and most important kind of evidence which lies

nearest to the prehistoric facts, consists of fossil bones and teeth. The evidence that these ancient fragments furnish is not always *directly* relevant because we cannot tell whether or not any particular fossil actually belonged to a human ancestor. Indeed, such a coincidence would be unlikely for any particular fossil individual (which may have died without issue), but whether it be the case or not can never be known for sure, even if the individual was mature. However, even if such fragments do not lie on the main and continuing stem of human evolution but are side branches, they can tell us something of the main stem from which they themselves evolved. Their relevance can, however, only be fully understood in the light of the other evidence that we have at our disposal.

Studies of fossil bones and teeth tell us directly of the skeleton and dentition of the animals of which they were a part because of our knowledge of the anatomy of living animals. We can, however, make a second inference from their structure. We can make deductions about the size and form of the muscles and nerves with which they formed a single functional unit. Muscles leave marks where they are attached to bones, and from such marks we can assess the size of the muscles. At the same time, such parts of the skeleton as the skull and vertebrae give us considerable evidence of the size and form of the brain and spinal cord.

Finally, knowledge of the general structure of the skeleton as a whole will tell us about the animal's mode of locomotion (swimming, running, jumping, climbing, burrowing, etc.) from which it is not a long step to infer the main features of its environment (marine, fresh water, terrestrial or arboreal).

2. *Dating.* The age of fossils is essential information needed to elucidate their relationships. Since we are concerned with reconstructing an evolving lineage of individuals and populations, knowledge of their relative age is vital if we are to build up an evolutionary or *phylogenetic* sequence. The advanced methods used in stratigraphy and radiochemistry have made it possible to establish both relative and absolute dates for many groups of fossils.

Relative dating methods are based on a thorough knowledge of stratigraphy—the study of the layers or strata which make up parts of the earth's crust where the rocks are sedimentary (as distinct from igneous) in their formation. Since sedimentary rocks are necessarily laid down with the younger on top of the older layers, fossils in upper layers of undisturbed deposits are younger than those in deeper layers.

Thus the relative age of fossils in a section of an excavation can easily be determined. The main problem arises when related fossils are obtained from different sites at some distance apart. In this case the stratigrapher has to correlate the sequence in the two sites to determine the age of one in relation to the other, which may introduce considerable uncertainties. The principle, however, is very important and allowed early paleontologists to develop a whole series of evolutionary sequences, long before the actual age in years of any fossil was known.

Chronometric or absolute dating depends on being able to determine the age in years of certain geological deposits which may contain fossils, or more often underly or cover fossil-bearing strata. The techniques have been developed as a result of the discovery that certain naturally occurring radioactive elements decay at constant, known and measureable rates into other known elements. Radioactive potassium (K_40) and radioactive carbon (^{14}C) are two such elements which decay into argon and nitrogen, respectively. These techniques can be used both directly and indirectly to date fossils in a number of ways, and form an essential basis for the construction of a reliable phylogenetic lineage.

3. *Environment.* Having created a picture of at least some part of our fossilized creature and its age, we can now make a further inference as to its whole biology and its way of life. In the first place, knowledge of its locomotion will enable us to infer its environment in a broad sense. Supporting evidence derived from a study of the geological context of the fossil may confirm the presence, in the prehistoric times when it lived, of seas, lakes, grassy plains, or forests. Another line of inference begins with the teeth. These may suggest the diet on which the animal fed (herbage, roots, flesh, etc.), and here again we find some indication of how the animal lived.

Next, and most important, other fossils accompanying the hominids can be used to establish a list of species that (in comparison with living forms) will tell us of the biological environment in which our ancestors lived. Were the accompanying species from mammals to mollusks arboreal, terrestrial, lacustrine, riverine or marine forms? The sorts of climates which then existed as well as the general environment of the fossil community can be determined. Finally, the plant remains can be examined to produce a list of flora for the site. Plant remains—usually pollen but sometimes seeds and rootcasts—will provide further indication of the climate and environment of the fossil-bearing (fossiliferous) land surface.

In this way, by rather extensive deductions, we can piece together a picture of the whole environment of the extinct animals we are studying (e.g., Andrews *et al.*, 1979).

4. *Living animals.* Our interpretations of prehistoric populations of animals are based on one further and essential line of evidence, that of living animals. This fourth kind of evidence, though seemingly remote, will prove of inestimable value. If we assume from the fact of evolution that all animals are related, it is reasonable to deduce that those most broadly similar are most recently descended from a common stock. We must therefore compare the anatomy and physiology of living animals—

and especially the monkeys and apes—with that of living humans. This method can also be used to assess the closeness of relationship of different fossils. Fossil remains most similar to a particular living animal may be postulated to be the remains of an ancestor of that animal, or a near relative of that ancestor (see Chapter 1).

In view of the methods of studying human evolution outlined above, it is not surprising that our concern will be mainly with bones and teeth, for they are the only parts preserved as fossils. However, that is not as limiting as might be supposed, since the skeleton is the most useful single structure in the body as an indicator of general body form and function. The teeth, too, are very valuable in assessing the relationships of animals because, apart from the effect of wear, their basic form is not altered by environmental influences during growth. It is a feature of the present study, to infer from the bones and teeth—by consideration of their *function*—the maximum possible information about the body as a whole and the way of life of the animal. In that way we shall attempt to trace the evolution of the whole human being as a social animal. We shall move from a study of the evolving human body to consider evolving human behavior and human society.

In synthesizing evidence to discover the course of human evolution, we therefore draw on four kinds of data; we study early human and fossil animal remains, their geological age, their environment, and related living animals. At one point, we shall be comparing the living primates—our nearest relatives—with ourselves, to gain insight into the differences between us; at another we shall compare their structure with that of fossil remains of our supposed ancestors. The structure of this approach is shown in the accompanying diagram.

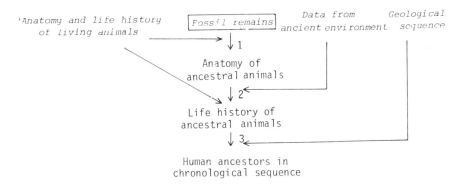

To describe the lineage of human ancestors in chronological sequence, we study fossil remains by comparing their anatomy with living and other fossil species (*1*); next, we consider their functional adaptations in light of their anatomy and the evidence of the environment in which they lived (*2*); and finally, we place them in chronological sequence on the basis of dating (*3*).

The only way to infer the life history of an animal from a few fragments of bone is to investigate the *function* of those fragments. For example, the form of fossil shoulder and arm bones will be informative only if we try to determine how the muscles were attached to the bones and compare them with those of living animals so as to deduce how the limbs actually worked. We may thus be able to determine the difference in function of our fossil bones from that of the bones of living animals. Was this fossil shoulder joint the type associated with animals that walk quadrupedally upon the ground or with animals that hang by their forelimbs from trees? An answer to that question gives us immense insight into the whole life history of the animal and such insight is gained from our knowledge of living animals.

This functional approach involves rather detailed study of the working of each party of the body—in particular the structural parts—and for that reason the method adopted in this book is to study, in turn, the evolution of the different parts. Such an approach presents considerable problems, however, since the body is a single complex mechanism and not merely a collection of discrete mechanisms. Any number of different "functional complexes" can be recognized, but every stage in the subdivision of the animal for descriptive purposes means a loss of truth. We must therefore examine a functional complex in its broadest possible interpretation.

The idea of a functional complex of characters is not new, and it is certainly the most informative and valuable way of analyzing the biology of an organism. A classic study of our origin along these lines was published in 1916 under the title *Arboreal Man* by F. Wood Jones. According to that approach, therefore, the body is not simply divided into skull, dentition, vertebral column, arm bones, leg bones, hands, and feet, but into the different parts involved in the different functions of posture, locomotion, manipulation, feeding, etc. This means, of course, that we must face a considerable overlap of subject matter in different parts of the book. It must also be stressed that the meaning of each functional complex itself cannot be understood alone; on the contrary, even consideration of the whole physical body as a single unit is ultimately meaningless without consideration of its psychological and social correlates.

Function, in fact, which describes the structure and operations of an organism, is only half the study of biology. We must also be concerned with *behavior,* which describes how the organism interacts with the environment (which is everything other than the organism): how the organism actually receives its sensory input, and how it delivers its motor output. The social and cultural behavior of humans, which is the correlate of our complex functions, is in essence no more than an extension and development of the simple behavior of the smallest and simplest protozoa: it is the means of interaction between the organism and the environment that makes possible the biological functioning of the individual and the survival of the species.

Our study of human evolution will therefore be set forth as follows: We begin with an introductory chapter on the nature of evolution, followed by a survey of the background to human evolution—that is, the mammals and our nearest relatives, the primates (Chapter 2 and 3). Then we proceed with a short review of some fossil evidence of early humans (Chapter 4). We then trace in detail the evolution of certain broad functional and behavioral complexes: posture, locomotion, and manipulation (Chapters 5, 6, and 7); sense reception, the head, feeding, and ecology (Chapters 8 and 9); reproduction and society, communication and culture (Chapters 10 and 11). Finally, Chapter 12 will review those different evolving complexes in a time sequence, and the last stages of the story of human evolution.

It should be stated that the text does not include a complete account of primate anatomy or taxonomy at even the simplest level; such accounts should be consulted elsewhere (Napier and Napier, 1967; Le Gros Clark, 1971; Szalay and Delson, 1979). Our consideration of anatomy will be limited to features that have evolved in such a way as to differentiate us from other primates and in so doing made us the creatures we are. Our aim at all times is to paint a picture that shows how primate biology was modified in the course of time into human biology. This study will show how the primate pattern of life associated with the forest was changed into a totally different pattern of life adapted to a different terrestrial environment—the human earth.

In such a task, and in an attempt to achieve consistency with recent research, an effort has been made to make this account of human evolution up to date by incorporating new ideas and data in their appropriate places. Yet we have attempted to maintain a properly balanced picture; new research has not been included simply because it is new, and the reader may be disappointed to find scant reference to topical and controversial issues. What is of central importance is not the controversial problem of classification of fossils or the evolution of altruism, fascinating though it may be, but the functional anatomy and ecology that we can learn from fossil remains. If famous sites like Sangiran, Choukoutien, Sterkfontein, and Olduvai, from which our fossil evidence has come, receive only occasional reference, it is because we are not attempting a historical review of the science of paleoanthropology but a study of the human paleobiology. We must not, however, lose sight of our debt to the men who have discovered the relics of our ancestry, for they make possible the preparation of a book of this kind.

In spite of the contribution of archeologists, geologists, and paleontologists, our knowledge is still limited. Crucial gaps in our account appear in every chapter, and we have not overlooked them. What we do not know is as exciting as what we do know, especially at a time like the present, when the subject is rapidly developing. Let us hope that in ten years this account of human evolution will be thoroughly superseded.

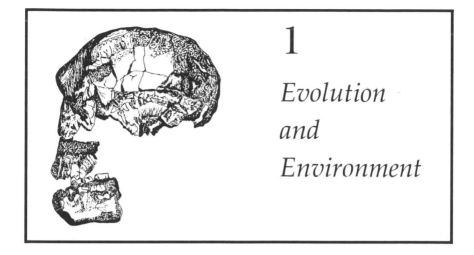

1

Evolution and Environment

I. Organic Evolution

When Charles Darwin and Alfred Russel Wallace published their theory of evolution by natural selection in 1858, they provided a rational and convincing explanation of the causes as well as evidence of the fact of evolution in both plants and animals (Fig. 1.1).

Both these naturalists had traveled widely and observed minutely the variation that clearly existed within each species. Members of species, they observed, are not identical, but show variation in size, strength, health, fertility, longevity, behavior, and countless other characteristics. Darwin in particular realized that natural variation was used by humankind in the selective breeding of plants and animals, for, by selection, farmers and gardeners would breed only from the particular individuals possessing the qualities desired by the breeder.

In due course, the key to how a similar kind of selection operated in nature to transform wild species of organisms came to both men, and it arose from the same source. The first edition of a book entitled *An Essay on the Principle of Population*, written by an English clergyman, T. R. Malthus, appeared as early as 1798. The author showed how the reproductive potential of mankind was far in excess of the natural resources available to nourish an expanding population. He showed that in practice the size of populations was limited by lethal factors such as disease, famine, and war, and that such factors alone appeared to check what would otherwise be an expanding population.

Both Darwin and Wallace read Malthus' essay independently, and, remarkably enough, recorded in their diaries how they realized (in 1838 and 1858, respectively) that therein lay the key to understanding the cause of the evolutionary process. It was clear to them that what Malthus had discovered for human populations was true for populations of

Charles Darwin

Figure 1.1. Charles Robert Darwin (*top*) and Alfred Russel Wallace (*bottom*): two great men in the history of biology, overturned our entrenched attitudes to the natural world and showed that humankind was part of the natural order, created by the processes of natural selection.

Alfred Russel Wallace

plants and animals: their reproductive potential was vastly in excess of that necessary to maintain a constant population size. They realized that the individuals which in fact survived must be in some way better equipped to live in their environment than those which did not survive. Thus it followed that in a natural interbreeding population any variation that increased the organism's ability to leave fertile offspring would most likely be preserved, while the variations that decreased that ability would most likely be eliminated.

The theory Darwin and Wallace formulated on that basis (at first, independently of each other) may be stated as four propositions and three deductions. Both propositions (P) and deductions (D) have since been well corroborated by careful observation.

P.1. Organisms produce a far greater number of reproductive cells and, indeed, young individuals than ever give rise to mature individuals.

P.2. The number of individuals in populations and species remains more or less constant over long periods of time.

D.1. *Therefore, there must be a high rate of mortality both among reproductive cells and among immature individuals.*

P.3. The individuals in a population are not identical but show variation in all characteristics, and the individuals that survive by reason of the particular sets of characteristics will become the parents of the next generation.

D.2. *Therefore, the characteristics of those surviving organisms will in some way have made them better adapted to survive in the conditions of their environment.*

P.4. Offspring resemble parents closely but not exactly.

D.3. *Therefore, subsequent generations will maintain and improve on the degree of adaptation realized by gradual changes in every generation.*

As a result of the weight of evidence presented by Darwin in his famous book of 1859, *On the Origin of Species by Means of Natural Selection, or the Preservation of Favoured Races in the Struggle for Life*, biologists became convinced of the value and truth of the theory of evolution that he and Wallace proposed. Since then, scientists have closely investigated the processes involved in the different propositions and deductions, and geneticists have come to understand the mechanism that accounts for the origin of variation and the transmission of characteristics (see, for example, Dobzhansky, 1962). Of direct interest to students of human evolution are the concepts involved in deductions 2 and 3, for, while human genetic processes are no different in kind from those of the rest of the animal kingdom, the selective factors that caused human evolution were unique.

Since Darwin and Wallace set forward their theory of the operation of natural selection, enormous advances have been made in many fields of biology that affect our understanding of the evolutionary process. Most important among these is the science of genetics, which has provided the necessary underpinning for Darwin's ideas. During the last 50 years (since the 1930's) advances in morphology, paleontology, systematics, cytology, classical genetics, and above all population genetics have been combined to form a basis for understanding the process of evolution, which has been called the *Synthetic Theory*. In spite of all this additional knowledge, it is believed that the directing force in evolution is still Darwinian natural selection, and the ideas Darwin developed have not so much been superseded, but rather thay have been confirmed, broadened, and given strong and well-tested foundations.

II. Natural Selection and Fitness

It is clear that only a proportion of individuals in a population survive long enough to reach maturity and, in their turn, bear offspring. The environment itself determines the fate of each and, in destroying a proportion, selects the remainder. Through its effect upon each individual the environment controls to a decisive extent the direction and rate of evolution, and for that reason it may be considered to be one creative factor in the process of evolutionary change.

Although natural selection acts on individuals, it is the population that evolves, since the genetic plan of an individual remains constant throughout its life. A novel genetic plan arises only in the production of germ cells (*gametes*) and in the fusion of male and female germ cells in sexual reproduction. Not only are successive generations therefore necessary for the introduction of new gene combinations, another creative factor, but they are in fact the source of variation on which natural selection acts. (This is not to deny the existence of evolution among animals and plants that reproduce asexually, but the sources of variation are more limited in them. In this book we shall consider only sexually reproducing animals.)

A series of successive generations reproducing sexually relates individuals not only through the dimension of time but also in the dimension of space. Animals must find a mate among their contemporaries, and if they mate more than once in a lifetime (as mammals do), sexual relationships will be widely spread. Thus, the unit of evolution, the breeding population (or Mendelian population), includes all the individuals able to mate with each other. The size of the population may vary, but it is the breeding unit with its network of sexual relations that evolves in the course of time.

The *fitness* of such a population requires not only the ability to cope with the existing environment and to reproduce, but also the potential to evolve in the future in response to environmental change. This potential requires not only genetic stability (see Section VIII), which reflects the broad stability of the environment, but also genetic variability (consequent upon sexual reproduction), which reflects the instability of the environment. That is to say, a population cannot afford to vary greatly in a stable, competitive, and hostile environment, for random variation may be lethal; the population must remain well adapted. At the same time, the population must be able to change, evolve, adapt to environmental change. This necessary genetic stability, accompanied by flexibility in the form of adaptability, is the basis of Darwinian fitness, and the balance struck between these two factors determines how fit a population is and the nature of its adaptation.

The dynamic stability of all the genetic components of a population (called the *gene pool*) makes possible adaptation to the environment as well as the modification of such adaptation in the presence of environmental change in the future. The absence of such modification can result in extinction; a proper balance must be found between stability and flexibility, and it is an alteration in the form of this balance that, among other things, characterizes human evolution.

It is clear that every gene, every characteristic, every complex of characteristics of the individual, its anatomy, physiology, and psychology, contributes to the biological fitness of the population, and it is in this sense and this sense only that a particular characteristic is of evolutionary significance. It is the contribution characteristics make to the population's fitness that results in their selection, in their survival. It is the population that evolves, not the individual.

At the same time, since all parts of an organism require energy for their maintenance, any part that ceases to have a useful function will be rapidly lost in the process of evolution. Not only any part but any process will also be lost. For example, color vision is believed to have evolved among reptiles and then been lost in the very early period of mammalian evolution, probably as an adaptation to nocturnal life; it appeared a second time in the evolution of the primates, but most other mammals cannot see the full range of color because it has not been selected again during their evolution. Thus, we do not often find characteristics without functions, a fact that may well apply to so-called vestigial characteristics; it seems probable that they have at least a reduced function, or are in the process of being rapidly lost.

The function of a characteristic can therefore be understood fully only as a process that is necessary and that contributes to the overall reproductive advantage of the population in which it has evolved. The function of any characteristic that cannot be interpreted from this perspective cannot be said to be properly understood. It follows that in order to understand human evolution it is desirable to consider the function of each new characteristic that evolved and to discover how it bestowed upon the population in which it became established a greater probability of survival in a changing environment.

How populations have survived by changing their nature is the story of evolution. The concept of fitness involves both adaptedness and adaptability, but, like evolution itself, though it may be elucidated in the past, it can only be surmised in the present.

We have said that natural selection operates only at the level of the individual, but it has been claimed by Darwin (1871) and others that under certain conditions it will also operate on social groups. This may possibly be the case among some higher primates and especially human groups, where social bonding and social interactions are so much closer and more vital than in any other species. This view was recently presented by Grant (1977), who suggests that if troops or bands of early humans came into competition for a foraging territory or some other essential resource, selection operated on the band as a group rather than on its individual members. The success of a band in competitive encounters is determined by, among other factors, the intelligence of its leaders and members, its technology, communications, and numerical strength. Grant suggests that group selection could come into play when competing groups differ genetically with respect to these characteristics. This might well be so, especially if the band itself is small and is totally destroyed either directly in warfare or indirectly through starvation. This was probably not a very common occurrence, but almost certainly a small degree of group selection would have operated in human evolution, which would have given significant advantage to groups with a high level of intelligence, good coordination and communication. Wilson (1975), whose theories of the biology of social groups make the concept of group selection redundant, nevertheless accepts the possibility of group selection as exclusive to human groups, particularly where warfare is practiced with a degree of genocide. Clearly, the operations of natural section have changed during human evolution and are still changing (Chapter 12).

III. Genotype, Phenotype, and the Environment

The *genotype* of an individual organism is its heredity—the factors that determine the path development may follow in different environments. The determinants are called *genes*, and they take the form of very complex chemical substances; in particular, that known as *DNA (deoxyribonucleic acid)*. This substance, which forms immensely long chainlike molecules, is found in the nucleus (the controlling center) of every living cell. The DNA of the male and female gametes (the sperm and egg), together with some other minor genetic factors outside the nucleus, when combined in the fertilized egg cell or zygote, determine the form and structure of the new individual into which the zygote will develop. From that time onward, as the zygote divides many times, the DNA molecules are copied exactly, and similar copies lie within every cell of the growing individual's body. Once the genotype is determined at fertilization, therefore, it is fixed throughout the life of the individual organism. It is a highly condensed coded bank of information, which during the development of the zygote determines the kind of organism that will grow.

We can with some certainty postulate that there are two kinds of genes within the genotype of each individual. The first, called *structural genes*, determine the structural ground plan of the organism. The second, called *regulatory genes*, determine the rate of growth of the different parts of the body, the different organs, including the time of the onset of sexual maturity, which signals the end of physical growth. It seems clear that alterations in a very few regulatory genes (affecting, for example, the size of certain structures or organs) can have quite a marked effect on adult structure without fatally altering the functional integrity of the organism.

The *phenotype* is the whole individual, the manifest characters of an organism, the discrete biological unit, the human being or the worm. It is formed as a result of the living interaction of the genotype and the environment, an interaction called assimilation or growth. The phenotype is neither constant in form and structure nor permanent. Through its interaction with the environment, the genotype determines not only the phenotype's mature form (its *morphology*) but also its total growth pattern: that is, its form at all ages, together with its growth rate, its capacities, and its lifespan. While the genotype is constantly copied in cell division, its form does not change, but the phenotype itself does change as it grows from a fertilized egg into old age.

Variation in the genotype occurs between generations as a result of sexual reproduction. In the process of fertilization, a single set of paternal genes is combined with a single set of maternal genes to produce a novel genotype. With the exception of identical twins, no two individuals of *Homo sapiens* are ever likely to be similar in spite of their vast numbers, so efficient is the shuffling of the pack of hundreds of thousands of genes that occurs in sexual reproduction. By combining different genotypes in fertilization, sexual reproduction is an essential source of genetic variation in evolution, and the process itself has been selected as a reproductive mechanism for that reason. It is through selective breeding that plant breeders can create such remarkable phenotypes as the hybrid roses we enjoy in our gardens. Originally created by sexual reproduction and selection, they are, in turn, preserved by asexual vegetative propagation in which no further variation of the genotype occurs.

However, although sexual shuffling is of great importance, it is not the fundamental source of genetic variation. There is also the phenomenon of so-called "spontaneous" change in the genotype (*mutation*), which results from "imperfections" in its natural and ordered reduplication during cell division. This valuable and inherent "flaw" in genetic replication is most common in the production of the paternal and maternal gametes, when the paired gene sets of the normal body cells are separated. In this process of division (*meiosis*), the DNA chains undergo stress, which very often results in chemical modifications. This kind of mutation is random and therefore usually lethal, for no highly complex mechanism is likely to benefit from random interference.

Thus, the mutations that allow survival of the gametes (let alone of the individual) and that are subject to natural selection are themselves already selected by the need for internal coordination in the gamete and in the zygote. This internal selective process limits the kinds of mutations available to be selected by the environment (Whyte, 1965). The nature of organic mechanisms is such, however, that slight changes that might improve adaptedness in a changing environment are possible. Organic characteristics (for example, growth rate, body temperature, size, and so forth) vary continuously (not discontinuously, in jumps), so small adjustments in the mean or average form of these characteristics may be advantageous at any time. It follows that spontaneous mutation, if it has very slight effect on the phenotype, can be of value to an individual, and if it is, it may spread by the aid of sexual reproduction to a whole population over a period of time. But the value of such mutations can be realized for the population only by natural selection, the selective interference of the environment upon the gene pool.

In practice, the environment can interfere with the genotype in two ways, which blend into each other:

1. Random interference from the environment is perhaps the least common. The most obvious example is the radiation that enters the earth's atmosphere from outer space and penetrates organic substance. It may be *mutagenic*; that is, it may cause mutations in the genotype of the cell that it penetrates. Should this cell be a gamete or newly fertilized egg cell, the radiation may either kill the cell or, by changing the chemical structure of one or more individual genes that do not lethally affect cell function, cause an increase in variation in the gene pool. Atomic radiation is powerfully active in this way, and although artificially created and localized, its effects on the genotype are random. Such influences are not likely to be a significant source of variation in organic evolution over the immense periods of geological time.

2. Most of the other ways that the environment acts upon individuals are more or less selective, and they may be either lethal or sublethal. Lethal interference of a particular type (for example, famine) may kill some kinds of individuals (for example, thin ones) more readily than others, and for this reason it will change the nature of the gene pool of the population (individuals with reserves of fat are selected). Sublethal interference will also affect the nature of the gene pool if it affects the reproductive capacity of any individuals as a result of a particular characteristic they possess (for example, disease or displacement of the uterus may result in abortion). Such selective interference is enough to result in evolution, even if all the individuals survive; it is only necessary for them to vary in the rate at which they reproduce themselves. The environment strongly affects the growth and development of individual phenotypes because they show a certain amount of developmental plasticity or *adaptability* (see Section V), yet this plasticity is itself adaptive and as such helps to ensure a high rate of reproduction in the face of environmental variation. The effect of the environment upon the gene pool will ultimately depend on the intensity of the interference, or *selec-*

tion pressure, as it is usually called, and the degree of adaptation developed by the species in response to this pressure.

It should be noted that the individual (and, of course, the population) will alter the environment to some small extent merely by its existence, but, while the individual genotype cannot change in the face of selection pressure, the gene pool can and does. Such change is evolution; failure to change may result in extinction of the population.

It must be recognized that as a correlate of the theory of evolution an element of competition is invariably associated with organic life. Because organisms reproduce at a rate much higher than that necessary to maintain a stable population, there is always competition within a single population for the available food and space. In this way, every individual changes the environment. Similarly, there is competition among populations of the same species, as well as between species. It is noteworthy, however, that members of a single species are in more immediate competition because they require the same food and habitat, and therefore they alter their own environment more than members of different species that are adapted to different parts of the environment or ecological niches (see Birch, 1957).

Natural selection resulting from competition within species (*intraspecific* competition) is therefore as much a cause of evolutionary change as selection resulting from competition between species (*interspecific*). Intraspecific competition is a permanent feature of organic life, as is natural selection, and a noncompetitive situation cannot exist for more than an extremely short time because the reproductive potential of species is always greater than that which is required to maintain even an expanding population in times of beneficial environmental change. This intraspecific competition means in effect that the evolution of further adaptation may occur even when the overall external environment is fairly constant. Thus, evolution within one population will, of course, come to change the environment of its neighbors.

It is clear that the total interaction among individuals, populations, and their environment is intricate, and the totality of living plants and animals upon the earth's surface forms a dynamic system of great complexity. This complexity is seen to be immense when we consider further that among animals the connecting link between the environment and the phenotype is the medium of behavior, since what an animal *does*, as well as what it *is*, determines its ability to survive.

IV. Variability, Speciation, and Taxonomy   The sources of variation operating in animal populations and described above result in a great variety of forms in any particular population. The fact that every individual has a different genotype (unless a monozygotic twin) and has been subject to slightly different environmental stress means that every individual, when mature, will be different in a large number of features. Beyond this we can expect variability due to age: babies and mature adults vary not only in size but in bodily proportion, and in the presence and absence of certain features. The two sexes also may vary considerably, not only in the generative organs but in other features as well—the secondary sexual characteristics, which may affect stature, weight, and even shape to a surprising extent. For example, male gorillas are almost twice as large as females, and the proportions of parts of the skull differ between them. In the past, paleontologists have made the understandable mistake of classifying rare fossils of different age and sex as different species or even genera.

Beyond the variation that we can expect within a population, there will also be variation between different populations of a single species. Populations living in different geographical areas will experience slightly different environments, for no two places on the earth's surface are exactly the same. These different *geographical races*, as they are called, vary in adaptation to the unique features of their differing environments: this is geographical or racial variation. Local races at the opposite extremes of the range of a species may well show very considerable morphological differences. In this case we can only be sure that they belong to a single species by tracing the continuity of populations over the area of land between them. If a single species is represented, then the continuity of individuals will be associated with a continuity of morphology, which implies that genes are being exchanged between neighboring populations. This exchange of genes between populations is termed *gene flow,* and it maintains the integrity of the species.

Should some barrier become interposed between parts of the species range, then gene flow will be interrupted, and different populations will become geographically isolated. This *geographical isolation* will allow the separated populations to vary independently, perhaps in different ways. If the isolation is complete and remains so for a long time, the independent adaptations of the two groups of populations may well result in morphological divergence to a point where they are very different. If the difference is so extensive that on coming into contact again at a later date no further gene flow occurs, then the populations will have *speciated:* that is, one species will have split into two. The necessary conditions for speciation are the variability that is present in all species, the operation of natural selection, which is equally a feature of all organic life, and geographical isolation. This last feature is the immediate cause of speciation.

The appearance of geographical barriers is not an everyday occurrence. Changes in climate, and particularly rainfall, may result in the increased size of rivers, lakes, and seas to form impenetrable barriers to certain animals. Tectonic movements may result in the merging or separation of continents, in new mountain ranges or deep rift valleys, with associated changes in climate. Far less extreme changes in climate and environment may result in what amount, in effect, to geographical barriers, depending on the mobility and behavioral plasticity of the species.

But the formation of the barrier is the reason one species splits into two, (called *allopatric speciation*), which is the way a single evolutionary lineage also splits into two, and so may bring about a whole new evolutionary radiation.

This discussion leads us to the definition of species as *groups of interbreeding natural populations that are reproductively isolated from other such groups* (Mayr, 1963). The critical feature of successful speciation is failure in interbreeding—the absence of gene flow and the buildup of morphological divergence. Biological species, or *biospecies*, are the groups that fulfill this definition, the species of plants and animals that surround us. The nature and recognition of fossil species, sometimes called *chronospecies*, present some special problems that are discussed in Chapter 4.

We have seen in this section that a species is a natural unit, called a *taxonomic unit*, that can be defined and recognized in nature, generally without much difficulty. Species are often divided into smaller units called *races*. The term race usually refers to a group of populations with a certain number of features in common that distinguish them from other such groups. However, since they are not isolated by genetic discontinuity, their boundaries are not easily recognized, and the definition of particular races is therefore always a matter of discussion and often of disagreement. Important and large racial groups are sometimes called *subspecies*, while minor ones are called *local races*. Almost without exception, animal and plant species can be divided into these lesser units according to their geographical range and variability. The presence of different races is a typical characteristic of living species and represents their potential for further evolution and speciation.

Finally, species themselves are grouped into units called *genera* (singular *genus*), which are groups of species with major adaptive features in common. Like races, these units are subjective and do not always represent easily defined natural categories.

Two Latin names, or *nomina*, are used to label each species; the first is the generic name, the second the species name. Thus, *Homo sapiens* is the name given by Linnaeus in 1758 to mankind: it indicates that we belong to the genus *Homo* and the species *sapiens*. A third name may sometimes be added to label a particular subspecies, as in *Homo sapiens afer*: the name given by Linnaeus to the African races of *Homo sapiens*.

The conventions of zoological classification require that at each level the taxonomic units, or *taxa* (singular: *taxon*), are grouped into taxa of a higher level. Thus, genera are grouped into *families*, families into *superfamilies*, superfamilies into *orders* (such as Primates), orders into *classes* (such as Mammalia), classes into *phyla* (such as Chordata), and so on. This system of classification is essential as an informative basis of communication among zoologists, and many of these terms will be used in the following pages. An indication of taxonomic practice is given in Table 1.1.

We must now consider how the taxonomist goes about classifying

TABLE 1.1. The Use of Taxa[a]

	Taxon	
Taxonomic category	Example 1	Example 2
Kingdom	Animalia	Animalia
Phylum	Chordata	Chordata
Class	Mammalia	Mammalia
Order	Primates	Primates
Suborder	Anthropoidea	Anthropoidea
Superfamily	Hominoidea	Hominoidea
Family	Hominidae	Gorillidae
Genus	*Homo*	*Pan*
Species	*H. sapiens*	*P. gorilla*
Subspecies	*H. sapiens afer*	*P. gorilla berengei*
(geographical race)		
Race (local race)	Nilo-Hamite	Virunga race

[a]Names of categories above the species begin with a capital letter. Generic, specific, and subspecific names are italicized.

populations of organisms. This is an important question, since it affects the resulting classification that he builds. The assessment of relationships by the study of comparative anatomy is a complex question. The simplest approach is that of *phenetics,* in which morphological relationships are analyzed by giving all characteristics compared equal weight. The analysis should be based on a large number of characteristics, and the results produced simply group like species together and separate them from those that are less alike. This technique produces a taxonomy which is useful but does not reflect evolutionary history, only present morphology. It is convenient when trying to classify forms such as bacteria, which have no fossil evidence that can help in understanding their actual evolutionary relationships (Sneath and Sokal, 1963).

The *phylogenetic* approach attempts to assess relationships as they actually exist, just as human genealogists do. Thus, the taxonomy will take into account fossils and their age; it will take into account similarities between groups, those that are shared with ancestral forms and those that are more recent developments; those that are specialized in the lineages being investigated and those that are primitive and widely shared. The phylogenetic taxonomist will take into account all characteristics, but will give greater weight to the specialized, recently evolved ones than to the ancient ones. Thus, he will weigh his characteristics to achieve a classification he hopes will reflect actual evolutionary relationships (see, for example, Simpson, 1961). For example, the distinctions between the different means of nourishing the embryo are given great weight in the classification of the two great subdivisions of the class Mammalia (marsupials and placentals), but these characteristics are given no weight in the classification of the order primates, or any other order of mammals, as they are widely shared and show no particular specializations in these orders. At this lower level, dentition, which is more variable, is given greater weight. It is important to note that in arriving at these judgments, taxonomists are not looking at characteristics separately, as in a phenetic analysis, but at the total morphological pattern of the population in its functional aspect.

Biologists using this method arrive at a generally acceptable consensus, but the process of classification requires both experience and subjective judgment, so that some scientists have described the process as not so much science as art!

A refinement of this approach, in which personal judgment and experience is believed to be reduced, takes the form of a method of analysis called *cladistics*. In this approach the taxonomist clearly divides the characteristics of the specimens he is studying into either ancestral (*plesiomorphic*) or derived (*apomorphic*) characteristics. Thus, all relationships within a particular taxon are determined by a study of individual derived characteristics only, because the ancestral characteristics are necessarily shared and therefore not taxonomically significant. The more derived characteristics that are shared between units of the group under study, the closer their relationship. Evolutionary change over time implies the substitution of ancestral by derived characteristics (Delson, 1977).

Occasionally, the process of change over time is inconstant; the rate of evolution varies, some characteristics evolve while others remain static, and in a few cases some reversal in the direction of change may occur. For these reasons, the conclusions about relationships reached by cladistic analysis are usually not believed to conform to actual evolutionary history, which is to say, they do not represent phylogenetic relationships, but practitioners claim that the results are the best that can be obtained and that they create the most useful classificatory schemes. However, it should be pointed out that cladistic analysis is not an entirely objective process. Human judgment is required in distinguishing ancestral from derived characteristics and in isolating those characteristics from the functional complexes in which they exist which are required for quantification. Thus, the data are used to establish the closeness of morphological relationships within major groups and only later used, if ever, to establish lineages. Some researchers never go beyond the establishment of morphological relationships and altogether avoid making deductions about evolutionary lineages.

Our view here is that cladistic analysis, through its rigid methodology, is a useful indicator of phenetic (morphological) relationships, and it will generate a branching sequence (Fig. 1.2). An evolutionary biologist, however, must take the step of formulating chronological sequences of fossils if he wishes to understand the evolution of a particular group of organisms. We shall be concerned with the overall pattern of

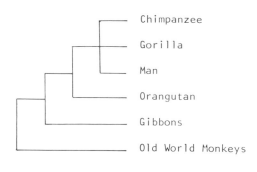

Figure 1.2. A study of shared and derived characteristics will generate a branching sequence such as this. This diagram, based on biochemical data, demonstrates the close relationship between humans and the African apes, and our more distant relationship with the Asian apes. The branching points are dated as follows: human–gorilla–chimpanzee 5 ± 1.5 mya; human–orangutan 10 ± 3 mya; human–gibbon 12 ± 3 mya (from Andrews and Cronin, 1982). For discussion, see Chapter 12.

the evolution of the hominids rather than with taxonomic procedures and conclusions. It should be pointed out, however, that different views on primate taxonomy can usually be attributed to the different theoretical approaches of the taxonomists who prepare them.

Using a phylogenetic taxonomy, the hierarchy of taxonomic terms (see Table 1.1) indicates to an evolutionary biologist a summary of deductions about the evolutionary history of a group. Cladistic analysis and the resulting taxonomy, which treats all groups as "contemporary" because the chronological data are not considered to be relevant, cannot reflect our fullest understanding of the evolutionary process. We shall therefore use the traditional phylogenetic approach in this book.

V. Homeostasis of the Individual

Before we examine further the interaction of the breeding population and the environment, it is necessary to consider the interaction of the Individual and the environment, in order to understand how one is related to the other. An individual organism is a very delicate, complex, and clearly very unstable living system, since it can stand only limited outside interference; a momentary electric shock or shortage of oxygen may destroy the living system entirely. Its survival depends upon its property of self-regulation.

A living organism is a self-regulating dynamic system that maintains a more or less steady state or equilibrium, both within itself and between itself and its environment. Cannon (1932) proposed the term *homeostasis* for the self-regulating property of organic systems. The particular character of an organic homeostatic system arises from the fact that it is an open system that depends on an energy supply obtained from the environment by chemical interchange for its continued existence, yet at the same time must maintain itself as an integrated and discrete mechanism separate from the environment. Furthermore, the system is maintained throughout the growth and development of each individual, and during this period the whole cellular structure is being continually broken down and replaced as it matures. The same body in infancy and old age has little in common beyond its genotype and a dynamic system of a particular kind. Human personality, the continuing identifiable nature of an individual, is an aspect of this system.

Cannon has described in his book, *The Wisdom of the Body* (1932), how the homeostatic physiological mechanisms of the human body maintain what has been termed its "internal environment" at a constant level. He has described how the body maintains constant (within narrow limits) its water content, salt concentration, level of sugar, fat, and protein in the blood, oxygen supply, temperature, and many other features of its organization. The efficiency of these systems is remarkable; for example, humans can survive dry-heat temperatures up to 128° C without an increase in body temperature above normal (37° C). Arctic mammals can similarly survive 35° C below zero without a drop in body temperature. This particular homeostatic mechanism is characteristic only of living mammals and birds, and it is of the utmost importance in the story of human evolution. It enables an animal to survive changes in the external environment that otherwise would destroy its delicate chemical systems. Animals with a limited range of homeostatic mechanisms are limited to more or less constant environments; animals with a wide range of homeostatic mechanisms can occupy unstable environments and survive external conditions that in no way approximate their internal environments.

Thus, marine organisms live in a relatively constant environment of water containing salt, food, and oxygen. For deep-sea animals, changes in temperature associated with the seasons are negligible. In the course of the evolution of marine creatures into terrestrial vertebrates, homeostatic mechanisms have been evolved that maintain their internal environment so that it will approximate that of their marine ancestors. Every living cell of the body of mammals is bathed in a fluid called lymph. The composition of this watery matrix is kept constant by the diffusion of salt, proteins, sugars, oxygen, and other substances from the blood vessels. Thus, we can survive dry heat and cold because the living cells of our bodies are preserved in a saline environment of constant temperature and unvarying composition.

The homeostatic mechanisms of mammals are numerous and complex; they may operate in varied ways such as sweating or suntanning (the former lowers the body temperature, the latter lowers the penetration of ultraviolet light). Both these adjustments to hot, sunny weather are methods of coping with environmental change and, as such, increase the chances of survival of their possessors. These, however, are not the only kinds of homeostatic adjustment that is made, for the maintenance of the organism depends on chemical interchange with the environment. Thus, a low water content will cause the sensation of thirst and lead to the consumption of water; a high carbon dioxide content will cause an acceleration in the respiratory rate to increase CO_2 loss and oxygen intake from the atmosphere. Food is required to maintain sugar, fat, and protein levels. The sensations of thirst, breathlessness, and hunger are therefore aspects of the homeostatic mechanisms involved, and so in turn are the behavioral responses necessary to satisfy these needs. If there is not food, water, and oxygen in the environment surrounding the organism (as is the case for a flourishing colony of bacteria), the organism must go to them. Behavior, then, is part of the homeostatic mechanism; it is the process whereby animals satisfy their need to maintain their internal equilibria. Indeed, the maintenance of this steady state is survival, and animal behavior exists simply to that end and that of reproduction.

An extension of the evolution of such behavior is the use of *artifacts* to increase a species' adaptive potential. Animals that build nests or dig burrows are artificially improving the immediate environment to make their survival more probable by avoiding extremes of heat, cold, or drought, or by gaining protection from predators. Human material culture, particularly in the development of *facilities* (see Chapter 9, IX and X), functions precisely to control body heat and energy flow. Fires permit adaptation to cold climates, and space suits make it possible to survive in a vacuum. Thus, one of the most significant features of human evolution is the development of cultural extensions of homeostatic mechanisms that allow adaptation to environmental extremes that a cultureless primate could not survive.

Physiological homeostatic mechanisms are not the only means by which an individual maintains a steady state in relationship with the environment. Another kind of homeostatic adjustment is discernible during individual growth and development; it is called, appropriately, *developmental homeostasis,* and is a mechanism, as yet not fully understood, whereby the development of the phenotype seems to involve some self-regulating capability that maintains its integrity in the face of environmental variation (Mayr, 1963). Like any self-regulating mechanism, developmental homeostasis is based on a factor of stability (due to the genotype) and a factor of variability, which is the *plasticity* of the phenotype during its development. Examples of such plasticity are found among lower animals and plants more easily than among vertebrates, which have other methods of dealing with environmental change (that is, versatility of behavior). Yet the plasticity of individual vertebrate development is clear. For example, our physique is affected by the kind of life we lead; muscles enlarge with use or atrophy as a result of disuse; fat deposition depends upon food, activity, and climate as well as upon genes. The digestive processes can become adjusted to different diets, and jaw strength and size respond to the demands made upon the masticatory apparatus by the food. This kind of plasticity is a valuable characteristic, which has evolved like any other and is under genetic control. The mechanisms of developmental homeostasis control the dynamic interaction between the genotype and the environment so that the phenotype is enabled to survive changes in the environment during its growth.

VI. Behavior

The study of behavior refers to what animals do in their interaction with the environment, and it is concerned to a great extent with the activity of locomotion. Broadly, it is concerned with how animals come into contact with their environment—how they breathe, how they touch and move upon the ground, how they eat the portion of the environment that constitutes food, how they escape from the portion that constitutes predators, and how they communicate with and copulate with the portion that constitutes their own species. Behavioral scientists describe the way that an animal is related to the environment, and much of the interaction between the phenotype and the environment occurs through the medium of behavior. An understanding of the behavior of an animal is therefore necessary for a full understanding of the interaction of a population and its environment; such interaction must be studied in an analysis of the process of evolution.

How is animal behavior determined? This is a difficult subject, but some broad generalizations may be made as a basis for discussion of the evolution of human behavior. In the first place, behavior, like morphology, arises as a result of the interaction between the genotype and the environment. The potentialities of all behavior patterns are genetically determined, but in the absence of a suitable environment a behavior pattern may not mature, just as an individual may not. Behavior that is derived from information coded in the genotype, with little contribution from the environment, is often called *innate* and is well exemplified among lower animals such as insects. A remarkable example involving a visual "language" is found in the honeybee. By a remarkably skillful piece of research conducted over many years, von Frisch (1950) has shown conclusively that a worker bee that has discovered a new source of food will, on returning to the hive, perform a dance on the face of the honeycomb. This dance has been shown to transmit information to other worker bees about the direction, distance, and nature of the food source and is a rare example of a descriptive language among animals. It is significant that it is found among one of the insects with a highly organized society. This descriptive information is in condensed, coded form and accurately enables other bees to find the flowers described. The bees' capacity for this complex behavior pattern is innate, but, since it records variable information, is determined in its details by the conditions of the external environment.

There is a great deal of variation among different species of animals as to the proportion of information input (which determines behavior) that comes from the genotype and the proportion that comes from the environment. There is an almost equally great variation in this factor within each individual species. The chart in Fig. 1.3 shows in simple form the situation in a species such as ourselves. Although some behaviors are described as innate, none can appear in the total absence of appropriate environmental input. For example, the nipple-searching and sucking reflexes of newborn babies are innate and genetically coded, but only appear as a response to contact with the mother's breast. At the other end of the scale, even behaviors that seem to depend solely on learning are ultimately based on potentials that are coded in the genotype. The ability to play a musical instrument, while learned, is based on a potential not evenly distributed among the population.

Thus, if we consider the genetic or innate component only, we see that it can vary from a detailed, programmed motor response (such as sucking) to a very generalized potential such as musical ability. An instance of even more generalized potential is car driving. Here the innate component lies in the physical capability of the individual and the possibility of rapid coordination between eye, hand, and foot; as we shall see, all these characteristics are associated with arboreal primates.

The environmental component also varies. At the left of Fig. 1.3, it merely consists of the minimal conditions for life and an appropriate stimulus: in our example, the skin of the mother's breast. At the right hand side of the chart, the information input may be not only from the general environment but from other members of the social group, and in this case it may take on the quality of *learning*. However, only a small proportion of the input from the environment can be called learning.

The environment influences each behavioral activity anew every time it occurs, obviously a necessity in the case of the bee dance in order that a correct report be given of each fresh discovery (see Fig. 1.4, *left*). But among many other animals it is clear that the effect of the environment upon each action may be *recorded* in some way in the central nervous system so as to influence future action. In the latter case some degree of learning has occurred (see Fig. 1.4, *right*).

In practice, therefore, it is impossible to draw a line between innate and learned behavior. Even the flatworms, with a much simpler internal organization than insects, have been shown to be capable of benfiting from experience; but the more complex the animal is, the greater the part that learning appears to play in determining behavior. Among birds, for example, there is wide variation in the relative contribution of learning and inheritance in the development of the mature song repertory. Some birds (the cuckoo is a classic example) develop their songs quite normally when reared away from their own species, while others develop abnormal songs when reared in this way. But, most commonly, broad characteristics are inherited and the details are learned.

Such data have been gathered by separating birds from their kind, an experiment that cannot be performed easily with mammals because so much behavior is learned that growth is deeply disturbed in the absence of a mother, and it is impossible to isolate the effect of deprivation upon

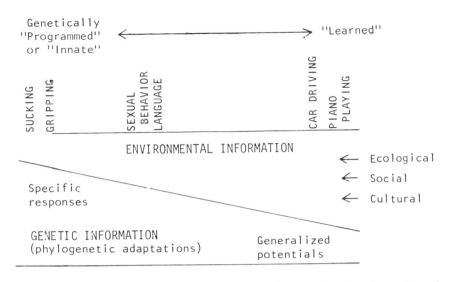

Figure 1.3. Diagrammatic representation of the contribution of genetic and environmental information to the development of behavior.

any particular behavior pattern. But, even in the absence of relevant experimental data, it is clear that in nearly all the behavior of the higher mammals there is an element of learning, and it may be very great. Genes supply the capacity for behavior of different kinds, together with some basic components that help form the full behavior pattern; learning makes possible the realization of this full pattern of activity. For example, studies of bird song suggest that in many species genes make possible the capacity to sing and, especially, to sing certain notes and trills; the actual pattern of the song is, however, usually learned.

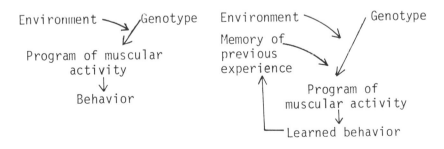

Figure 1.4. Behavior is always a product of both genotype and environment. The memory of previous experiences is not a component in the production of innate behavior (*left*), but is an essential factor in learning (*right*).

Clearly, behaviors that are primarily determined by information from the genotype and those primarily determined by information from the environment each have certain advantages. Behavior that is mainly innately generated obviates the need for learning processes that take time and may involve danger, and saves energy on the part of the animal. On the other hand, it is not flexible, and in a fast-changing environment or a long-lived species could be disastrous, since it could be changed only by the slow processes of variation and natural selection. Learned behavior, on the other hand, is expensive in time and danger, but it is much more flexible and allows readaptation to occur throughout a lifetime and in every generation. Just as evolution results from the interaction of factors causing variability and stability, so behavior has this dual character. Heredity can make innate behavior patterns more or less stable (though subject to evolution), while learning introduces a flexibility that may be of great survival value. A different balance between the two characterizes each species of animal (and see Chapter 2, V).

There are three different ways of learning: by trial and error, by observation and imitation, and by instruction. The trial-and-error method is found in all animals that can learn and is the sole means of learning in invertebrate animals. Compared with other methods, it is lengthy and dangerous. It is the method usually referred to in discussions of learning and in experiments testing learning ability. For example, rats can learn their way through a maze by trial and error (just as humans can) so long as some reward is offered. The rate at which learning can take place by this method has been used as a rough indication of the degree of complexity evolved in the central nervous system. Learning by trial and error reaches its highest development in mammals. As a method it is highly flexible and adaptable to environmental change, and therefore it has been evolved to a remarkable extent among some groups of mammals, particularly among the primates.

Learning by observation and imitation is, however, a speedier and more sophisticated means of building up behavior patterns. Imitation depends in the first place on the ability of an animal to recognize and copy another member of its species, usually its mother. Imitation is a shortcut to learned behavior, but it introduces a certain inflexibility in the behavior pattern that is not found in trial-and-error learning and is therefore more valuable when the behavior relates to the more constant features of the environment. Rats can learn to pass through mazes more quickly than by trial and error alone if they are allowed into the maze with other rats already trained in the art. Many birds learn their song repertories to a great extent by imitation. Monkeys and apes also learn a great deal that way, and children learn still more and faster. During the early stages of human development, imitation was probably the most important factor in developing behavior patterns.

Finally, instruction is a uniquely human way of creating behavior patterns, for it involves conscious thought and intent. It is the means of learning that is essential to the development of culture, yet is dependent upon culture for its existence.

Another and very different determinant of behavior that has not yet been mentioned may be termed "intelligent thought." It is in all probability a uniquely human characteristic and so is a novelty of great importance in human evolution. Action determined by intelligent thought may draw on learning but is itself not the result of a learned behavior pattern; it is original. (It is discussed in more detail in Chapter 11, IV.)

With its three determinants—heredity, learning, and intelligence—the study of behavior is highly relevant to the study of human evolution, for the evolution of behavior itself has played a vital role in human evolution. In the following sections we shall see more clearly the importance of the less easily detected changes in learning and behavior; we must study them as well as the more obvious changes in anatomy and physiology that characterize our evolution.

VII. Sociobiology and Human Behavior
Sociobiology, the science of the biology of social behavior, is an important and growing branch of biology that is of particular interest to us since we are a very social species. The science, with its roots in the study of social insects, is concerned with the behavior of all social animals from insects to humans. Its origin as a new science can be traced to the concept of *inclusive fitness*, first discussed by W. D. Hamilton in 1964.

The concept implies that the "Darwinian fitness" of an individual not only includes its own reproductive success but that of its relatives, which share the individual's genes. It is, as it were, the particular packages of genes that are being selected rather than particular individuals. This became clear from the study of social insects such as bees. In this group of insects, because of their unusual genetics with haploid males (i.e., males carry only one-half the normal diploid gene complement, so that the nucleus of the cells does not undergo meiosis in the production of gametes), the daughters inherit a full set of their father's genes as compared with the usual 50% of their diploid mother's genes. This means that since each worker bee shares (on average) 75% of its genes with its siblings (50% from having the same father and 25% from having the same mother), it pays it to work for the survival of its siblings as much as (in fact more than) for its own putative offspring, were it to have any, who would carry only 50% of its genes.

Similarly, if a bird giving a warning cry can thus protect a number of individuals that share its genes, the gain in survival of its genes may be greater than the loss that would occur if it was killed by its predator. In practice, the selective advantage of an individual assisting a relative in this or any other way will depend on three factors: the risk to its own fitness, the benefit to the relative or relatives, and the degree of related-

ness of each relative. We find that on this basis we can account for the
selection of what looks like altruistic behavior in close-knit social groups
where individuals are closely related.

In reality, however, the behavior is not truly altruistic insofar as an
individual is protecting its "own" genes. The phenomenon has been
termed *kin selection* or *gene selection*. It is not, however, an alternative to
natural selection but an extension of it. The term *gene selection* is appropriate in one sense, but it should be recalled that selection always operates on the phenotype and not on the genotype. The best way to
describe this phenomenon, therefore, is to see it as a development of the
concept of Darwinian fitness that is now extended to include not only
the reproductive success of an individual and its offspring but the reproductive success of all individuals in a social group that carry a particular
individual's genes. We shall follow the suggestion of Hamilton and refer
to the phenomenon as "inclusive fitness."

The concept is extremely important in understanding the behavior of
individuals in the kinds of social groups that are found among the
higher primates. We must conclude that natural selection produces individuals that are adapted to maximize the spread of their own genes
through both their own offspring and their relatives.

This limited kind of *bioaltruism*, which is effectively an extension of
maternal care, finds a sound genetic basis in the concept of inclusive
fitness. The theory can be extended to include a considerable range of
behavior, including behavior that has always been considered typically
and uniquely human and that would not appear to be based in traditional human genetics. Such behavior would include human altruism,
art, love at its highest, religion, and so on. E. O. Wilson, who has most
strikingly developed the concepts of sociobiology, has attempted to
apply them to human evolution and behavior in some detail (Wilson,
1978). The basis for Wilson's view lies in the fact of the ultimate genetic
control of human populations—a control that we cannot escape. He
points out that if the brain evolved by natural selection, and we believe
that it did, then its capacities to develop and select particular esthetic
judgments and religious beliefs must have arisen by the same mechanis-
tic process. They are either direct adaptations to past environments, or
at most constructions thrown up secondarily by deeper, less visible
activities that were once, long ago, adaptive in a strictly biological sense.
"The brain exists in its present form because it promotes the survival
and multiplication of the genes that direct its assembly. The human
mind is a device for survival and reproduction, and reason is just one of
its various techniques" (Wilson 1978, p. 2). As evolutionary biologists
we must accept this general statement, but we are entitled to ask
whether it precludes the possibility of a sizable amount of nonadaptive
behavior. We need to be clear as to whether all humankind's achievements are part of our biological adaptation, with no other goal, or
whether we are now in a position to develop behavior with another

(higher?) goal in view—a purpose beyond the imperatives created by our genetic history.

Wilson answers this question negatively. He writes: "Genes hold culture on a leash. The leash is very long, but inevitably values will be constrained in accordance with their effects on the human gene pool. . . . Human behavior—like the deepest capacities for emotional response which drive and guide it—is the circuitous technique by which human genetic material has been and will be kept intact. Morality has no other demonstrable ultimate function." (Wilson 1978, p. 167).

"The leash is very long." This is surely the critical statement. In fact we now know just how long it is. It is long enough for the human individual to choose to have no children, and for the species to destroy itself entirely, with nuclear weapons. If this were to occur, even in error, and the possibility exists, then we should have escaped the genetic leash completely and our brains would finally have destroyed our genes.

If this is even a possibility, and it is, it demonstrates that human behavior can indeed escape the biological constraints a biologist might predict would limit its wilder excesses. The possibility of self-destruction by a rational species suggests that the genetic leash is so long that humans can and often do act in a nonadaptive manner, and it suggests that they may well be capable of setting goals for themselves that are not adaptive in any biological sense and moving toward them.

Among animals, pleasure is a reward for biologically adaptive behavior (e.g., feeding, sex). Humans have discovered how to generate pleasure without necessarily producing adaptive behavior. Sex with the pill is one example, while listening to Mozart is another. Humans can also set goals for themselves (zero population growth, nirvana, and so on) that seem to fly in the face of biological imperatives. Perhaps they do not, perhaps they themselves are quite complex adaptations, and zero-population growth is a possible example of that. However, the conclusion we must face is that we are not driven to look for the adaptive value of every part of human behavior. While we may be satisfied to find it (for example, if homosexuality is a birth control mechanism that arises in conditions of overcrowding), we should not be surprised if we do not. The leash is long enough to permit the most startling freedom for humans, who, if they understand what they are doing, can, with their technology, cut the leash that binds them to their past.

This is possible because the species has evolved the capability of rational thought, which, though it gave great power of adaptation to a technological species, also brought an understanding and made possible a freedom of action that animals do not possess. The brain has given us freedom from the control of the lower brain centers, the limbic system (see Chapter 2, V), which makes animals do what they have to do. It has enabled us to escape from the prison of our biology, our past, and given us increasing freedom to explore ourselves and our world. This development, which has done as much as anything else to make humankind the extraordinary species that it is, has been possible because of the success of the human adaptation. Because we are many and widespread, we can afford, as a species, to take risks; we can afford to try out our newfound freedom, and risk our genes in living out our dreams. The ultimate mystery lies, perhaps, in the origin of those dreams: the new goals that give meaning to the lives of so many people.

VIII. Homeostasis of the Population

Just as the physiological and behavioral mechanisms that we have discussed help to maintain a steady state in each individual organism during its lifetime, so are there mechanisms that maintain the steady state of the whole population over long periods of time, keeping an equilibrium between it and the environment.

The environment is always oscillating as well as undergoing long-term change. At the same time, natural variation in individuals from generation to generation will tend to cause random changes in the average characteristics of a population. Living populations are homeostatic in the face of both these kinds of variation, insofar as they remain in a relatively steady state in spite of minor environmental and genetic variations. This phenomenon has been termed *genetic homeostasis* (Lerner, 1954). Such a mechanism will maintain an equilibrium between the population and the environment and will thus tend to ensure the survival of the population. Long-term environmental change, however, will result in a homeostatic readjustment (evolution) on the part of successive generations of the population.

For example, the fittest mean weight and size of a species in a more or less constant environment will be maintained in the face of individual variation. Natural selection will cut down the reproductive rate of the more extreme variants and will thus operate to regulate this characteristic (Fig. 1.5). On the other hand, a gross change in external conditions (such as greater food supply resulting from climatic change) may alter what is the fittest weight, and so a homeostatic adjustment will be made in the population to a new, increased mean weight and size. Such an adjustment constitutes evolution and fulfills the definition of evolution as *a systematic change in the genetic structure of the population* (Fig. 1.5). Such a change is termed an *adaptation*—an adaptation to changed environmental conditions. This, then, is evolution by natural selection: shifts in the mean of existing variable characteristics (adaptations), that help to maintain the continuing survival of the population. New mutations are not necessary for such changes, but existing variability is essential for change and so for survival.

There are rare circumstances in which the stabilizing pressure of natural selection fails and permits random variations in the genetic structure of the population to become fixed. This is called *genetic drift* (though it is random and nondirectional), and it may result in the appearance of

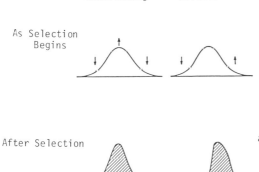

Stabilizing   Directional

As Selection
Begins

After Selection

Figure 1.5. Homeostatic action of natural selection.

neutral or even maladaptive characteristics in the phenotype. Drift may occur when a small population is isolated in a relatively rich environment. Under these conditions novel, but random, characteristics may survive for a time because selection pressures are low (see Section X).

We see, then, that at the level of both the individual phenotype and the population, homeostatic mechanisms are operating to maintain a steady state. In both cases the steady state is the key to survival, and the maintenance by evolution of such effective homeostatic systems is the unique property of living organisms.

We can also see that the individual phenotype is maintained by the mechanism of genetic homeostasis—i.e., by the properties of the genes, and of course the function of the phenotype is to ensure survival of the genes. As Dawkins (1976) put it: phenotypes are survival machines for genes.

Many different kinds of phenotype have in fact survived, and it is of interest to consider whether any one form of survival machine is in any way better or more progressive than another.

IX. Evolution and Progress

Evolution, as indicated by the foregoing discussion, is an adjustment in the mean characteristics of a population due to natural selection, which usually results in an increased likelihood of the population's survival in its particular environment. There is no force operating within individuals that causes evolution, nor are systematic changes in the genetic structure of a population necessarily "advances" on the *status quo;* they are changes which maintain that status. Some changes may prove retrospectively to result in more success than others, depending entirely on the definition of success in a particular context.

The idea of success and progress and the idea of evolution are not unrelated. The idea of progress was hardly in existence 300 years ago for, like evolution, it was, above all, a 19th-century concept. Evolution has formed humankind from the primeval ooze, so it is clearly progressive, at least from our viewpoint. But the view *sub specie aeternitatis* is very different. Here is a process, operating in time, which has produced a whole range of living organisms, from viruses to humans. While there has been, on the one hand, a trend toward size and complexity, the smaller and simpler forms of life have at the same time survived and evolved in great variety and numbers. The plant kingdom as a whole has achieved a dominant position without the evolution of a nervous system or musculature. Our greatest competitors in the world today are not other mammals but minute and relatively simple viruses, bacteria, and protozoa. As Inge has written:

> The progressive species have in many cases flourished for a while and then paid the supreme penalty. The living dreadnoughts of the Saurian age (the ruling reptiles) have left us their bones, but no progeny. But the microbes, one of which had the honour of killing Alexander the Great at the age of thirty-two, and so changing the whole course of history, survive and flourish (1922, p. 166).

Evolution as a process can hardly be described as progressive in the usual sense of that word; it is characterized by change (not necessarily involving improvement), the sole positive result of which is survival. It is in that case reasonable and perhaps more valuable to examine the process from the human viewpoint in order to discover whether the human species is in any sense progressive compared with the rest of the animal world. In what sense are we more progressive than other animals?

The success of a species could be measured in terms of its numbers, or the gross weight of its *biomass* (the total mass of all of its individual members); it could be judged in terms of its stability in time, its variety of adaptations, or its geographical range. We humans are in fact actually increasing our population and our total biomass faster than any other organism, yet that can hardly be called a sign of real progress, since we know it to be both undesirable and dangerous. Other species have had population explosions; so even if we excel in such characteristics, they do not explain our special evolutionary position. If we look at *Homo sapiens,* we see, for example, that the species is very complex (especially the nervous system) and self-conscious (which is probably unique in the organic world). It has also intentionally altered its environment. Again, we cannot be sure that it is necessarily progressive to be complex, or to be self-conscious, or to alter the environment. Any of those characteristics could prove disastrous to their possessor, and they may do so to humankind. But their combination does reflect one single evolutionary trend in which humans appear to be progressive: they have an improved homeostatic response to environmental change, or, as Herrick puts it, progressive evolutionary change is "increase in the range and variety of adjustments of the organism to its environment." (1946, p. 469)

Let us examine the mechanism of homeostasis further in order to discover how such a progressive improvement may have come about. A homeostatic system consists basically of three parts: (1) the receptors, which measure the status of critical variables in the internal or external environment (for example, the *hypothalamus* at the base of the brain senses the temperature of the blood, as described in Chapter 2); (2) the controller (regulator or mediating mechanism), which acts when the signal from the receptor indicates a need for an adjustment (again, in our example, it is the hypothalamus); and (3) the effector, which the controller uses to bring about correction of a deficiency or excess (in our example, the effectors are various and cause shivering, sweating, etc., as described in Chapter 2). Their interaction can be summarized as shown in Fig. 1.6.

Receptors that sense the status of the internal environment are of course internal, and they are usually found in the brain itself. The survival value of response to changes in the external environment (for example, absence of food) has resulted in the evolution of the sense receptors (which in our example can detect food chemically or visually). The evolution of such receptors makes possible response to the external environment, but by themselves the sense organs may not be sufficiently accurate as generators of input to allow very precise homeostatic adjustment. For example, the very simple eyespot of certain single-celled animals is able to detect the direction of light sources, but it required not just the eye but the vertebrate visual system as a whole to recognize distinct shapes within the overall visual pattern of the environment. Thus, the evolution of the sense organs must be accompanied by the evolution of analytic mechanisms in the brain that can differentiate between different intensities and patterns of stimulus. The evolution of such a refined input mechanism has been accompanied by the evolution of more sophisticated controllers and effectors that generate a wide range of behavioral response.

With the early mammals we find not only the full range of sense organs that we ourselves have, which respond to changes in the external environment, but also the full range of internal homeostatic mechanisms already mentioned (see Section V) and described by Cannon (1932). But during the more recent stages of human evolution, during the primate stage, we find the development of a general-purpose controller—the cerebral cortex of the brain—and a general-purpose effector—the versatile trunk and limbs—which make possible a wide range of behavior. In this study of human evolution we shall place great emphasis on this new kind of homeostatic mechanism, which is less specific and more generalized and versatile. It takes the outward form of learned and intelligent behavior.

Thus, we shall see in primate and human evolution that increased

Figure 1.6. Diagrammatic illustration of a homeostatic circuit.

homeostatic response is based on (1) broad sense reception with very detailed analysis of input, (2) an advanced general-purpose computer and controller, the cerebral cortex, and (3) much increased versatility of the effectors, the nerves, muscles, and bones by which the limbs and body move.

It is in this respect that humans can claim to be more progressive than other animals: our sensory inputs supply far more data about the environment that do those of other animals, and our effector organs make possible a greater range of behavior. The evolution of the large brain is the necessary concomitant of this extended interaction with the environment.

How do we reconcile the recognition of progress in human evolution with our earlier definition of evolution as no more than adaptive change? The answer is that evolution involves two kinds of adaptation. The first and simplest is an adjustment by existing homeostatic mechanisms (different mean weight, different salt concentration, different coloration). The second, and the more complex, is an alteration in the functional homeostatic systems themselves. For example, in the evolution of many internal parasites, such as the tapeworm, various functional homeostatic mechanisms have been lost because the present external environment (the body of another animal) is more constant than that in which the tapeworm originally evolved as a free-living organism. In this example evolution is not progressive, but the animal has become very highly specialized to exist in a very limited environment.

On the other hand, the evolution of the temperature-regulating mechanism in primitive mammals was something altogether new and certainly progressive, since it gave mammals improved homeostasis and a greater independence from environmental change. Since the appearance of mammals, further improvement in homeostasis has arisen from behavioral rather than physiological mechanisms. Man's body in its structure and function is very similar to that of other mammals; his difference from them lies in his behavior and the mechanisms determining it. Such differences are far-reaching and constitute a progressive change in an evolutionary lineage.

X. Environmental Change and Evolution Rate

Environmental change is the prime determinant of homeostatic adjustment in the gene pool of organic populations. Environmental change may widely vary in form, but its ultimate cause is change in the earth's climate. A long-term climatic change in a geographical region directly affects every single living organism within that region. In addition, every organism will be indirectly affected by the change in every other organism that the change in climate causes. For example, a steady lowering of rainfall in a particular region will change the vegetation from plants adapted to a relatively high rainfall to plants adapted to a relatively low rainfall. Such a change will immediately affect all herbivorous animals, which may be unable to find enough of their usual food plants. If they survive, such animals will have slowly adapted their behavior and digestive processes to the new food plants that become dominant. In turn, their carnivorous predators will also be affected. Change in rainfall may also, for example, affect the transmission and occurrence of parasites, for they may depend on damp conditions to survive the period of transmission between their host animals or plants. Drinking water will perhaps become scarce, and animals may survive only by an adaptation in their water-storage capacity so that they require only one visit to a water hole per day. Ultimately, the climate may become subject to occasional extremes of drought, which may be responsible for the extinction of certain species—herbivores, carnivores, or their parasites.

It is clear that a single change in one component of a community of plants and animals may have very complex and far-reaching effects on the whole nature of the environment. Homeostatic adjustment among all the different populations of organisms may be a lengthy process, for every change in one species will affect the environment and in turn effect change in every other. In stating, therefore, that the ultimate cause of environmental change is a change in climate, we refer to the whole causal chain of adjustments that result from a change in the prime determinants of the earth's environment: the level of solar radiation, the nature of the earth's crust and atmosphere, and the earth's rotation.

Evolution, which is the resultant of genetic variation and environmental change, can therefore occur in the following ways:

1. In a more or less constant environment, a population may evolve a better adaptation to it. Such evolution may, however, be limited in extent.

2. From such a more or less constant environment, a population (usually part of an existing population) may expand into a neighboring and different environment to which it may prove to be already adapted to some extent. New adaptations may then follow.

3. The environment may change, and the population may adapt to the change. That is of course the most common situation; it results in situation 1 and very often causes the changes in distribution that result in situation 2. It is, however, worth noting that environmental change may not necessarily result in evolution, since, if the change takes the form of movement of climatic belts on the earth's surface, the fauna may move with the belts, so no new adaptation is necessary. Such a situation would be particularly likely in animals like monkeys or the herbivores of the plains, which are highly mobile and would move within shifting climatic belts without any modification in their way of life.

It has been claimed that a change in gene frequency resulting not from selection by the environment but from random genetic variation in small populations temporarily relieved of selection pressures, known as genetic drift (Section VIII), can result in significant evolutionary change. Evidence that this phenomenon of drift has been a significant factor in primate or human evolution is not at present impressive. It may, however, have played an important part in the evolution of other groups of plants and animals. The matter does not require further discussion here (see Mayr, 1963). This does not mean, however, that we understand the selective factors that have caused the evolution of every characteristic of living species.

Although Darwin did not have the knowledge of genetics that we have today, he did visualize the process of evolutionary change to be gradual and the product of a succession of small changes in the germplasm being subjected to natural selection. It was not until the turn of the century that the Dutch botanist de Vries drew attention to the importance of major mutations in the evolution of new species. In his studies of the evening primrose, *Oenothera*, he observed such mutations and recorded the appearance (within a few generations) of many entirely novel varieties and species, which bred true.

De Vries believed that such mutations must be the basis of evolutionary change, and that natural selection operating on minor variations could never be sufficient to bring about the varied evolutionary radiations of plants and animals that today cover the earth.

Since then our knowledge of the evolutionary process has grown enormously. The modern version of Darwin's original ideas, the *Synthetic Theory*, which incorporates modern genetics, recognizes both gradualism in evolutionary change and the possibility of episodes of very rapid evolution. The hypothesis that small isolated populations, in particular, may undergo rapid change has been part of the synthetic theory for 50 years.

This is important because a lively controversy exists today between those who believe that the synthetic theory still represents an adequate, reasoned explanation for the process of evolution, and those who believe that changes are needed in it and that a much greater weight must be given to the phenomenon of rapid change interspersed by periods of adaptive stability (stasis). This concept of *punctuated equilibrium* has been developed as constituting a central process in organic evolution by some contemporary scholars (e.g., Gould and Eldredge, 1977; Stanley, 1979; Gould, 1982). They believe that steady, gradual change is a rare phenomenon, and imply that the synthetic theory as it stands cannot account for very rapid change.

It is the view of this writer, and probably still the consensus among biologists, that there is no need to modify the synthetic theory or to postulate major and highly improbable adaptive mutations to account for periods of rapid change. Simpson discussed what he termed *quantum evolution* in a number of books (e.g., 1944, 1953). He described the means by which natural selection can bring about a very rapid phase of evolution between two adaptive zones that are more or less discontinuous. The evidence for such events is well manifested in the evolution of many groups. Such a quantum change may have occurred at a number of points in human evolution, but most probably between the earlier adaptation to the forest woodland zone and the later adaptation to the savanna. It is clear, however, that the synthetic theory as generally presented does allow for apparent discontinuities in the fossil record, which represent periods of rapid evolutionary change; and that this rapid change can be understood without modification of the synthetic theory. We can conclude that periods of gradual change and periods of rapid change may have often occurred and that the latter may have played an important part in the evolutionary process. We may find that a lineage passes from a period of almost complete stasis through gradual change of a quantum jump, or through any succession of these differing rates. A very detailed study of a complete and very well-dated fossil record would, however, be required to identify each part of the process.

The rate of evolution is ultimately related to its two causative factors: genetic variation and environmental change. Environmental change is the prime determinant of the evolution rate, but in times of fast change the maximum rate will depend upon the amount of genetic variation available from mutation and from the sexual shuffling of genes. How is the mutation rate itself determined? It varies considerably among different animals and plants, and it seems most probable that the rate is selected in evolution like any other characteristic. A fast rate will allow fast evolution, yet will produce a higher ratio of ill-adapted members in a stable population. On the other hand, in times of rapid environmental change a slow rate might block sufficiently fast adaptation to permit survival. Like every other characteristic, the mutation rate of a population is adapted to the environment, and in particular to its overall rate of change. Genetic stability and variability, as we have seen, are important features of biological fitness.

**XI.
Specialization
and
Pedomorphosis**

The extent to which a population is committed to a particular environment is termed its degree of *specialization*. As we shall see, some primates are more highly adapted to arboreal life, are more specialized in this respect than others. Thus, their degree of specialization can be seen as their degree of irreversible commitment to their environment: highly specialized species may not survive environmental changes that less specialized species might. Specialization in fact seems to reduce the possibility of further evolution, and, in the long view, highly specialized groups (such as the tapeworm) can be seen to have been all too often doomed to ultimate extinction when gross changes in the earth's environment have occurred. From the point of view of survival, therefore, a species must remain fairly *generalized* in its overall character; extreme specialization may ultimately involve extinction—and indeed it usually has.

There is, however, one remarkable way out of the narrow corner of overspecialization, and from our position of hindsight we can see that it has occurred on a number of occasions. The process, called *progenesis*, can effect despecialization and operates somewhat as follows: we can suppose, as we have seen (see Section III), that the adult form of an individual is determined by two sets of genes; structural genes determine the adult characteristics themselves, and regulatory genes determine the rate of growth and time of onset of sexual maturity. Since specializations must be present in the adult stage, they usually appear during growth, and their structure at all stages is determined by the first set of genes. The second set of genes, however, establishes the permanent adult form of the individual by their power to time the onset of sexual maturity. By speeding up the onset of sexual maturity, the adult form may become established and growth may stop before all the specializations can be fully developed. The animal becomes sexually mature when in the young stage; the result is a very remarkable evolutionary change, which is due to a very simple alteration in growth pattern. The phenomenon, recently discussed by Gould (1977), has been an important factor in evolution. Perhaps the most famous instance is the evolution of the chordates (the ancestral group of the vertebrates) from the echinoderms (the sea urchins and starfish). We have evidence that the primitive chordate evolved from a creature like the larva of a sea urchin that had become sexually mature. There is no doubt that the adult echinoderms are a very unusual and specialized group that could not possibly evolve into chordates; yet by the process of progenesis that specialization is believed to have been cast off, with the result that there appeared what amounted to an altogether different kind of animal of unspecialized form—the sexually mature echinoderm larva.

Thus, progenesis gives evolving groups a way out of the dead end of overspecialization, and, with limited mutations affecting mainly growth rate and time of maturity, it can result in a saving generalization of form.

A somewhat similar process, called *neoteny*, results in the retardation of somatic development of certain organs or structures. As might be expected, there is evidence that such a development is a frequent evolutionary adaptation and that it has played a part in the evolution of *Homo sapiens*. While the process in no sense accounts for human evolution as a whole, there is no doubt that many characteristics considered typically human are present in the early stages of the development of other primates and that they have been retained in humans by retardation in growth rate of certain parts of the body. For example, the limited body hair of *Homo sapiens* and the fair skin of certain races may be the result of the arrested development of the hair follicles and melanin cells, respectively. A list of human neotenous characteristics is given by Montagu (1962), who discusses this interesting phenomenon. The results of both progenesis and neoteny are termed *pedomorphosis*—the retention into adult life of juvenile characteristics.

All evolution is due to developmental and genetic changes, and the phenomenon of pedomorphosis is of particular interest because it is a shortcut to drastic change through change in growth rate, without change in structural genes. Thus, it may have come about that although we are in many ways profoundly different from chimpanzees, we nevertheless share with them some 99% of our DNA coding. That 1% in which we differ includes, no doubt, not only structural genes but many regulatory genes that control rates of development of the components of the body.

It is also likely that mutations affecting growth rate and the timing of the onset of sexual maturity are less likely to encounter the effect of internal selection (see Section III) than are those affecting structure, because the former threaten less fundamentally the harmony of the whole organism. Though the phenomena of progenesis and neoteny both give evolving species a way out of the trap of overspecialization, they do not require us to change our ideas about the action of natural selection and the way that it brings about evolutionary change.

XII. Organism and Evolution— A Summary

The organism can be described as a self-regulating, self-reproducing open system that passes vital information (coded in DNA molecules) to copies of itself. It can be considered to be of two parts: the genotype, which consists of the informational nuclear molecules of DNA within each cell of the individual, and the phenotype, which is the whole individual formed by the genotype through its interaction with the environment.

The genotype interacts with the environment through a causal chain of which the phenotype and its behavior are the connecting links. This interaction maintains the homeostatic equilibria of the individual and population that is necessary for survival. The individual phenotype system maintains the physiological and developmental equilibrium of the internal environment for the preservation of the genotype.

Reproduction by the fusion of gametes from different individuals of two sexes creates over a period of time a network of sexual relations between individuals that constitutes the interbreeding population. In the formation of the gametes, variation arises in the information store of the DNA, and thus gives variability to the population and makes possible changes in it. Both the extent of variation at a given moment (which is ultimately generated by mutation) and the long-term alterations in the gene pool (which constitute evolution) are controlled by natural selection, which is the process whereby the reproductive capacity of individuals not well adapted to the environment is depleted by environmental interference. The change in the population brought about by changes in the environment (which constitute selection pressure) can be described as adjustments made in the population that serve to maintain a homeostatic equilibrium between it and the environment. Because living processes are so unstable and delicate, such equilibria must be maintained at all times for the survival of life.

Variability between populations in different geographical regions may lead to speciation if gene flow is interrupted by geographical barriers for a certain period of time. Isolation allows for the evolution of morphological and behavioral divergence, which may eventually replace the geographical barriers as an effective isolating mechanism so that genetically discrete species are evolved.

Progress in human evolution is seen as the appearance of new homeostatic mechanisms that make possible survival in the face of ever greater environmental variation. The mechanisms themselves increase in turn the complexity of the genotype and phenotype. Progressive evolution in any but an anthropocentric sense cannot be recognized. Evolution is merely a homeostatic adjustment in the mean or average characteristics of a living system in response to environmental change and can be the outcome only of such adjustment and nothing more. By a study of the anatomy, physiology, behavior, and environment of living organisms related to ourselves, it is possible to discuss how human nature was formed and to attempt to understand how each characteristic of our species serves its survival.

Suggestions for Further Reading

There is a vast literature on all aspects of evolution, although the concept of homeostasis has not been greatly stressed until recently. For general reading see G. G. Simpson, *The major features of evolution* (New York: Columbia Univ. Press, 1953) and S. M. Stanley, *Macroevolution: Pattern and process* (San Francisco, Calif.: Freeman, 1979). For a more technical treatment see E. Mayr, *Animal species and evolution* (Cambridge: Harvard Univ. Press; London: Oxford University Press, 1963). For a treatment of genetics and the general biology of human evolution see T. Dobzhansky, *Evolution, genetics and man* (New York: Wiley; London: Champman and Hall, 1955), and *Mankind evolving* (New Haven, Conn., and London: Yale Univ. Press, 1962). For a treatment of classification and nomenclature see G. G. Simpson, *Principles of animal taxonomy* (New York: Columbia Univ. Press, 1961). For a text on the relationship of social behavior and evolution see E. O. Wilson, *Sociobiology: The new synthesis* (Cambridge: Harvard Univ. Press, 1975).

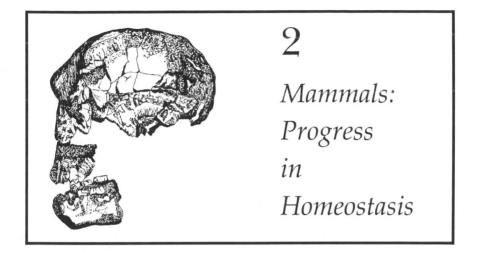

2

Mammals: Progress in Homeostasis

I. The Origin and Evolution of Mammals

In tracing the origin of humankind it is not our concern here to describe in detail the part of our evolutionary history that we share with other mammals. Our interest lies in the most recent stage of human evolution, in which our own peculiar nature was evolved. It would, however, be a serious omission to neglect to mention the features that, although shared with other mammals, are the essential basis for the evolution of an organism as complex as a human being. Human evolution cannot be fully understood without some consideration of the main features of our zoological class, the Mammalia.

Mammals are separated by zoologists into three subclasses, the monotremes (such as the platypus), the marsupials (such as the kangaroo and opossum), and the placentals. The first two groups are limited today both in species and in range, and, although the marsupials had a period of dominance in the past, they were superseded by the placental mammals—the group to which we belong (see Table 1.1). Although what follows applies in part to the two primitive subclasses, we shall concern ourselves from now on with the great *radiation*—the evolutionary divergence and dispersal—of the placental mammals.

While the most important characteristics of mammals are poorly recorded in the fossil record, consisting as they do for the most part of soft tissues, the fact that certain characteristics are common to such a large and varied group of organisms suggests that they evolved in a common ancestral lineage. Paleontologists see the origin of the mammals from reptiles of the Triassic period; by Cretaceous times, the main orders of the placental mammals were established (Table 2.1 and Fig. 2.1).

Mammals are characterized by evolutionary novelties of the utmost

TABLE 2.1. Events in Human Evolution (Approximately Dated)

			Millions of years	
Stratigraphical phases of the earth's history		Extent	Beginning before present (BP)	Event
Quaternary	Pleistocene	1.6	1.6	Rise of humans
Tertiary	Pliocene	3.4	5	Rise of hominids
	Miocene	20	25	Rise of monkeys
	Oligocene	10	35	and apes
	Eocene	23	58	
	Paleocene	5	63	Rise of placental mammals including primates
Mesozoic	Cretaceous	72	135	First placental mammals; extinction of ruling reptiles
	Jurassic	46	181	First mammals, the pantotheres
Paleozoic	Triassic	49	230	
	Permian	50	280	
	Carboniferous	65	345	First reptiles
	Devonian	60	405	First amphibians
	Silurian	20	425	
	Ordovician	75	500	
	Cambrian	?100	?600	First fossilized animals

importance to the story of human evolution. The characteristics that relate mammals to each other and separate them from other groups, such as birds and reptiles, are very numerous indeed, yet they arise from the evolution of only four great complexes, all of a homeostatic nature. These complexes can be seen in turn to contribute to a single adaptive development, which made it possible for this most remarkable class of animals to succeed the mighty reptiles that had previously dominated the earth, yet which shared with the mammals a common origin. This all-important adaptation was the ability to maintain a steady level of activity in the face of basic changes in the external environment. From being dependent, like reptiles, upon the external temperature with its daily and seasonal rhythms, mammals became able to remain active at any time due to the evolution of new homeostatic mechanisms.

The four adaptive complexes that made this constant activity possible

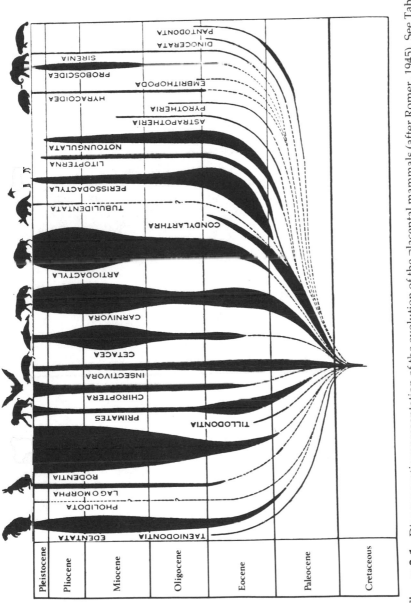

Figure 2.1. Diagrammatic representation of the evolution of the placental mammals (after Romer, 1945). See Table 2.1 for estimates of the duration of the geological periods.

will be considered in turn: homoiothermy, mastication, improved reproductive economy, and new ways of determining behavior.

**II.
Homoiothermy**

The chemical reactions that underlie homeostasis
and constitute the process of living, like any other
chemical reactions, can generally occur more
quickly at a higher than at a lower temperature,
but such a temperature should not be so high as to cause breakdown in
the relatively unstable organic molecules. At the same time, a constant
body temperature makes possible the regularity and reliability of the
highly complex chemical reactions that underlie the activity of higher
organisms. With the highly elaborate patterns of activity in the brain,
constant internal temperatures also make possible the very rapid and
precise control of motor activity and the storage of an extensive cerebral
memory. The evolution, dating perhaps from 150 million years ago, of
homoiothermy—the phenomenon of constant and appropriate body
temperature—enabled the early mammals to maintain a high level of
activity at times when reptiles and amphibians became sluggish (at night
and in cold weather). During the Cretaceous period an animal had for
the first time gained a tactical advantage over the great ruling reptiles of
that age, such as dinosaurs. At the same time, this constantly heated
animal could move north from the tropical regions to which terrestrial
life was then more or less limited and could exploit the vast forests of the
temperate zones.

The maintenance of body temperature, at first approximate but later
exact, involved a complex regulating mechanism under control of the
hypothalamus, a small structure at the base of the brain (see Fig. 8.16).
As has been explained (Chapter 1, V), the hypothalamus is the thermo-
stat of the mammalian body, and it brings into action the effectors of
temperature regulation when the blood departs from that set point to
which the species has become adapted. Temperature regulation involves physiological and behavioral adjustments of many kinds. To
lower the temperature of the body during hot weather or great exertion,
the usual rate of heat loss is increased as follows:

1. An increase in effective surface area is effected by spreading the
limbs (and the ears in some mammals); increasing the ratio of surface
area to volume increases heat loss.

2. Fur or hair, which evolved as an insulating layer, is laid flat upon
the skin to reduce its effective thickness.

3. Respiration rate increases to allow the loss of heat by the vaporiza-
tion of water in the lungs, throat, and mouth. Panting is most commonly
seen in dogs, though it is a human characteristic as well in times of great
overheating.

4. The *capillaries* (small blood vessels) under the skin dilate and so
increase the circulation of the blood near the surface; blood supply to the
internal organs is reduced. Heat loss through the skin is thus increased.

5. Sweat glands produce water on the skin, which vaporizes, thereby lowering skin temperature. Blood temperature is in turn lowered.

To raise the body temperature, the usual rate of heat loss is decreased as follows:

1. A reduction is made in effective surface area by curling up roughly into a ball; a sphere is the shape with the minimum ratio of surface area to volume.
2. The fur is raised to form a thick insulating layer.
3. The capillaries in the skin are constricted and reduce circulation near the surface of the body.

There are two positive reactions to a lowered body temperature:

1. The generation of body heat is increased by an increased rate of metabolic oxidation of sugar.
2. Involuntary muscular activity occurs in the form of shivering.

Thus, the maintenance of constant body temperature is made possible by the evolution of certain new organs common to mammals. The most important are the fur or hair and its *follicles*, with their erector muscles, which raise and lower the animals' coat, together with the sweat glands in the skin. Mammals are also the bearers of a layer of subcutaneous fat below their outer layer of skin, which helps to insulate them from the external environment and acts as a food store. The whole

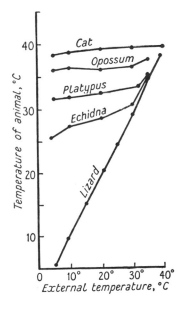

Figure 2.2. The evolution of homoiothermy is demonstrated in the diagram. Shown are changes in body temperature of different animals during changes in the external temperature. The body temperature of the lizard varies with the environment; the three primitive mammals have made some headway toward controlling body temperature, which is almost constant in the cat (from Martin, 1902).

thermostatic adjustment is effected by the novel functioning of the hypothalamus as a heat receptor. Both it and the eye (the light receptor) develop from the same tissues at the base of the brain.

Finally, mammals, like reptiles, have a behavioral response to temperature change in the skin, for they will seek out a warm place during cold weather and a cool, shady spot in the heat of the day. The effectiveness of the whole mechanism has increased during the evolution of mammals, and in Fig. 2.2 we can get some idea of this progress by noting the response of different animals to change in the external temperature. Although in some mammals and a few primates the regulation of body temperature is subject to activity rhythms (DeVore, 1965), this pattern is superimposed on the homoiothermic mechanism.

III. Heterodontism and Mastication

The maintenance of a constant body temperature requires a reliable source of food (fuel) to develop not only muscular power but also heat, which is carried around the body by both the deep and the superficial blood vessels. Mammals therefore need constant nourishment not altogether dependent on season. The evolution of a new kind of jaw function and tooth structure (*dentition*) made possible a more complete and effective exploitation of the environment's resources.

The mammal's *heterodont* dentition, in which different teeth have different shapes and functions, arose from the reptilian *homodont* dentition, in which the teeth were similar in shape and function (Fig. 2.3). The reptile jaw and homodont dentition acted mainly as a trap, and food was swallowed whole, as is commonly seen in snakes and alligators. The teeth were often incurved and served only to grasp the prey or, in the case of reptilian herbivores, to tear the foliage. In contrast, the mammal jaw and heterodont dentition achieved much more. The jaw was shortened and strengthened and different kinds of teeth evolved specially for chewing, grinding, and cutting, with the result that the entrance to the alimentary canal was functionally a more useful organ than a mere food trap. In particular, more kinds of food could be exploited, and the effectiveness of the digestive juices was increased by mastication of the food (as well as by the higher body temperature). A new digestive enzyme (*ptyalin*) was introduced to the food in the mouth as saliva, so that starch digestion could begin during mastication, even before the food was swallowed. The heterodont dentition has evolved within the different orders of mammals in a remarkable variety of ways, and not least in the primates, though in that group the more extreme specializations (such as tusks) have been avoided. A reversal to a homodont condition can be seen in toothed whales, while in other members of the Cetacea, such as the whalebone whales, the teeth have been lost altogether.

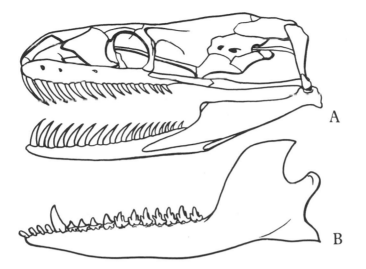

Figure 2.3. Skull of python (a reptile) (*A*) and lower jawbone of primitive Jurassic mammal (*Amphitherium*) (*B*) drawn the same size. Note that in the python all teeth are identical in shape and function, whereas in the mammal the teeth are differentiated into incisors, canines, premolars, and molars (Python after Romer, 1956; *Amphitherium* after Simpson, 1935).

The evolution of mastication and the resulting heterodontism (to be considered further in Chapter 9) is of great importance to mammalian evolution. Because of it, mammals were enabled to exploit food resources that were not available to their reptile predecessors. The teeth of mammals were able to release nutriment from food that the digestive juices alone could not penetrate. Just as the gizzard performed this function among birds, the heterodont dentition released to mammals the full food value locked up within the hard outer skeleton of insects and in plant-food storage organs (such as nuts and other seeds). The densely packed carbohydrates and fats of seeds, the sugars and starches of underground tubers and roots, and all the plant proteins locked up in tough vegetation were released to mammals by the chewing and grinding action of their teeth, thus making possible the maintenance of constant body temperature and more or less continuous activity.

Another development, of less importance, that affected the dentition was the loss of continuous tooth replacement and its evolution into the appearance of only two successive series of teeth during the growth of the individual: the so-called "milk" teeth and the later permanent dentition. Changes in diet and tooth action probably reduced tooth loss, which in turn removed the need for perpetual tooth replacement. The milk teeth or deciduous dentition that develop during the period of breast feeding make mastication possible even when the jaws are still small and unable to carry the permanent dentition. The permanent dentition erupts during the second half of the growth period and forms an extremely strong and powerful masticatory apparatus.

IV.
Reproductive
Economy

A. Prenatal. The early reptiles had become fully adapted land animals by the evolution of internal fertilization by copulation, thus removing dependence on fresh or saltwater for the external fertilization of their eggs. The reptiles also evolved the *amniote* egg (an egg for dry land, sealed against dehydration by a shell and membranes), which allowed the growing embryo to develop in its own little aqueous environment. Reptiles, however, maintained their numbers, like more primitive animals, by the mass production of eggs rather than by the successful establishment of small numbers of offspring. Their numerous eggs were usually hidden and left to hatch alone, and the young animals had to fend for themselves from hatching. Both eggs and young were easy prey for small carnivores, especially the little mammals that were then evolving.

These evolving mammals developed a new kind of reproductive system, which, by cutting down losses of eggs and young, improved their overall reproductive economy. It consisted of two basic developments, each involving a whole range of new physiological and anatomical characteristics. The first was *viviparity* (which has occurred as a parallel adaptation in some other classes)—the retention and nourishment of the fertilized egg within the mother until a relatively advanced stage of growth. The second, to be considered in the following section, was the evolution of parental care of the young after birth, consisting of nourishment and protection.

While such behavior as feeding affects the future of every individual separately, the reproductive activity of animals affects future generations rather than the fate of the individuals involved. In this sense, reproductive mechanisms are above all the property of the species rather than the individual. As a result, reproductive behavior tends to be innate and stereotyped and much less flexible, for too much variation in reproductive behavior might endanger the survival of the species. The first requirement of biological fitness is the maintenance of the reproductive rate by regular and successful reproductive activity.

Partly for historical biological reasons and partly for the reasons discussed above, we find that the reproductive processes of mammals are primarily under control of the *endocrine system* rather than the nervous system. The endocrine system evolved early as a method of communication between the cells and organs of an animal's body. It operates by means of *hormones*. Hormones, sometimes described as "chemical messengers," are chemical substances secreted by one group of cells and transported in the bloodstream and by diffusion around the body. According to their chemical structure, they evoke a specific response in another group of cells, tuned, as it were, to receive them. In mammals the endocrine system operates in cooperation with the nervous system: the former tends to effect response in long-term life processes, such as the rate of basic metabolism and the seasonal activity of the reproductive system; the latter effects more immediate changes in the body, such as day-to-day behavioral responses.

The part of the endocrine system that controls seasonal egg-laying and copulation in reptiles and amphibians has been expanded in mammals to control a regular cycle of sexual activity throughout the mating season and, in some species, throughout the whole year. Known as the *estrous cycle*, it arises with the evolution of viviparity and is part of the mammal's reproductive adaptations. It results from the need to follow up the growth and production of the egg or eggs (*ovulation*) within the *follicles* (egg-producing bodies) of one or both *ovaries* with changes in the walls of the ducts through which they pass (the *oviducts*). In the higher mammals these ducts are extensively modified and become the *fallopian tubes* and *uterus* (Fig. 2.4). These changes allow the *implantation* of the egg or eggs after their fertilization. Implantation is the means by which the developing egg becomes attached to the specially modified wall of the uterus—the *endometrium*; the egg derives nourishment for its growth from the maternal blood vessels within the endometrium in an almost parasitic fashion. If implantation occurs successfully, it constitutes the beginning of pregnancy or *gestation*. If not, the changes in the wall of the uterus are arrested, and new eggs develop in the ovary, later to be shed in turn into the uterus. The complete estrous cycle can be observed only in the absence of fertilization, for implantation arrests the cycle. The complete cycle is as follows:

1 Egg develops in an ovarian follicle.
2 Egg is shed, passes into fallopian tube and into the uterus.
3 Endometrial lining of uterus is prepared for implantation.
4 If no implantation occurs, endometrial lining of uterus returns to former condition.
2.1 New egg develops in an ovarian follicle.

Since in most mammals the egg remains in the female for only about three days after ovulation, fertilization, if it is to occur, must take place during this period. Sexual interest of the male, culminating in copulation, is induced at this time by the particular physiological changes in the female known as *estrus*, commonly called "heat." (The word estrus comes from the Greek word for gadfly or its sting, which sends an individual into a frenzied condition.) By means of scent signals emitted during estrus (and occasionally visual signals), the female attracts the male to copulate. It is characteristic of mammals generally that only during the period of estrus, and at no other time in the estrous cycle, is the female receptive or the male sexually stimulated.

The estrous cycle is controlled by a system of hormones produced by

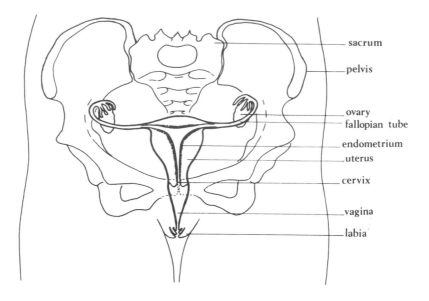

Figure 2.4. The oviducts in the female mammal have been modified to supply nourishment to the developing embryo. The arrangement of the reproductive organs of a woman is shown relative to the pelvic girdle; the front of the
pelvis (the pubic region) is cut away where it would obscure the cervix. The
arrangement is not strikingly different in other higher primates.

the *pituitary body,* a small endocrine gland that lies close under the base
of the brain (see Fig. 8.16). In spite of intensive research, it is still not
fully understood how the whole system of the estrous cycle is maintained, but at least part of the mechanism is known. Probably under the
control of the hypothalamus (see later, Fig. 8.11), the pituitary, during
each estrous cycle, produces two hormones successively, which have
been termed the *follicle-stimulating hormone (FSH)* and the *luteinizing hormone (LH).* Their production is shown graphically in Fig. 2.5A. Following
secretion into the bloodstream, the two hormones stimulate, respectively, the growth and production of one or more eggs in the follicles of one
or both ovaries (FSH) and what has been called the "yellow body," or
corpus luteum (LH), which develops in the ovary in place of the egg after
it has been shed (Fig. 2.5B). The developing ovarian follicle and the
corpus luteum each produce hormones when they are activated by the
pituitary: the well-known female sex hormones *estrogen* and *progesterone*
(Fig. 2.5C). Estrogen induces estrus by activating the scent glands and
causing enlargement of the external female genitalia; progesterone initiates the special changes in the endometrium that prepare it for implantation.

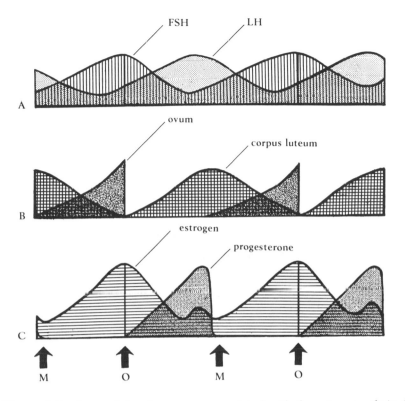

Figure 2.5. Some of the phenomena associated with the estrous cycle in the absence of fertilization and implantation: *A,* pituitary hormone levels; *B,* development of ovarian products; *C,* ovarian sex hormones; *M,* menstruation, *O,* ovulation and peak of estrus. *FSH,* follicle-stimulating hormone; *LH,* luteinizing hormone (adapted from Mason, 1960).

If fertilization and implantation take place, a hormone produced by the developing zygote arrests the cyclic activity of the pituitary; the production of progesterone continues and in turn brings about all the accessory changes associated with pregnancy. If fertilization or implantation do not occur, the pituitary produces a new secretion of FSH, which causes the development of new ovarian follicles and new eggs. The corpus luteum in this case degenerates, and the progesterone level falls. *Menstruation* is the loss of blood seen in certain mammals when the preparatory changes in the endometrium are reversed at the end of the cycle, and the spongy, blood-filled tissues that have developed are shed. [The estrous cycle of mammals has been described by Eckstein and Zuckerman (1956) and Mason (1960). It is shown diagrammatically in Fig. 2.6.]

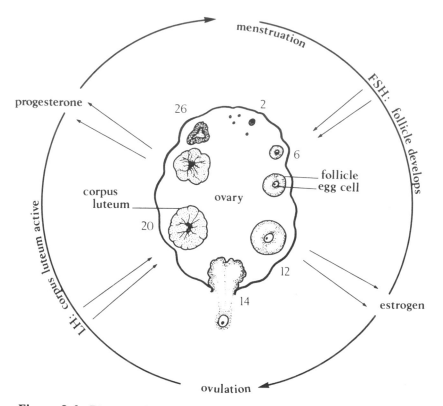

Figure 2.6. Diagram showing the changes in the ovary that take place during the estrous cycle. The cycle in higher primates lasts approximately 28 days, and the numbers shown around the ovary give some indication of the timing of the events, counting from the first day of menstruation, when the follicle starts to grow. Menstruation has been recorded as characteristic only of the Old World monkeys, apes, and man, although it is known in a few other primates.

This remarkable series of changes in the ovary and uterus is brought about by the activity of the pituitary hormones. While egg production in lower vertebrates is also under simple endocrine control, mammals have evolved the additional functions of the luteinizing hormone, the corpus luteum, and the endometrium, together with all the other changes that occur in pregnancy. Mammalian viviparity also involves other changes in the mother, some of which are also under the endocrine control of the pituitary gland. For example, the hormone *oxytocin* contracts the muscular wall of the uterus at the end of pregnancy, and together with *prolactin* promotes milk secretion after birth. We see, then, that the evolution of mammalian viviparity involves the evolution of new endocrine activity, as well as anatomical modifications such as those seen in the oviduct.

Correlates of these evolutionary developments are the adaptations of the amniotic membranes of the *fetus* (the developing young in the womb) to form a *placenta*, a blood-filled organ that penetrates the endometrium and, by diffusion, receives nourishment for the growing fetus from the maternal bloodstream (see Chapter 10, I). Waste products from the fetus also pass into the maternal bloodstream by this route.

The intimate, almost parasitic, association of mother and fetus typical of placental mammals means that the egg is fully protected during its development and that the young stand a far greater chance of survival than do the young of reptiles, which are left to the mercy of predators. However, the prenatal relationship between mother and fetus is hardly more important as a means of reproductive economy than is the postnatal relationship between mother and young.

B. Postnatal Parental Care.   Care of the newborn by one or both parents is not peculiar to mammals but has evolved extensively among birds and is found occasionally among reptiles and a few invertebrates. What is unique to mammals is that the "parasitic" mode of life of the offspring is continued after birth, in that offspring are fed by the special milk glands (*mammae*) of the mother, which are subcutaneous in origin and probably evolved from sweat glands (another mammalian novelty). Milk contains all the substances necessary for the growth and development of the young (including antibodies) (Table 2.2) and may continue to be consumed for long periods (the record for all mammals is probably four to five years among Alaskan Eskimos). The relative safety of the postnatal environment, in which the young mammal remains close to its mother, is also important in the reduction of mortality and in allowing, secondarily, a slower growth rate in some orders.

As a correlate of this lowered mortality rate, we may note a reduction in the number of eggs produced and fertilized by an individual female, from some millions in fishes to dozens in reptiles and no more than one per year among humans. This is the general spectrum of life histories— of reproductive strategies—designated by MacArthur and Wilson (1967) to range from the "*r*" strategy, in which the reproductive rate and number of offspring is maximized, to the "*K*" strategy, in which environments are exploited to their maximum carrying capacity, in which parental care is maximized and the number of offspring minimized. MacArthur and Wilson suggested that in an environment with abundant food and little crowding, selection would favor "opportunistic" species— those animals which, on entering a new environment, harvested the most food and rapidly raised the largest number of offspring. On the other hand, where food was limited, selection would favor those animals which could replace themselves with the minimum of food requirements. Thus, in the *K*-strategy, generated by *K-selection*, efficiency for the conversion of food into offspring is selected, because they will reach the carrying capacity of the environment more quickly. *K*-selection

TABLE 2.2. Growth Rate and Milk Protein[a]

Animal	Days to double birth weight	Protein in milk (parts per 100)
Rabbit	6	10.4
Dog	8	8.3
Sheep	10	7.0
Pig	18	6.9
Cow	47	4.0
Horse	60	2.3
Man	180	1.9

[a]After Sivertsen, 1941.

is an adaptation to stable environments with consistent food supplies; more seasonal and unstable environments would favor *r-selection*. The tropical rain forest is one of the most stable environments on earth, with its enormous diversity of species; predictably it harbors many large *K*-selected mammals up to its carrying capacity; all primates fall into this group. It is expressed most strikingly among the higher primates—the monkeys, apes, and humans. It is correlated with large body size, large relative brain size, a longer gestation period, a longer life span, and slower growth. The strategy can be summarized as a bid for quality rather than quantity of offspring (Table 2.3).

It is interesting to note that in human ontogeny we find a record of a change from a high reproductive rate to a *K*-strategy. The 5-month-old human female embryo contains two million egg follicles; at birth, less than one million (Soules and Bremer, 1982). An adult, unfertilized female may produce no more than about 400 eggs during her lifetime, though she continues to carry thousands (Table 2.4).

We shall see (in Chapter 10, VII) that the exceptionally slow growth rate is a characteristic of great importance in human evolution, and it is interesting that the quantity of protein in mother's milk (the food most essential for growth) is precisely related to the growth rate of the infant. Table 2.2, which indicates the early growth rate of different mammals, also shows how the protein content of the milk is correlated with this feature. In addition, the female primate carries only two teats rather than the large number found in *r*-strategy mammals such as mice and cats. She will suckle the young with great frequency throughout the day and night, and for a relatively long period of time, correlated with the slower growth rate.

The intimate relationship between generations, which results from suckling and caring for the young, makes possible the transmission of learned behavior by observation and imitation. Behavior patterns learned by trial and error have been recognized among lower vertebrates and invertebrates, but the simplicity and origin of such behavior puts it into a different class from that of mammals. It is the transmission of

TABLE 2.3. Some of the Correlates of r Selection and K Selection[a]

Correlate	r Selection	K Selection
Climate	Variable and/or unpredictable: uncertain	Fairly constant and/or predictable: more certain
Mortality	Often catastrophic, nondirected, density-independent	More directed, density-dependent
Population size	Variable in time, nonequilibrium, usually well below carrying capacity of environment; unsaturated communities or portions thereof, ecological vacuums, recolonization each year	Fairly constant in time, equilibrium, at or near carrying capacity of the environment; saturated communities; no recolonization necessary
Intraspecific and interspecific competition	Variable, often lax	Usually keen
Attributes favored by selection	1. Rapid development 2. High reproduction rate 3. Early reproduction 4. Small body size 5. Single reproduction	1. Slower development, greater competitive ability 2. Lower resource thresholds 3. Delayed reproduction 4. Larger body size 5. Repeated reproductions
Length of life	Short, usually less than 1 year	Longer, usually more than 1 year
Emphasis in energy utilization	Productivity	Efficiency
Colonizing ability	Large	Small
Social behavior	Weak, mostly schools, herds, aggregations	Frequently well developed

[a]Modified from Pianka, 1970.

complex learned behavior by means of imitation that makes it unnecessary to learn solely from direct experience—a dangerous process (see Chapter 1, VI). Instead, among mammals the experience of generations can be assimilated in a short time with reduced danger to the young. Thus, mammals have the great advantage (shared to a minor extent by birds) of being able, as it were, to "inherit" (by observational learning from parents) "acquired" (i.e., learned) behavioral characters.

TABLE 2.4 Reduction in Egg-Bearing Follicles during the Human Life Span[a]

5-Month embryo	2 million
At birth	< 1 million
At menarche	400,000
25 years	160,000
35 years	60,000
45 years	35,000
Postmenopause	Very few

[a]Data from Soules and Bremer, 1982.

The inheritance of learned behavior patterns makes possible a high rate of behavioral evolution. Thus, the behavior patterns of mammals have increased in complexity in such a way as to put them at a different level from those of all other living organisms. Their learned behavior repertory is immense, complex, and flexible (because it is still subject to trial-and-error learning). It makes possible the maintenance of the population in the face of even extensive and sudden environmental changes.

V. Determinants of Behavior

A. Emotions. Possibly the most important developments that occurred in the evolution of the early mammals lay in the brain. The mammalian brain has evolved far beyond that of other classes of vertebrate, and its characteristics and potential comprise probably the most significant single feature of the class. The human brain, the most complex organ in the organic realm, is a direct development of the modified early mammalian brain.

The mammalian brain is characterized most obviously by the development of the *cerebrum*: a new layer of brain cells arranged in two hemispheres, which come to overlie the primitive reptilian brain (Fig. 2.7). The first parts of the *cerebral hemispheres* to evolve among the early mammals, which now lie underneath its later and more highly evolved areas, are usually referred to as the *limbic lobes,* one on each side. From recent experimental work, it is now established that their primary function is the generation of emotional responses. Together with those lower brain centers with which the lobes are associated, they constitute the *limbic system* (McClean, 1980) and generate the emotions of fear, anxiety, anger, sexuality, and so on (Fig. 2.8). These we know are consciously experienced as intense feelings, and we understand them because we share these brain centers with all other mammals. In the face of a threat from a predator, for example, the limbic system generates impulses that make the mammal want to do what it needs to do for its survival and for the perpetuation of the species. Although reptiles do, of course, re-

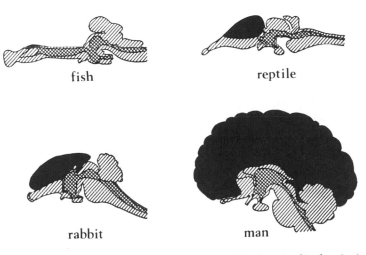

fish   reptile

rabbit   man

Figure 2.7. Sections (not to scale) through the longitudinal axis (termed "medial" or "sagittal") of the brain of a fish, reptile, rabbit, and human showing the difference in size of the cerebral hemispheres in relation to the older parts of the brain. For names of the different parts see Figs. 8.7 and 8.12 (from Magoon, 1960).

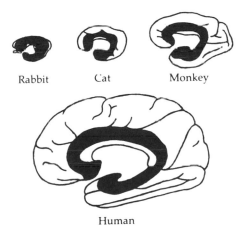

Rabbit   Cat   Monkey

Figure 2.8. Sagittal sections of the brain show the extent of the limbic system (black) in different mammals. (Drawn approximately to scale.)

Human

spond to their environment in appropriate ways, their range of responses is very limited, and they are not visibly accompanied by any signs of what we consider to be the hallmarks of emotion (rise in heart rate, increase in blood sugar level, feelings of anxiety, etc.).

The evolution of emotional responses is of profound importance to the life of mammals, and especially significant for primates and human beings. One of the most important emotions, which is a key to the evolution of parental care, is the response of separation anxiety found in both mother and infant when they are apart from each other. As a contrast to the bonding we see in mammals, the giant 10-foot lizard from Southeast Asia known as the Komodo Dragon provides an instructive example. As soon as the eggs hatch, the young are programmed to climb rapidly up the nearest trees, where they remain for several months. If they linger on the ground near their mother she may well catch and eat them, for she has no means of distinguishing them from the other small animals that normally constitute her prey (Auffenberg, 1972).

Mammals have effectively overcome this problem by evolving vocal signals and a complex emotional response to them. Thus, among most mammals we find precisely tuned signals, usually varieties of squeak, which are generated by the limbic lobes; the evolution of the mammalian ear, a structure vastly more complex than the reptilian ear; and cerebral structures that bring about feelings of anxiety and appropriate searching behavior when parents and young are separated. When the vocalizations, which are more or less continuously produced, are interrupted for a certain interval, the mother will become anxious and search for the young. Later, the young will respond to her absence in a somewhat similar way. Among humans, the baby's cry is such a response; it is an extraordinarily effective means of gaining the mother's attention.

Separation anxiety, just one of the emotions that became of central importance in the evolution of mammalian behavior, plays a vital role in the evolution of effective parental care. It is this emotion of anxiety and loss that bonds the parents to their young. The limbic system as a whole characterizes and generates much of the basic behavioral repertoire of mammals and clearly plays a central role in the roots of human behavior. We shall return to this topic in Chapter 8, VI.

B. The Exploratory Drive.   Behavior has been described as both innate and learned (see Chapter 1, VI), and we have seen that learned behavior assumes great importance among mammals. The intimate relationship between different generations enables mammals to build up complex behavior patterns by imitation, without the need to learn by trial and error in each generation. Such learning is confined to those animals (birds and mammals) in which parental care has evolved.

Mammals also depend on learning by trial and error, which must always be the original source of learned behavior. While in most animals activity and learning result from the need to satisfy the primary biological drives of hunger, sex, and self-preservation, a new determinant of behavior is found among mammals that does not address these immediate needs. This new factor is a drive to investigate and explore the environment.

The *exploratory drive* is seen in the playful and exploratory activity of young mammals with the parts of the environment that provide changing and interesting feedback in connection with effort expended. This activity leads individuals to discover how the environment can be changed and what consequences flow from the changes. Such moderate, but persistent, activity is possible only under the protection of parents, because its value to the individual and the species is not immediate, as are the activities based on primary drives, but is of a long-term nature. This type of behavior forms part of the process whereby the young animal learns to interact effectively with the environment and build up its perception of its surroundings (see Chapter 11, II).

The value of a computer in the solution of a very complex problem (such as climate prediction) depends not only on the computer's size and complexity but also on the amount and quality of data fed into it. For reliable results, a large input is a prerequisite. It is equally necessary in the evolution of the brain as a mechanism for prediction and problem-solving; complex problem-solving depends on sufficient input data. This explains the evolution of an exploratory drive. During mammalian evolution the motor activity that results in the satisfaction of the primary drives of hunger, sex, and self-preservation was supplemented by the increased input that was derived from exploratory activity. This was an essential development of the evolution of mammals into organisms with brains much larger (in relation to body size), much more complicated, and much more effective than had been known before in the organic world.

There were two developments within the brain itself concomitant with the increased input. The first was the evolution of a greatly expanded memory store (that is, an unconscious record of experience); the second was the evolution of an expanded "computer" (that is, the part of the brain that actually makes predictions and solves problems).

As we have seen, when we compare the mammalian brain with that of lower vertebrates, we see that it is typified by the relatively great development of the cerebral hemispheres (Fig. 2.7). It is in these structures, and particularly in their surface layers or *cortex*, that we believe the memory store and most complex computing activities are to be found. We know the function of some of the lower parts of the brain, and they seem to be of a simpler order. For example, we have recorded how the hypothalamus controls body temperature and the estrous cycle, and it has other relatively simple functions, such as the control of sugar metabolism, heart rate, blood pressure, salt and water balance, and sleep. The *brainstem* and *cerebellum* (see Chapter 8, VI) control many innate reflexes relating to balance and muscular control. It is the highly evolved cerebral hemispheres, however, that are concerned not so much with executing a program of repetitive bodily activity and maintaining it, but with initiating activity with a "decision" as to what should be undertaken next, on the basis of the sensory input and accumulated experience that is memorized.

We have seen that the lower limbic lobes of the cerebral hemispheres are concerned with the generation of emotions. Just which parts of the hemispheres are concerned with memory will be discussed later (Chapter 8, VI). Investigations into which parts are concerned with decision-making have achieved poor results, which has led to the theory that it is the activity of certain relatively large areas of the cerebral hemispheres, rather than that of a clearly defined part, that determines most mammalian behavior (the so-called "law of mass action"). Extensive damage to the human brain may be sustained in some parts without any serious impairment of function, leading us to deduce that the problem-solving and predicting activities of the brain are not truly comparable to those of an electronic computer, which cannot function after the mutilation of any part.

In summary, we see the evolution of the cerebral hemispheres of mammals into a mechanism somewhat but not exactly similar in function to a large computer with an extensive memory store. Unlike a computer, the brain operates to some extent with nonlocalized systems of activity; like a computer, its value to its user in complex prediction is directly related to the information content of the input. The mammalian brain functions, therefore, not only to control the mechanisms that maintain the steady state of the body, but also to store the memory of an enormous range of experiences and interpret a wide range of inputs. Input is maintained by the exploration of the environment by highly evolved sense organs, which bring to the animal an enlargement of experience. On the basis of such experience the brain is able to make predictions about the likely course of external events, and on this basis to initiate novel patterns of adaptive behavior.

VI.
The Human
Mammal

We have discussed six evolutionary novelties that characterize the placental mammals, each one a whole complex of characteristics. Two of them, homoiothermy and heterodontism, contribute in an obvious way to make possible the constant metabolic level of the organism in the face of environmental change. Following our discussion of evolutionary progress (Chapter 1, IX), we see that insofar as they bring improvements in the internal homeostatic state of the organism, they are progressive in the direction of human-kind and in the sense defined by Herrick (1946). We have already indicated their importance to the early mammals, in the advantages that these new characteristics gave to their bearers in the reptile-dominated environment of the Cretaceous period.

Clearly, any advances in internal homeostatic mechanisms increase the chances of survival, not only of the individual but of the whole species, because each individual is more likely to survive and reproduce itself. The advantages to the species of the second two character complexes are perhaps more obvious. The improvements in reproductive economy that result in a much lower wastage of eggs and young mean an increase in biological efficiency as a whole for the species: fewer eggs need to be produced, and less energy and living substance is devoted to that end. The trend toward a K-strategy with a lowered rate of egg production reaches its apogee in *Homo sapiens*.

Our fifth evolutionary novelty, the appearance of emotion, brings into mammalian life mother–infant bonding and an awareness of internal emotional states, which were to become of great importance in the evolution of society. The evolution of the limbic system is closely tied to the evolution of vocalization and the complex mammalian ear.

Finally, the exploratory drive is a behavioral development concomitant with the evolution of a larger brain, which, by making possible a better knowledge of the environment and a more effective response to it, increases the chances of survival of the individual and so of the species. The ability to assimilate and utilize this knowledge is selected in the genotype, but its transmission between generations is made possible by imitation within the close mother–infant relationship. This transmission of behavior patterns in an extragenetic manner is the seed from which human culture was eventually to grow (see Chapter 11).

It is not necessary to stress the importance to humankind of these complexes and all their structural correlates. With our remarkable ability to survive every sort of climate, to live off almost every sort of food (Chapter 9, II), we have a wonderfully efficient reproductive system (Chapter 10) and a body of learned and transmitted behavior (Chapter 11) that is far beyond anything known in the rest of the animal kingdom. We owe this vast body of transmitted data of different kinds mostly to the extraordinary knowledge of the environment that we have gained by experience. We owe our perception of the environment, which gives us that enriched experience, to the highly evolved sense organs and exploratory drive that we share with other mammals. Improved homeostatic mechanisms and a new behavioral relationship with the environment made mammals the potential ancestors of *Homo sapiens*. Humans are first and foremost mammals and could belong to no other class of vertebrates.

Suggestions for Further Reading

For an outstanding text on the biology of mammals see J. Z. Young, *The life of mammals* (Oxford: Oxford Univ. Press, 1957). For a simple taxonomic survey of the Mammalia see M. Burton, *Systematic dictionary of mammals of the world* (London: Museum Press, 1962). Useful introductions to the social behavior of animals include W. Etkin, *Social behavior and organization among vertebrates* (Chicago: Univ. of Chicago Press, 1964) and E. O. Wilson, *Sociobiology: The new synthesis* (Cambridge: Harvard Univ. Press, 1975).

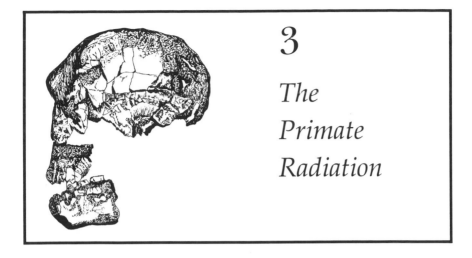

3

The Primate Radiation

Anyone who is familiar with the anatomy of Man and the Apes must admit that no hypothesis other than than of close kinship affords a reasonable or credible explanation of the extraordinarily exact identity of structure that obtains in most parts of the bodies of Man and Gorilla.

G. Elliot Smith
The Evolution of Man, 1924

I. The Origin of Primates

A glance at Fig. 2.1 will show that the mammals that gave rise to the primates (and indeed to all other placental mammals) are believed to have been of a kind that would now be classified as Insectivora. The central position of the order Insectivora is intended to indicate that they have diverged least from the ancestral mammalian stock that gave rise to the whole radiation of mammals.

Some of the fossil insectivores from the Paleocene period are very similar to some living genera of this order, and in particular to the genus *Sorex*, the little shrew (Fig. 3.1). Though many of the Cretaceous and Paleocene insectivores were rat-sized, the shrew and its relatives are very small animals; the one illustrated is only 4.5 inches long, including tail. The insectivores are insect-eaters, as their name implies, but, like the shrew, the majority supplement their diet with other small animals, seeds, and buds; they are in fact omnivorous. The shrew is an active, nervous, nocturnal animal that eats its own weight in food daily. A large food supply is necessary to maintain a creature so active and so small, since small animals have a much larger surface area in relation to their weight, through which heat may be lost, than larger animals have, as well as a high metabolic rate.

Figure 3.1. Common European shrew (*Sorex araneus*), with a body about 3 inches in length, is very similar to the North American species *S. cinereus*.

The shrews and other ground-living genera have exploited the dense environment of the forest floor, where, living almost exclusively upon the ground, they are able to find the practically continuous food supply they require. As can be seen in Fig. 3.1, shrews have a long snout and sensitive tactile *rhinarium* (the sensitive area of naked skin at the end of the muzzle, familiar in dogs). They snuffle through the grass after seeds and minute animals. They owe their obscurity to their small size and dense habitat, which has certainly contributed to their survival by protecting them from the larger carnivorous reptiles and mammals. The adaptations of small size and nocturnal activity have obviously given these lively and omnivorous creatures an immense advantage over their reptilian competitors. Yet their adaptations did more than this, for not all of them remained in obscurity. The moles went underground; the African otter shrews and American marsh shrews (among others) have become aquatic; other insectivores took to the trees and evolved into tree shrews, and from some of the arboreal shrews evolved the bats. Still other ground-living forms evolved at the same time into the whole range of terrestrial mammals (Fig. 3.2).

Primatologists are especially interested in the tree shrews [Fig. 3.3 (1)], a group that invaded the trees, climbing with their sharp claws and using their long tails for balance. From animals not unlike them the order Primates is believed to have evolved, and within it the human species.

II. The Primates   The diagnostic characteristics of the order Primates, whose origins it is claimed can be traced to the Late Cretaceous period of North America (Van Valen and Sloan, 1965), some 65 million years ago, are not easy to define and describe in a simple list. But their characteristics, though numerous, can be understood as adaptations to insect predation in an arboreal

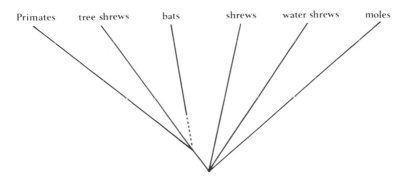

Figure 3.2. Simplified diagram showing part of the radiation of the primitive ground-living shrewlike placental mammals.

environment. We will consider first the primate adaptations as a whole and then the characteristics that evolved with them.

In the face of competition from other forest-living terrestrial creatures, we may suppose that the earliest primates evolved the ability to climb shrubs and trees. Tree-climbing was advantageous to fully exploit the forest environment—to hunt for insects and later to gather fruit and vegetation, as well as to escape from predators on the ground. The environment was a difficult one, densely filled with objects of different kinds and demanding accurate mobility and keen sensory awareness. Since scents were unreliable in identifying the exact direction of food or enemies and were of little help in negotiating trees and undergrowth, efficient vision and the ability to grip branches became the great need of these forest dwellers. It is supposed, therefore, that the primates became characterized in the first place by the evolution of forward-looking eyes for insect predation and soon after by limbs adapted for tree-climbing.

There are nine orders of mammals with arboreal species, but not one has evolved a range of adaptations truly analogous to those of the primate order. Therefore, it seems unlikely that primate specializations can be accounted for merely as an arboreal adaptation. For this reason, Cartmill (1974) has suggested the possibility that the early primates first became adapted as active insect predators operating in the shrub layer of woodland, and in this way developed first-class binocular vision with forward-looking eyes and good hand–eye coordination, together with grasping fingers and toes. This constitutes a character complex such as we find today fully evolved in the lorises. Only when this adaptation was established did these creatures come to occupy the forest habitat more fully and evolve the complete range of primate arboreal adaptations. At present, the fossil record does not present good evidence for this or any other hypothesis of primate origins, so this scenario remains an untested hypothesis. It is, however, a reasonably likely explanation

Figure 3.3. The tree shrew (1) together with some living members of the order Primates: a lemur (2); a tarsier (3); an Old World monkey (4); a chimpanzee (5); and an Australian (6) (not drawn to scale). (From Le Gros Clark, 1971.)

for the origin of the primate character complex and accounts well for the distinctive nature of the order.

Almost every characteristic we normally associate with primates is part of one or the other of these two functional complexes, and from this point of view they fall into place very neatly. Let us consider each complex in turn.

1. The following sensory adaptations were selected: (a) Eyes that look forward, with an overlapping visual field that gives three-dimensional perception of the environment, accompanied by analysis of input to the brain and mechanisms for the perception of depth (stereoscopic vision). (b) Enlargement of the eyes, which increased the amount of light and detail received. (c) Development of a bony protection for the large eyes. (d) Evolution of an improved retina with increased sensitivity to low levels of illumination and, in some groups, to different frequencies (that is, to color). (e) Reduction (as a corollary of enhanced visual sense), especially in the higher primates, of the apparatus involved in the sense of smell, the olfactory lobes, and protective bony snout or muzzle.

2. Insect predation and tree-climbing involved the following adaptations of body and limbs: (a) Retention of a generalized limb structure and *pentadactyly* (five fingers or toes), with free mobility of limbs and digits, giving support to the body from any direction, and with limb movement not restricted to one plane. (b) Development of long mobile digits capable of grasping, with sensitive pads supported by nails (which replaced claws in primate evolution) and palmar surface with extensive friction skin. (c) Retention of a tail as an organ of balance (or as a "fifth limb" in some New World monkeys). (d) Evolution of upright body posture and extensive head rotation. (e) Increase in body size—ultimately limited by arboreal environment. (f) Evolution of the nervous system to give precise and rapid control of musculature.

The foregoing two functional complexes reflect only part of the whole pattern of primate biology, but the most important part. One complex, the developed visual sense, is concerned with input—with receiving information about the environment; the other is concerned with output—the active exploitation of the environment. The brain processes information from the sensory input and instructs the motor output. This complex process is an example of the intensification of total metabolic and nervous activity typical of the primates. The brain, the mechanism for analysis, prediction, and programming, is the necessary correlate of the evolution of either sensory or motor activity and its coordination. Development of both these functions goes hand in hand with development of the brain.

The sight of gibbons moving through a forest canopy is a beautiful instance of incredibly rapid and precise motor action under accurate visual control. The speed and accuracy of their movements depend on a brain that can assimilate many factors of weight, distance, and wind and process them to give immediate and precise motor instruction. It is to the exploitation of this dense, complex, and demanding environment that we owe the development of a brain that is unique in the extent of its memory bank, in the quantity of the data it processes, and in the speed of its operation. It is accompanied by nervous and muscular mechanisms that are accurate and flexible in operation.

The structural characteristics listed here will be referred to in the following chapters, when we consider how they have contributed to the human adaptation. In this chapter we shall consider first the classification of the order Primates, then review the way that the various groups within the order have developed different locomotor adaptations. Lastly, we shall consider some aspects of the evolution of the nervous system that have accompanied these important adaptations.

III. Primate Classification
Among living primates we find a wide range of 57 genera of different appearance, size, and habitat. This varied order, which extends from a prosimian the size of a mouse to ourselves, constitutes a taxonomic order of great variety and extraordinary zoological interest. Seven main terminal groups of living primates, the lemurs, aye-ayes, lorises, tarsiers, monkeys—New World and Old World—apes and humans are classified as superfamilies. These particular types have survived because they became successfully adapted to particular ecological niches; intermediate forms have been lost.

It is a remarkable, rare, and happy fact that we have this variety still living, for they can help us interpret some of the multitudinous fossil forms. Of course, we can never be certain of the interrelationships of living forms in the past, but, *on the basis of the evidence available at present*, a working hypothesis of the type shown in Fig. 3.4 forms a popular interpretation of the evidence.

A formal classification of the order Primates is given in Table 3.1. While there is a great variety of classifications in the literature, this rather traditional taxonomy is widely accepted. Five of the seven superfamilies listed have been illustrated in Fig. 3.3; we shall be concerned mainly with the so-called "higher" primates or Anthropoidea: the monkeys, apes, and humans. The monkeys and apes are the groups most closely related to ourselves (see Tables 3.2 and 3.3). The tarsiers (one genus only) were almost fully evolved 50 million years ago in the Eocene period. They have some characteristics, especially in the optical system, which have caused them to be classified with the Anthropoidea (Martin, 1979), but in the past have more often been considered unusual prosimians. The "lower" primates (Prosimii) are of less immediate relevance, though of immense interest.

We see from the classification that there are two groups of monkeys: the Ceboidea, or New World monkeys, confined to Central and South America, and the Cercopithecoidea, or Old World monkeys, which are spread throughout the tropical regions of Africa and Asia. (These two groups, which have striking and important differences, are also sometimes termed the *platyrrhine* and *catarrhine* monkeys, respectively.) The

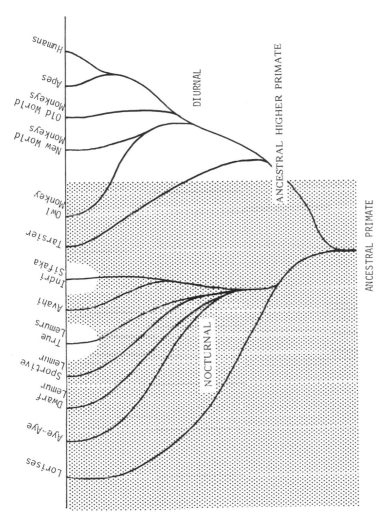

Figure 3.4. Evolutionary radiation of the primates. This hypothetical evolutionary tree indicates the supposed relationships between some important primate taxa that are, or are believed to have been nocturnal or diurnal. Two anthropoids (tarsier and owl monkey) are secondarily nocturnal, and two taxa of prosimians (true lemurs and indris) are exceptional in being diurnal. The ancestral primate is believed to have been nocturnal (from Martin, 1979).

TABLE 3.1. Classification of the Primates Showing Numbers of Living Genera

Order: Primates		
Suborder: Prosimii		
Superfamily		
Lemuroidea	Lemurs (9 genera)	Madagascar
Daubentonioidea	Aye-ayes (1 genus)	Madagascar
Lorisoidea	Lorises (5 genera)	Africa and Asia
Suborder: Anthropoidea		
Superfamily		
Tarsioidea	Tarsiers (1 genus)	Southeast Asia
Ceboidea	New World monkeys (16 genera)	Central and South America
Cercopithecoidea	Old World monkeys (14 genera)	Africa and Asia
Hominoidea	Apes and men (5 genera)	Worldwide

third superfamily of Anthropoidea, the Hominoidea (apes and humans), was also confined to the Old World until humans entered America and Australia in geologically recent time. Geographical and anatomical evidence make clear (Fig. 3.5) that the Hominoidea are much more closely related to the Old World than to the New World monkeys. The New World monkeys became isolated at an early stage in primate evolution, and may even have had a separate prosimian ancestry (Ciochon and Chiarelli, 1980). Therefore, we shall refer very little to them; we shall be mainly concerned with the Old World forms, which lie nearer to our ancestral lineage.

Of all the characteristics deserving consideration in any study of primate evolution, the most important, as we have seen, are their locomotor adaptations to tree-living. We shall now discuss the different ways that the primates have become adapted to the forest.

IV. Primate Locomotion

A. Prosimians and Monkeys. The mode of locomotion of a group of animals is by necessity closely adapted to the ecological niche they occupy. Different species of primates occupy different niches, and this factor, coupled with the differences in body size of the various species, results in considerable divergences in locomotor behavior and anatomical adaptations. The observation of locomotor behavior in wild primates is difficult, but its description is even more demanding, ideally concerning itself with total locomotor and postural behavior, including all kinds of movement, feeding, and resting. Such complete descriptions have only been made of a few species (e.g., Ripley, 1967; Fleagle, 1976,

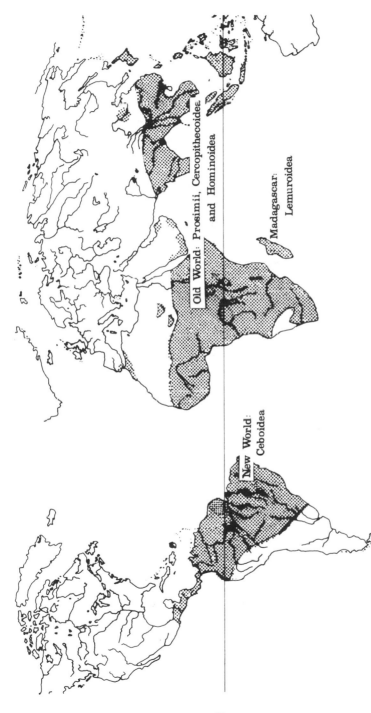

Old World: Prosimii, Cercopithecoidea and Hominoidea

Madagascar: Lemuroidea

New World: Ceboidea

Figure 3.5. World map showing the distribution of the three great radiations of nonhuman primates.

TABLE 3.2. Genera of Living Monkeys[a]

Ceboidea: New World monkeys			
Callithrix	Marmosets	*Callicebus*	Titis
Cebuella	Pygmy Marmosets	*Pithecia*	Sakis
Saguinus	Tamarins	*Chiropotes*	Bearded sakis
Leontideus	Golden lion tamarins	*Cacajao*	Uakaris
Callimico	Goeldi's marmosets	*Alouatta*	Howlers
Cebus	Capuchins	*Ateles*	Spider monkeys
Saimiri	Squirrel monkeys	*Lagothrix*	Woolly monkeys
Aotus	Douroucoulis	*Brachyteles*	Woolly spider monkeys

Cercopithecoidea: Old World monkeys			
Cercopithecus	Guenons	*Cynopithecus*	Celebes black ape
Erythrocebus	Patas monkeys	*Colobus*	Guerezas
Mandrillus	Mandrills	*Presbytis*	Langurs
Cercocebus	Mangabeys	*Nasalis*	Proboscis monkeys
Papio	Baboons	*Simias*	Pagai Island langurs
Theropithecus	Geladas	*Rhinopithecus*	Snub-nosed langurs
Macaca	Macaques	*Pygathrix*	Douc langurs

[a]From Napier and Napier, 1967.

1980). For the present, we must be content with a very general description of what has been called *positional behavior* (Prost, 1965) in different primate groups. Differences in positional and locomotor adaptations are broadly reflected by the taxonomic groupings of the primates, so they can conveniently be considered under taxonomic headings.

Since the primates evolved from ground-living forms, insectivore-like creatures, we can be sure that the earliest mode of locomotion in the group was running up the trunks of trees and along the branches

TABLE 3.3. Classification of the Hominoidea[a]

Family	Genus	Common name
Oreopithecidae	*Oreopithecus*	Swamp apes
Hylobatidae	*Hylobates*	Gibbons, Siamangs
Pongidae	*Pongo*	Orangutan
	Sivapithecus	Indian apes
Gorillidae	*Pan*	Chimpanzee, gorilla
	Dryopithecus	Oak apes
	Proconsul	African apes
Hominidae	*Homo*	Humans
	Australopithecus	Ape-men

[a]Excludes Oligocene species.

quadrupedally—on all four limbs—gripping with their claws, as can be observed in the living tree shrews. This mode of locomotion would have involved minimal arboreal specialization in the first place, but was evidently accompanied by the development of the hindlimbs so that they became far more powerful than the forelimbs. This is sometimes called *hindlimb dominance,* and it characterizes all primates, distinguishing them clearly from most other quadrupedal terrestrial mammals in which the forelimb is dominant, i.e., better developed than the hindlimb (Fig. 5.3). Kimura *et al.* (1979) have shown that forelimb dominance is more efficient in a terrestrial quadruped in terms of energy consumption; hindlimb dominance is more efficient in an arboreal form.

One aspect of hindlimb dominance can be demonstrated in all primates by consideration of the form of the *calcaneus* (the heel bone) and the part it plays as a lever in the foot. This characteristic will be discussed more fully in Chapter 6, III.

Hindlimb dominance was accompanied by other characteristics of the limbs; in particular, all the primates evolved grasping feet at an early stage. This development involved the evolution of a divergent first digit, the *hallux* in the hindlimb, the *pollex* in the forelimb. Such grasping feet would seem to imply that the early primates moved from running up the trunks of trees and along the heavy boughs to exploring the smaller branches of the forest, where claws were less effective and where much of the insect life was to be found. Grasping feet evolved in adaptation to the small-branch zone of the rain forests. Of all primates, only *Homo sapiens* has lost the divergent hallux.

1. *Prosimians.* From our knowledge of the living prosimians, we can now identify two further early adaptations to arboreal life.

i. Vertical clinging and leaping. Further elongation and development of the hindlimbs is characteristic of many small prosimians (some lorises such as *Galago*) as well as some larger forms (such as the lemur *Indri*). It is also to be found in the tarsier, and has given these creatures the ability to leap from tree to tree like small kangaroos, and over enormous distances in relation to their body size. Thus, they can avoid predators, both by avoiding the ground altogether and by leaping if they are attacked in the trees. This development is accompanied by a movement of the center of gravity to a position near the hindlimbs, and by the adoption of a resting posture with the trunk held vertically—a common feature of primates. They usually cling to a vertical rather than a horizontal branch. This locomotor adaptation has been called *vertical clinging and leaping* (VCL). It is an arboreal specialization, but it has the added result that if such an animal rapidly crosses the ground it proceeds by hopping, like a kangaroo, rather than running quadrupedally.

Although today the monkeys and apes do not carry this extreme locomotor adaptation, all Old World higher primates are characterized by the hindlimb dominance and the erect trunk that we see in the prosimians, at least when at rest. The specializations of the VCL complex may be summarized as follows (Napier and Walker, 1967):

a. A considerable lengthening of the hindlimbs in relation to the trunk. This lengthening includes the femur, tibia, and fibula, and in tarsius and galago the tarsal bones are also lengthened (see Fig. 4.3 for the bones of the skeleton).

b. Powerful muscles are associated with the hindlimbs. These developments bring the center of gravity back over the hindquarters.

c. A large foot carries a divergent big toe and reduction of digits II and III. This allows the animals to grasp bigger branches.

d. The vertebral column and its relation to the head reflect the vertical posture.

e. The tail is usually long and carries thick fur (it is absent in *Indri*). Where present, its function is probably to give aerodynamic control during a jump to maintain the animal's attitude during flight, as does the tail of an aeroplane.

ii. A second specialized adaptation seen among prosimians contrasts strikingly with VCL. A group of four genera move by very *slow quadrupedal climbing* (SQC), with neither leaping nor running. Typical of this adaptation is the slow loris (*Nycticebus*), which moves without shaking a branch and relies on its stillness both for stalking its prey of insects, lizards, and birds, and for protection from predators. SQC specializations are summarized as follows (Walker, 1969):

a. The lorises and other slow climbers have very robust limb bones and musculature adapted for slow, powerful, and prolonged contraction. The hindlimbs are only slightly longer than the forelimbs.

b. The digits are long, but the tarsal footbones are short.

c. Hip and ankle joints show great mobility, allowing suspension in strange postures by any or all the limbs. Wrist joints are also specialized to give great flexibility (Cartmill and Milton, 1977).

d. Tail absent or very short.

In both VCL and SQC adaptations the size of the individuals and mode of locomotion closely reflect the animal's adaptation for feeding and for predator avoidance in a particular type of arboreal habitat.

2. *Old World monkeys.* Old World monkeys are larger than living prosimians. Some are terrestrial, others arboreal, and their greater size allows them to run quadrupedally along the larger branches and to leap from tree to tree without the extreme specializations we find among prosimians. Their large size gives them not only increased mobility, but also protection from the smaller animals that prey upon the prosimians.

There is great variation in behavior within genera, within species, and even within populations and social groups. Locomotion may also vary with the age and sex of monkeys as well as with environmental variables. For example, Fleagle (1978) reports that two related species of leaf monkey (*Presbytis*) living in the same area of Malaya have very distinct locomotor patterns: one moves mainly by leaping and the second mainly by quadrupedal walking. Any generalizations about primate locomotion must be considered in this light. In consideration of their total locomotor and postural behavior, however, the Old World monkeys are classified by Rose (1973) as *quadrupedal*; Andrews and Aiello (1984) describe them as *above-branch feeders* (to distinguish them from those primates that hang from their forelimbs to feed and so are classified as below-branch feeders). Within the group we can recognize two main kinds of locomotor adaptation:

i. Arboreal quadrupedalism—Typical of this group are the monkeys of the genera *Cercocebus, Cercopithecus,* and some species of *Macaca.* These genera have a number of anatomical adaptations in common: (a) A relatively longer trunk, especially in the lumbar region, which gives them greater power in leaping. (b) A high brachial index, which means that the forearm is longer in relation to the upper arm than in other primates. This gives them rapidity rather than power of movement at the elbow joint (Aiello, 1981a). (c) The shoulder girdle is built for stability and allows only restricted movement (Corruccini and Ciochon, 1978). (d) The elbow joint also shows restricted mobility (Napier and Davis, 1959). (e) The hand and foot have long flexible digits, with independently mobile thumb and big toe that are opposable for gripping branches. (f) The tail is retained as an aid in balance when leaping. (g) The characteristic primate backward shift in the center of gravity is associated with a sitting position when resting and feeding, with *ischial callosities,* and with the trunk held more or less vertical.

In some genera of arboreal quadrupeds there is substantial use of *forelimb suspension* in climbing and feeding. (These genera were classified as semi-brachiators by Napier and Napier, 1967). The most notable are *Colobus, Nasalis,* and *Presbytis.* These genera carry the following specializations: (a) The fingers are relatively longer and the thumb is rarely used in opposition to the other fingers, but alongside them. It is somewhat reduced in length, and in *Colobus* it is lacking or reduced to a stump. (b) The scapula and associated musculature show modifications associated with increasing the power of the arm to lift the body from above, as well as to support it from below. The freedom of movement of the arm above the head is also improved.

ii. Terrestrial quadrupedalism—All primates can cross open ground and many often do so. A number of genera of Old World monkeys are to varying extents adapted to living on the ground, such as the smaller vervets (a species of *Cercopithecus*), which spend some time on the ground, and those which are most fully terrestrial and have increased substantially in body size, such as the mandrill, *Papio,* and *Theropithecus.* In general, those species most fully adapted to terrestrial life have evolved further in the direction of forelimb stability. Certain characteristics are especially notable: (a) More stable shoulder girdles (Corruccini and Ciochon, 1978). (b) More stable elbow joint with wide epicondyles of the humerus, more cresting and backward reflexion of the olecranon process of the ulna. (c) More robust vertebrae, producing an increasingly short vertebral column as body size increases (Schultz and Straus, 1945; Aiello, 1981a). (d) A higher intermembral index than is found in the arboreal quadrupeds (Aiello, 1981a); that is, their forelimbs are longer than their hindlimbs, promoting greater efficiency in terrestrial locomotion. This lengthening of the forelimb is increased in some genera, such as the Patas monkey, by walking on the fingertips (*digitigrade*) rather than on the palms of the hands (*plantigrade*). The longer forelimb will also improve vision over grass in these terrestrial species, which is vital for protection against predators. In the patas monkey all the limbs are long in relation to trunk length; an adaptation for speed. (e) The fingers in the digitigrade species are shorter and more robust than those of arboreal primates, but the thumb is well developed and in baboons permits a precise finger–thumb grip.

The terrestrial primates have proved of great importance to students of human evolution, since we too are a terrestrial species, and it is useful to explore some of them in a little more detail. The baboon (*Papio*) is classified into two species. The common savanna baboon, which extends from West Africa to Ethiopia and to the Cape of South Africa (and is called the chacma, yellow, olive, and Guinea baboon in different regions), is best considered a single species (*P. cynocephalus*) with a variety of geographical subspecies. This baboon is the most extensively and intensively studied monkey in the wild. It is large and powerful and spends most of the day on the ground in the sparsely wooded savanna areas of its range. It feeds on grass shoots, seeds, and fruit, often in a sitting position. At night it climbs into trees (or occasionally cliffs) for protection from predators. It walks quadrupedally and takes its weight on the fingers (digitigrade) but on the toes and distal soles of its feet (plantigrade) (Fig. 3.6).

The second species is the hamadryas baboon (*P. hamadryas*). This monkey is closely related to the savanna baboon and differs from it only in relatively superficial respects. It lives in the even drier environment of eastern Ethiopia and parts of the Arabian peninsula and is adapted to subdesert conditions where trees may be absent. At night it sleeps on cliffs for protection from predators.

Besides the baboons, two other ground-living monkeys should be mentioned. The patas monkey (*Erythrocebus patas*) is probably the terrestrial monkey with the most striking locomotor adaptations. It is distributed along the subdesert steppe country between the southern edge of the Sahara and the tree savanna to the south. The patas sleeps in trees but finds its food on the ground during daylight, like the baboons. Its main difference lies in its lighter build and its adaptation for speed. The

Figure 3.6. The baboon (*above*) is a large terrestrial quadrupedal monkey; digitigrade and plantigrade. The gorilla (*below*) is a semiterrestrial ape with an arboreal climbing ancestry; when walking quadrupedally it takes its weight on its knuckles. Note that the gorilla's arms are much longer than its legs (not drawn to scale). [By permission of the trustees of the British Museum (Natural History).]

monkey has long legs in relation to trunk length, and this gives it great speed with which to escape from predators in its relatively treeless environment. It runs digitigrade, like the baboon, but its fingers are shortened and strengthened even further. Like the baboon, it often stands bipedally to improve vision, but it uses a powerful tail to form a tripod with the hind legs, like a kangaroo.

A fourth monkey, the gelada (*Theropithecus gelada*), probably carries more adaptations to terrestrial life than the other forms, and they do not lie only in its locomotor organs. These monkeys live on mountain slopes in Ethiopia, at altitudes of between 6,000 and 16,000 feet. Like the baboons, they are powerful, digitigrade animals that normally remain on the ground at all times. They feed continuously in the sitting posture, picking up their food from the ground.

All four species of monkey have particular anatomical and social adaptations to terrestrial life to which we shall return. Here we have been concerned only with their locomotor and postural adaptations.

The fossil evidence for monkey evolution is limited. Remains of *Victoriapithecus* from the early and middle Miocene and of *Mesopithecus* from the late Miocene and Pliocene periods are associated with open savanna and not with forest habitats. They also show anatomical adaptations to terrestrial bipedalism. Many other fossil monkeys from the Pliocene and Pleistocene are similar in both respects. Andrews and Aiello (1984) present evidence that the ancestral monkeys evolved from a relatively small-bodied arboreal ancestor that increasingly engaged in terrestrial locomotion. As body size increased, the form of the skeleton was determined by the requirements of *terrestrial* adaptations at the expense of forelimb mobility and the acrobatic postures that are associated with arboreal movement and feeding behavior. Thus, as the monkeys adapted to a terrestrial lifestyle, various changes took place: (a) body size increased; (b) intermembral index increased—forelimbs lengthened; (c) more stable forelimb skeleton and musculature evolved to bear increased weight.

Some of the monkeys then returned to an arboreal environment. As a result, the following changes appear to have occurred: (a) body size increase stopped; (b) intermembral index dropped—forelimbs shortened; (c) more flexible shoulders evolved to raise forelimbs above the head for climbing and picking fruit, leaves, etc.

This scenario requires further investigation, but suggests that the monkeys were originally adapted to a terrestrial life as compared with the apes, which were arboreal. Further fossil evidence should help to confirm or dispose of this hypothesis.

3. *New World monkeys.* Although the New World monkeys have also been classified by Rose (1973) as arboreal quadrupeds, they show a wide range of adaptations depending on size, and the larger species use a variety of suspended postures. While the smaller species, including the marmosets, have few extreme arboreal adaptations, the larger species have adapted in a different direction to those of the Old World monkeys. In place of the adaptations of terrestrial quadrupedalism that seem to characterize the Old World monkeys, New World monkeys have remained fully arboreal and have evolved advanced adaptations to suspensory activities by modifications of the arms, somewhat reminiscent of those we find in Old World monkeys that make use of forelimb

suspension. The larger New World primates are characterized by features associated with great mobility of the forelimb and stabilization of the trunk in climbing postures (Cartmill and Milton, 1977), together with the evolution of a fully prehensile tail in four genera. They are able to spread their weight with five clinging appendages that enable them to enter the small-branch zone of trees where food is most abundant.

The spider monkey (*Ateles*) is one of the most highly adapted New World monkeys, with its long limbs and extraordinarily long and powerful prehensile tail. The tail carries special friction skin on the ventral surface and a well-developed sense of touch, as well as a good sense of position. The monkeys can hang by the tail alone while feeding, and almost all suspensory activities involve use of the tail. The New World monkeys depend far less than Old World monkeys on leaping between trees: instead, they tend to climb and swing.

The difference between Old and New World monkeys is to some extent related to their history. While it seems very possible that the Old World group were originally a terrestrial adaptation, there is no reason to believe that New World monkeys ever passed through this stage. A continuous process of arboreal adaptation has given them a wide range of specialized anatomical adaptations, and no single species has ever become adapted to ground living as far as we know.

B. Apes and Humans. 1. *Asian apes.* The suspensory postures and locomotion we find among the larger South American monkeys reach their most advanced form among the Asian apes. In Southeast Asia there are two groups: the smaller gibbons (the gibbon *Hylobates* and the closely related Siamang *Symphalangus*) and the large orangutan (*Pongo pygmaeus*). The gibbons have the most evolved suspensory locomotion among the primates and most commonly move through the trees hanging by their forelimbs alone, with no support from below. When they are feeding, however, they tend to move by climbing (Fleagle, 1976). When traveling suspended by their arms, they hold their feet tucked up beneath their bodies. Because of the degree of dependence on the arms, this type of locomotion has been called *brachiation*. The range of adaptations that we have traced through the order of primates, from almost total dependence on the hindlimbs among VCL prosimians to the larger New World monkeys with their suspensory feeding postures, reaches its end point with almost total dependence on the long and powerful forelimbs of the gibbons (Fig. 3.7). This adaptation includes the following features: (a) The forelimbs are lengthened relative to trunk length so that they are one quarter again as long as the hindlimbs. (b) The hand is long (especially the fingers), but the thumb has not shared this increase in length. (c) Structural modifications of the shoulder and arm give even greater power and flexibility than we find among any Old World monkeys. (d) The tail is lost. (e) The trunk has reduced flexibility: it no longer adds to the power of a leap. The arms have taken over almost the entire

Figure 3.7. The gibbon (*Hylobates*) is a skilled brachiator. When he is travel-ing rapidly from tree to tree, his legs are fully retracted; under these circum-stances they have no locomotor function. Note the very long arms and long hands.

locomotor role in the trees. There is almost no leaping activity propelled by the hind legs.

It is noteworthy that the gibbon is probably the only nonhuman pri-mate that occasionally adopts a bipedal run along a branch, most com-monly on large branches. It will also run bipedally when crossing open spaces. We shall return to this observation when we discuss the origin of human bipedalism.

The heavily built orangutan has many of the anatomical characteris-tics we associate with brachiation, but it is perhaps better described as a climber. While it regularly uses its arms above its head, it usually takes weight on its feet too, which grip branches below and to its side. Although the forest habitat of the orangutan (now found only in Borneo and Sumatra) is somewhat different from that of the gibbon, it appears that the differences in their locomotor behavior and adaptations are due more to their difference in size than any other single factor. The gibbon is a small and fast-moving animal, while the orangutan is heavy and usually slow; it has no predators but the tiger, the clouded leopard, and man. When resting in the trees, it usually holds onto a branch above its head; although it rarely brachiates fully, its long arms probably supply more locomotor power than its legs. Though it sometimes walks bi-pedally on the ground, it usually walks quadrupedally, with clenched fists and feet, and occasionally its hands are used palmigrade (Tuttle, 1967). The orangutan is primarily an arboreal animal and has no stereotyped pattern of terrestrial locomotion. Its arboreal adaptations include: (a) Some reduction in hindlimb length so that forelimbs are much longer than hindlimbs. (b) Great reduction of hallux (great toe) and elongation of fingers, but reduction of the thumb to a short, almost vestigial, digit (see Fig. 7.13E, Tuttle, 1970). (c) Great mobility of the shoulder and wrist, with independent finger control. (d) Powerful de-velopment of the shoulder, arm, and finger musculature (Tuttle, 1969). (e) Loss of tail.

2. *African apes*. The Asian apes are less closely related to us than the African apes, so we must discuss African ape locomotor adaptations in some detail. The chimpanzee and gorilla (*Pan troglodytes* and *P. gorilla*) are closely related and have many features in common. The most striking difference is one of size: gorilla males may weigh up to 400 lb with a mean of about 340 lb, while male chimpanzees average only 110 lb (the orangutan male averages 160 lb). They live in neighboring territories of the rain forests of central Africa, though there is some overlap northwest of the Congo River (Fig. 3.8). There are three subspecies of gorilla, the best known of which are the mountain gorilla and lowland gorilla; the former is adapted to high altitude woodland with small trees, the latter to dense tropical rain forest. The chimpanzee is usually divided into two species, *Pan troglodytes*, the common chimpanzee, and *Pan paniscus*, the pygmy chimpanzee or bonobo. *Pan troglodytes* is divided into three subspecies, the western, central, and eastern types, though their morphological differences are very slight.

Variations in anatomy, locomotor behavior, diet, and social behavior throughout these species of the genus *Pan* are not yet as well known as they should be, because only limited field work has been possible in western and central Africa. It is, however, permissible to make some generalizations based on what we know.

The genus *Pan*, as we have said, occurs in species of three different sizes, and locomotor behavior, feeding, and diet vary accordingly. The larger the animal, the more time it appears to spend on the ground. When arboreal, these species show a rather varied repertoire of locomotor patterns. The smaller species, as well as juvenile chimpanzees and gorillas, are sometimes seen to brachiate though less rapidly than the gibbon. For the most part they and the larger animals climb the trees slowly, using all four limbs, as we ourselves climb trees.

Unlike the Asian apes, however, these species have very well-defined terrestrial adaptations. They walk quadrupedally on the soles of their feet and backs of the middle segments of their fingers (Fig. 3.6). Appropriately, the skin on the *outside* of the middle finger bones is tough, hairless, ridged friction skin, such as we find on the palms of our hands, and the wrists are strengthened by alterations in the structure of the short wrist bones (the carpal bones) to provide better support. This curious adaptation gives extra height to the animal's head when it is moving quadrupedally, an advantage where the undergrowth or grasses are high, as it allows more effective vision. It also brings the center of gravity forward and increases the animal's efficiency in terms of energy use (Stern, 1976; Andrews and Aiello, 1984). All species will stand bipedally to display or to get a better view, and the common chimpanzee, in particular (the best-known species in both captivity and the wild) will often walk bipedally over short distances. But bipedalism has never become a major locomotor adaptation among the apes.

It seems clear that the terrestrial adaptations of the African apes are secondary, in the sense that they probably evolved following a long period in which tree climbing was the primary mode of locomotion.

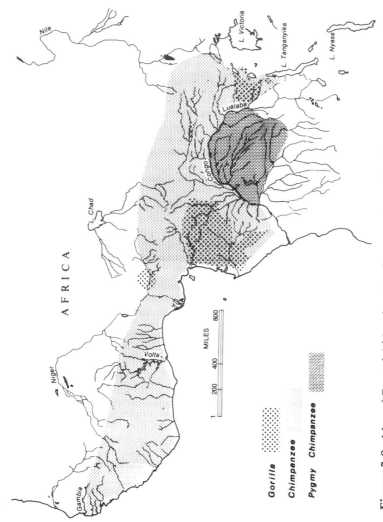

Figure 3.8. Map of Central Africa showing the approximate distribution of the three species of *Pan*. The data for much of the area are very limited and unreliable, and these rare species may today occupy a much smaller range than that indicated. The apparent continuity of distribution shown on the map is ill-founded, due to increased deforestation, cultivation, and hunting.

78

The brachiating gibbons and slow-climbing orangutan have lost their tails (as have the slow-climbing prosimians) during the course of their evolution, and so have the African apes. This development seems, therefore, to be associated with a climbing mode of locomotion where the function of the tail as an aerodynamic organ of balance is no longer as appropriate as it is for the running and leaping quadrupedal monkeys and prosimians. In a mainly terrestrial form like the gorilla, the tail is even less necessary.

We can summarize the adaptations of the African apes as follows: (a) Short, rather inflexible trunk. (b) Highly mobile and powerful forelimbs. (c) Long hands, short thumbs. (d) Powerful, opposable, great toe. (e) No tail.

These features are adaptations to arboreal climbing and below-branch feeding. Superimposed upon them are clear-cut terrestrial adaptations: (a) Modifications of the wrist bones to give greater stability. (b) Friction skin on the backs of the middle phalanges of the hand. (c) Plantigrade feet.

The differences in locomotion and general anatomy between the terrestrial monkeys, on the one hand, and the more or less terrestrial African apes, on the other, can be accounted for by considering their past. As we have seen, the terrestrial monkeys may have a long history of quadrupedalism on the ground. The modern African apes, on the other hand, have evolved, we believe, from creatures that were somewhat smaller than the existing great apes and that were adapted to arboreal climbing from an early date. They have only relatively recently become adapted to terrestrial quadrupedalism, and this was accompanied by a substantial increase in size. We shall see that humans also find their beginnings in a climbing arboreal adaptation.

3. *Humans.*   Although birds are bipedal, no mammals but humans have evolved our own peculiar kind of alternating bipedal gait. Many mammals move on two legs but, like the kangaroo, they move by hopping. Probably humans alone share with some of the dinosaurs the symmetrical alternating walk, but both dinosaurs and kangaroos have brought the tail into use as a prop, counterweight or propulsive organ, and neither group shows any equivalence to the human erect posture. Among the whole animal kingdom, we are unique in our combination of an erect trunk and a bipedal walk. In this respect, the human mode of locomotion, which looks like a permanent balancing act and which involved important changes in structure and function, is a remarkable and unique evolutionary development. We know the two hominid genera *Australopithecus* and *Homo* to have been plains-living and bipedal. It will be our purpose to investigate the changes that must have taken place during the evolution of *Australopithecus* from a tree-living form, and to trace from this time the final evolution of modern humans.

**V. Neural
Correlates
of Arboreal
Locomotion**

In the functional analysis of the evolution of the human body, it has not yet proved possible, except in a very few instances, to determine in detail the neural correlates of a functional change. Studies of the cortex of the cerebral hemispheres, however, and in particular of that part concerned with motor outputs, show clear variations in different groups of animals. Electrical stimulation of points on the "motor area" of the cortex produces movements in different body parts, so that a map of the function of the cortex can be constructed on the basis of the parts that move when each point is stimulated (Fig. 3.9). The larger the area concerned with a particular part of the body, the more precise the control exerted in the cortex. In primates, as we might predict, the limbs occupy a large proportion of the total motor area; in particular, the hand and foot are very well represented. However, it is noteworthy that in humans the area representing the foot is relatively smaller than that found in the rhesus monkey (Penfield and Rasmussen, 1957), presumably because the human foot is no longer a grasping organ. We see that among the primates as a whole the areas concerned with the hands and feet in the motor cortex are relatively large.

Studies of the descending motor nerve pathways from the brain to the muscles also show that some important changes have occurred in the evolution of primates that are probably correlated with the evolution of primate arboreal locomotion. Some significant differences between primates and other mammals, on the one hand, and between the lower and higher primates, on the other, have been demonstrated in these descending nerve tracts.

Among mammals generally we find a relatively direct system of nerve fibers running from the enlarged cerebral cortex of the brain to the motor *neurons* (nerve cells) of the spinal cord, which supply nerves to the muscles and effect their contraction. These *corticospinal fibers* (also called the *pyramidal system*) are exclusive to mammals and reach their greatest development in the higher primates. This phylogenetically new system (that is, one relatively recently evolved) is complementary to the phylogenetically ancient and indirect extrapyramidal tracts, the fibers of which arise in the brainstem rather than the cortex and reach the motor neurons via a sequence of relay neurons. The evolution of the corticospinal tracts is correlated with the evolution of the cerebral cortex; recent experiments have shown that they are responsible for precise motor control and that they add both speed and a capacity for very precise movement to muscular activity. This refinement is superimposed upon that of the indirect brainstem systems of the extrapyramidal tracts that direct movement in a more general way. Figure 3.10 shows in diagrammatic form the difference between the arrangement of these two kinds of nerve tracts.

In all mammals the contraction of the muscles is brought about by nerves that find their origin in the motor neurons within the spinal cord.

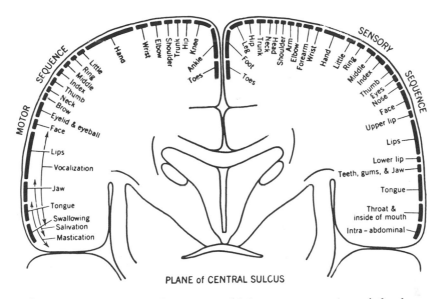

Figure 3.9. Schematic diagram combining a cross section of the human motor cortex (*left*) and the somatic sensory cortex (*right*), in the plane of the central sulcus. The diagram shows the motor and sensory sequence of representation determined by electrical stimulation of conscious patients. With stronger stimulation wider zones for a given part might be found, and there is considerable individual variation. Note that the sensory area devoted to the fingers and hand is greater than that devoted to the much larger skin area lying between forearm and shoulder (from Rasmussen and Penfield, 1947).

As has been said, these motor neurons are activated by connections from the descending nerve fibers from either the brainstem or the cortex. The connection is made by a relay neuron, the so-called *internuncial neuron*. However, it has been shown through studies on monkeys, chimpanzees, and humans, that in primate evolution an increasing number of corticospinal fibers make direct connection with the motor neurons and bypass the internuncial cells (Kuypers, 1964). This is a most important primate evolutionary development; it is especially interesting that the motor neurons of *distal* muscles (farthest from the trunk) receive more direct cortical fibers than do those of *proximal* muscles (nearest to the trunk), which correlates nicely with the evolution of the delicately controlled primate hand and foot, and especially with the evolution of the human hand. Kuypers has shown that these direct connections from the cortex to the motor neurons are not established at birth but develop during early life with the coming of precise motor activity.

In the evolution of the human motor-nerve pathways, we see, therefore, (1) a large increase in the number of direct connections between the cortex and the spinal cord that are shown to affect muscular speed and

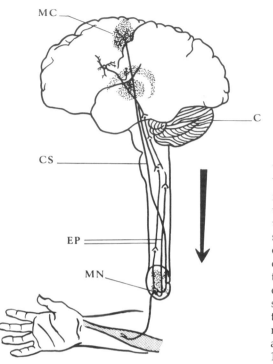

Figure 3.10. Diagram of motor (descending) nerve fibers. *CS* is a corticospinal fiber that in higher primates forms a direct link between the motor cortex (*MC*) and the motor neuron (*MN*) of the spinal cord. *EP* are two of the extrapyramidal fibers that connect the cerebral cortex through a complex sequence of relay neurons in the brainstem to the motor neurons of the spinal cord (internuncial neurons omitted). (Redrawn after Noback and Moskowitz, 1963.)

precision, and (2) the appearance of a direct connection between these fibers and the motor neurons, especially in connections to the distal musculature of the hand and foot. Here we have a clear-cut neurological correlate of the evolution of that extraordinary precision instrument—the human hand; a striking and significant organizational improvement in motor control.

Precise muscle control is not, however, achieved only by fast muscular contraction. The extent of such contraction is controlled by information returned to the brain by special sensory nerve endings within every muscle, the *proprioceptors,* which cause adjustment in the frequency of the motor-nerve impulses. The precise control of muscular contraction by proprioceptors is believed to be carried out in the cerebellum (Fig. 3.11, C) as well as in the cerebral cortex.

Ascending fibers from the proprioceptors and the sensory nerve endings in the skin (the somatic sensory nerve tracts) are therefore of almost equal importance with the descending fibers in the evolution of primate locomotion. While studies of these ascending tracts in mammals have not yet revealed such a striking novelty as that which we have recorded in the descending pathways, we do have evidence of two equivalent and complementary systems of nerve paths like those described above. The *lemniscal system* is prominent in mammals and reaches its highest development in the primates, especially in humans. It is a relatively direct

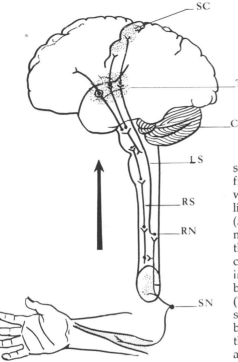

Figure 3.11. Diagram of sensory (ascending) nerve fibers. *LS*, fiber of the lemniscal system, which forms a relatively direct link between the sensory neuron (*SN*) (from proprioceptors in the muscle shown here, but also from the skin) and the somatic sensory cortex (*SC*) via relay neurons (*RN*) in the brainstem near the cerebellum (*C*) and in the thalamus (*T*). *RS*, a reticular fiber that constitutes a relatively indirect link between the sensory neuron and the cortex (redrawn after Noback and Moskowitz, 1963).

system of fibers linking the peripheral sensory nerve cells with the cerebral cortex. The phylogenetically older *reticular system* is, as its name implies, more indirect and diffuse and is the more primitive of the two. Nerve tracts of the two systems are shown diagrammatically in Fig. 3.11. Here the sensory neuron, which lies in a small *ganglion* or knot of nervous tissue near the spinal cord, can transmit an impulse from the proprioceptors in the muscle of the forearm to both the lemniscal and reticular systems. According to Herrick (1956), the lemniscal system is fast, specific, and analytical, the reticular system nonspecific and integrative. The latter has been shown experimentally to be responsible for arousal and wakefulness and to bring about a general condition of alertness. Both systems are well developed in primates, particularly in humans.

When we look, therefore, at the ascending and descending nerve pathways in primate evolution, we see a general increase in relatively fast direct connections between the cerebral cortex, the sense receptors, and the muscles. The precision of primate muscle activity is due not only to the evolution of the direct systems of nerve fibers, however, but also to a very complex feedback control mechanism, which depends on both descending and ascending nerve tracts as well as upon an increasingly complex organization in the brain.

TABLE 3.4. Approximate Extent of Visual Fields of Certain Mammals (Without Head Movement)

Mammal	Angle between optical axis of eye and midline of head (°)	Total visual field[a] (°)	Greatest width of binocular field (°)
Horse	40	360	57
Dog	20	250	116
Cat	8	287	130
Monkey, ape, and man	0	180	120

[a]Note that, since in the primates the eyes come to point straight ahead, the total visual field is reduced; this reduction is compensated for by free movement of the head. Data from Walls, 1963.

VI. Neural Correlates of Binocular Vision

In order to understand the evolution of primate vision, it is necessary to understand in some detail the structure and operation of the visual system as a whole.

The brain is a spatially organized and oriented organ, the two sides of which are connected by nerve tracts to the opposite sides of the body. Through these motor and sensory connections, the two sides of the brain are, as it were, concerned with the opposite two halves of the environment, the left hemisphere with the right half and the right hemisphere with the left half. The optic nerves, which carry the input to the brain from the eyes, therefore cross, and they do so at a structure called the optic *chiasma*. The evolution of stereoscopic vision has, however, involved an overlap in the field of vision of each eye, so that light from a single object can enter both eyes at the same time (Table 3.4). There is some overlap— some binocular vision—in many mammals, especially cats (whose eyes face forward), but this feature reaches its full development in the primates.

In primate evolution the eyes have moved toward the midline of the face, so in many lower and all higher primates the two fields of view overlap almost entirely. As sense organs, the eyes are no longer concerned individually and exclusively with the two halves of the environment; instead, they are each receiving information from both halves of the environment, which involves a new distribution of the optic nerve tracts. In animals with stereoscopic sight the nerve fibers are so arranged at the chiasma that those from the left side of *both* eyes pass to the left side of the brain, and similarly for the right side (Fig. 3.12). (We have to bear in mind that the image on the retina is reversed by the lens so that it is upside down and left to right.) As a result, a portion of the nerve

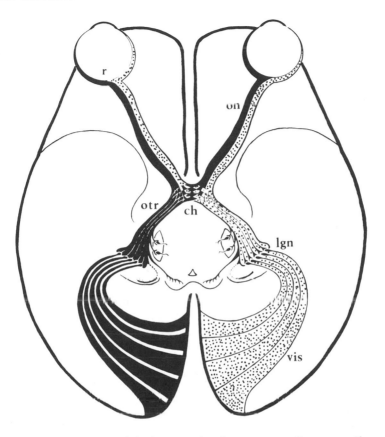

Figure 3.12. Diagram of the human visual system: *r*, retina; *on*, optic nerve; *ch*, optic chiasma; *otr*, optic tracts; *lgn*, lateral geniculate body; and *vis*, tract to visual cortex. Fibers originating in the inner or nasal halves of the retinae cross at the chiasma so the visual cortex of each side receives signals from the same half of the field of view, that is, the opposite half. The signals, however, represent two slightly different images, and on the basis of these differences depth and distance are perceived (after Polyak, 1957).

fibers do not cross at the chiasma, and, as the extent of the binocular field increases over that of the monocular field, the noncrossed fibers increase over the crossed fibers until in humans almost 50% of the fibers from each eye pass to the same side of the brain. Information from each eye is thus more or less equally distributed to the two sides of the brain, which receives a double image of the environment.

It is mainly on the basis of the slight differences in this double image that the brain is enabled to assess the relative and, with experience, the absolute distance of different objects. There are also a number of other cues of less importance (Gregory, 1966). Visual knowledge of the relative distance from the animal of different parts of the environment makes possible three-dimensional perception. Depth is added to breadth and height.

The estimation of distance on the basis of a difference in the analysis of form (that is, a slightly different view of an object) involves a comparison of the images of the two eyes. The difference between the two images of a near object will be much greater than that between the images of a distant object. It follows that to interpret depth at a distance involves greater visual discrimination than such interpretation at close range. This fact no doubt goes some way to explain the degree of visual discrimination available to primates, for a high level of discrimination is essential for an accurate analysis of the difference between the two images received by the brain. The eyes, it appears, have evolved as precision range-finders; the ability of humans to read fine print may be a byproduct of this adaptation.

We are only beginning to understand the process of visual analysis in the brain. Experiments with the cat (Hubel, 1963) have shown that analysis begins in the *retina* and continues in the *lateral geniculate body* and then in the *visual cortex* (Fig. 3.12). Hubel has shown that the pattern of impulses from the retina is analyzed by banks of neurons for individual components of pattern, color, and movement. We must assume that the result of the analysis is then compared in some way with visual patterns stored within the brain, and that such comparison may result in "recognition" of certain environmental phenomena.

In animals with stereoscopic vision the analysis must also include a simultaneous comparison of images, and the differences between them are interpreted as distance and depth in accordance with memory traces of previous visual experience. Though the complexity of this process is immense, it seems clear that to turn an image into a view of the environment (that is, perception), it is essential to compare the image with recorded memory of other images with their context of sensory and motor experience. Perception involves memory, and perception of the detailed kind available to higher primates (and eventually humans) involves increasingly extensive analysis and increasingly extensive memory banks. As we shall see (Chapter 8,VI), the parts of the brain believed to function as memory banks (in particular, the *temporal lobes*) have expanded in primate evolution. Investigations by Kuypers *et al.* (1965) show the tremendous extent to which nerve tracts from the visual cortex pass to other areas of the cerebral cortex. These intracortical connections are very typical of advanced mammals and the higher primates, and they are surely correlates of the immense importance of vision in almost every aspect of the animal's life history (Fig. 3.13).

We are visual animals. Three-dimensional vision gives a wide range of precise information about the environment that cannot be obtained in any other way. Because light travels in straight lines and because the nature of reflected light is determined by the chemical and physical structure of objects, precise data about the nature of objects can be

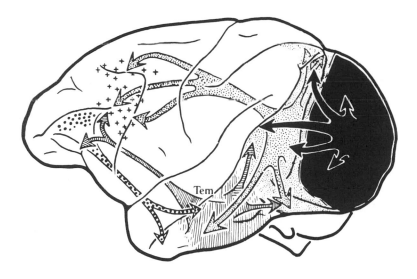

Figure 3.13. Diagram of some of the intracortical connections in the brain of the monkey *Macaca* recently revealed by Kuypers. Note the dense connections between the visual cortex (black), its neighboring association cortex (shaded), and the temporal lobe (*Tem*) (after Kuypers, 1965). See also Chapter 8, VI.

obtained at a distance. Stereoscopic vision and competent motor investigation (in particular, manipulation) make the world view of the higher primates unique both in quality and in kind. Primates alone have come to know the structure of the environment in terms of both pattern and composition. They see the environment as a collection of objects rather than merely as a pattern, and the recognition of objects is, as we shall see (Chapter 11, II), the beginning of conceptual thought. With stereoscopic vision, the primate can begin to perceive nonspatial abstractions, and this advanced perception finds its foundation in combining the analytic functions of the two sides of the brain. The evolution of vision is one essential basis for the evolution of an animal that came to understand its environment so well that it could control it for its own direct benefit.

VII. Summary

We ourselves are primates, and the origin of our nature may be traced precisely to the adaptations of the primates that evolved in response to the forest environment. The special problems associated with that environment demanded sensory awareness, precise mobility, and a brain able to make accurate and immediate prediction. In the following chapters, an attempt will be made to show how each aspect of our physiology and behavior developed in response to the challenge either of the forest or of the totally different, more open environment of the plains, in which our ancestors later evolved. The majority of mammals have evolved for tens of millions of years in one environment alone: antelopes on the grass-lands, whales in the sea, apes in the forests; that fateful move of a primate from the forest to more open country made us finally into what we are, and that is the story of our own evolution.

Suggestions for Further Reading

The most convenient reference book on the living primates is by J. R. and P. H. Napier, *Handbook of living primates* (New York and London: Academic Press, 1967). For a simple account of primate anatomy, see A. H. Schultz, *The life of primates* (New York: Universe Books, 1969). On the general features of the nervous systems of mammals, consult J. Z. Young, *The life of mammals* (Oxford: Oxford Univ. Press, 1957). For a comparative review of some aspects of higher primate behavior and anatomy, see R. Passingham, *The human primate* (Oxford and San Francisco, Calif.: Freeman, 1982). For the evolution of the nervous system, see H. B. Sarnat, *Evolution of the nervous system* (New York: O.U.P., 1974), and of the vertebrate visual system, consult S. Polyak, *The vertebrate visual system* (Chicago: Univ. of Chicago Press, 1957). For papers on the primate brain, see E. Armstrong and D. Falk, *Primate brain evolution* (New York: Plenum, 1982).

Those interested in prosimians may consult C. A. Doyle and R. D. Martin, *The study of prosimian behavior* (London and New York: Academic Press, 1979). A recent publication on New World monkeys is also recommended: R. L. Ciochon and A. B. Chiarelli, *Evolutionary biology of the New World monkeys and continental drift* (New York and London: Plenum, 1980).

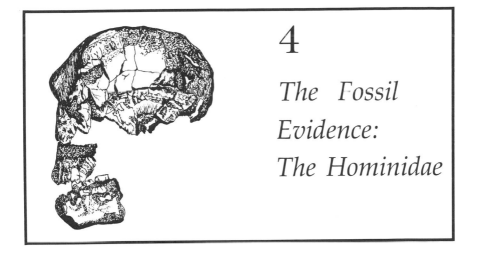

4

The Fossil
Evidence:
The Hominidae

In each great region of the world the living mammals are closely related to
the extinct species of the same region. It is, therefore, probable that Africa
was formerly inhabited by extinct apes closely allied to the gorilla and chimpanzee; and as these two species are now man's nearest allies, it is somewhat more probable that our early progenitors lived on the African continent than elsewhere.

Charles Darwin
The Descent of Man, 1871

**I. Fossil
Hominids:
*Australopithecus***

We have already seen that, while the greatest
amount of evidence for the nature of our ancestors
comes from the study of living primates, the most
direct evidence is to be found among fossil forms
that bear close resemblance to us. Our relationship
to both fossil and living Old World higher primates is illustrated very
approximately in Fig. 4.1. The diagram represents no finally accepted
fact of evolution, but rather a hypothesis that is supported by the
present evidence.

Some important fossil groups are shown in their approximate relationships (Fig. 4.1): *Aegyptopithecus* (E), *Propliopithecus* (P), *Proconsul* (Pr),
Dryopithecus (D), *Sivapithecus* (S), genera of fossil apes; and the two
genera of the Hominidae, *Australopithecus (A)* and *Homo (H)*. The figure
also shows the most probable route for the evolutionary lineage of the
Hominidae. In the past many alternatives were considered. For example, it was suggested that human evolution from a prosimian was
independent from that of any other anthropoid primate (Kurten, 1972).
Straus (1949) argued that we had more in common with monkeys than

89

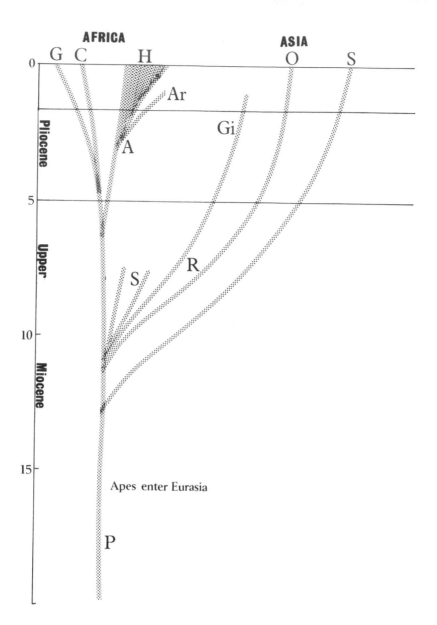

Figure 4.1. Diagrammatic representation of lineages suggesting the course of the evolution of the Old World Anthropoidea. The origin of the different groups is not known with certainty, and the dates of origin are based on biochemical and fossil evidence. *Homo* (H), *Australopithecus* (A), *Australopithecus boisei* (AR), gorilla (G), chimpanzee (C), orangutan (O), the gibbons (S), *Gigantopithecus* (Gi), *Ramapithecus* (R), *Sivapithecus* (A), *Dryopithecus* (D), *Proconsul* (Pr), *Aegypotithecus* (E), and *Pliopithecus* (P).

with apes. The evidence to be considered in the following chapters will show, however, that any lineage basically different from that suggested here is improbable.

The oldest specimens which can be recognized as higher primates (Anthropoidea) are probably those grouped in the genus *Propliopithecus* and known since 1910 from the Fayum Depression in northern Egypt. They come from the Early Oligocene and are dated to about 35 mya. They are clearly apelike in their dentition, but in their little-known cranial and skeletal anatomy they are very primitive and cannot be grouped with any of the living higher primates. For the present they can be considered possible ancestors to monkeys, apes, and humans.

From a slightly later date (c.32 mya) we have the better known genus *Aegyptopithecus* which has been collected from the same area by Simons since 1964. It shows considerable similarities in its dentition to the later Miocene apes, but in the postcranial skeleton is quite distinct and can be interpreted as a heavily muscled, slow moving, arboreal, climbing quadruped (Fleagle, 1983). On the basis of the present evidence it has been considered by Simons and others to be a possible ancestor to all later apes.

From the Middle Miocene (c.18 mya) of Kenya (mainly from Rusinga Island) we have specimens of the fossil ape named *Proconsul*. This was first discovered by Mary Leakey in 1948 and recently a number of new specimens have been collected. *Proconsul* is now recognized as an important genus which stands clearly at the base of the ape-human lineage and can be taken to represent an early ancestral ape. There are three species known which differ mainly in size; the smallest, *P. africanus*, is best represented in the fossil record. Although the new material recently found (5 specimens) has yet to be described, we already have a good idea of the nature of this species. The skull, forelimb, and hindlimb have been studied, and most of the other parts of the skeleton have been discovered. The creature appears to have been a small, robust, primitive, and unspecialized ape. It weighed perhaps 20–25 lb and stood 18 inches high at the shoulder when standing on all four limbs. It was probably a powerful quadrupedal climber, not obviously adapted to either brachiation, knuckle-walking or fast terrestrial quadrupedalism (Fleagle, 1983; Walker and Pickford, 1983).

During Miocene times the Tethys Sea, which had separated Africa and Eurasia for so long, retreated, and around 18 million years ago the continents became joined together, at first intermittently, as a single land mass. Both were tropical with high rainfall, and they became connected by richly forested corridors, a narrow one through Gibraltar in the west and a wider one through Arabia in the east. Animals of many kinds passed between the continents by the eastern corridor. But at the same time, a cooling in temperature and a lowering in the rainfall began in the more northerly latitudes, producing a more arid climate. The Alps, Urals, and Himalayan mountains were uplifted. Africa underwent tectonic changes too; rifting and volcanic activity began, which lasted well into the Pleistocene. A fundamental change in the environment of the Old World was under way; by the late Miocene, grassland areas began to replace forest in both Africa and Eurasia, and became even more extensive in the Pliocene.

In late Miocene deposits we find a number of genera of apes including the well-known *Sivapithecus* (from Africa and Asia) and *Dryopithecus* (from Europe and Asia). On the basis of the known fossil remains, these genera seem slightly more like modern apes than the earlier *Proconsul*, but nevertheless retain much in common with that genus, with highly flexible limbs and a lack of the kinds of specializations that we find in the living apes (Rose, 1983).

These apes were generally still confined to forest, and as the forests retreated in Eurasia, they disappeared from Europe and only the immense *Gigantopithecus* survived in the dryer regions of Asia (especially China) until about 1 million years ago. The lesser apes and the orangutan remained in the forests of China and Southeast Asia.

The new grasslands of Africa and Eurasia became occupied by newly evolved grazing quadrupeds and their predators. Among the latter, early in the Pliocene, we can identify the first evidence of a hominid, ancestor *Australopithecus*.

Although claims have been made that *Australopithecus* is represented in the Chinese Pliocene fauna, these claims require further investigation; the genus is best known at present by a wealth of fossil bones and teeth from Africa. *Australopithecus* is invariably associated with grassland and savanna animals, but often with some evidence of woodland as well. The first find of *Australopithecus* was made in 1924 and was recognized as a missing link in our ancestry by the anatomist Raymond Dart. The skull in question came from the site named Taung, now within the Republic of South Africa. Further finds in the Transvaal followed in the 1930's, and since 1945 many more fossils have been added from this very fossiliferous region. More recently (1960–1980) a large number have been discovered in East and Northeast Africa (Tanzania, Kenya, Ethiopia).

Unfortunately, the taxonomy of the group presents a number of problems, and qualified workers disagree widely as to how many and which species* are represented at any particular site. However, as it is not the purpose of this book to discuss their taxonomy but rather to assess the functional anatomical significance of the fossils, we shall merely summarize the differing views of their taxonomy. The most important sites are as follows:

---

*When we use the term species here, we have in mind something different from the biospecies described in Chapter 1, IV. Fossil species are properly called chronospecies, which implies that they have a dimension of time as well as spatial and morphological variability. They represent a segment of an evolving lineage, a biological species extended through a long period, with the variation

Transvaal. Fossils from this region are usually divided into two species: (1) *Australopithecus africanus,* from Taung, Sterkfontein, and Makapansgat. Age, 3–2.5 million years ago (mya). (2) *A. robustus,* from Swartkrans and Kromdraai. Age, 2–1.5 mya (Fig. 4.3).

Eastern Africa. The taxonomy is very complex, but we can distinguish the following main groups: (1) *A. afarensis,* from Laetoli. Age, 3.75 mya. (2) *A. boisei,* from Olduvai, Koobi Fora, and Peninj. Age, 2.5–1.2 mya.

Ethiopia. (1) *A. afarensis,* from Omo and Hadar. Age, 3.3–2.8 mya. Some authors also recognize fossils related to *A. robustus* among those from this site (Fig. 4.3).

Thus, the fossils that fall into this genus represent four commonly recognized species. Although their nature and occurrence may be an issue for debate, most taxonomists agree that they can be seen to form two contemporary and related lineages. One lineage, of gracile lightly built forms, leads to *Homo,* while the second, which becomes increasingly robust in build, evolves for a time and then becomes extinct (Figs. 4.1 and 4.4). Fossils from the first lineage range from about 4.0 to 2.3 mya; those from the second, from around 3 to 1.2 mya. The exact relationship of the different species is not entirely clear. In one scheme (Tobias, 1980) *A. africanus* is ancestral to both lineages and includes *A. afarensis.* In a second scheme (Johanson and White, 1979), *A. afarensis* is the ancestral form and *A. africanus* leads only to the robust *A. robustus* (which includes *A. boisei*). See Fig. 4.4.

The two older species, the gracile taxa *Australopithecus afarensis* and *A. africanus**** are well represented in the fossil record by parts of some 500 individuals, being better known than any hominid species except *Homo sapiens.* By taking all the evidence into account we can get a good idea of the variation present in the different species in both space and time. Spatial variation is due to environmental differences in different ecological zones (such as we find among modern human races), while temporal variation represents the evolution of these species from an earlier ancestral form (which also gave rise to the African apes) to a creature that eventually became *Homo.*

---

due to evolution added to that which we would expect of any population at a given point in time (Fig. 4.2). The main difficulty that arises from the concept of the chronospecies is that *successive chronospecies* in a single lineage are continuous and not discrete; there is no real natural boundary between them. Those that represent two contemporary lineages can easily be distinguished by the methods usually employed to recognize biospecies, but where they succeed each other in a single lineage, the boundary must be artificially drawn by the taxonomist. The result is that many different opinions exist as to the best place to draw these boundaries; the same fossils are given different names by different authors and their nomenclature becomes confusing. The classification and nomenclature used here (Table 4.1) are the simplest and least controversial.

*Under the term *Australopithecus africanus* I am including fossils previously designated as *Plesianthropus transvaalensis.*

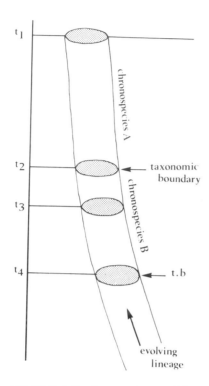

Figure 4.2 The difference between biospecies and chronospecies is here shown diagramatically. An evolving lineage is divided by time lines that represent taxonomic boundaries, at time *t*2 and *t*4, into two chronospecies, A and B. Biospecies can be recognized at any point in time (e.g., *t*3) and the term is used to describe the present status of the lineage (*t*1).

TABLE 4.1. Approximate Chronological Range of Hominid Species

Hominid species	Range (in thousands of years)
Homo sapiens	300–0
Homo erectus	1500–300
Homo habilis	2500–1500
Australopithecus africanus	3000–2500
A. afarensis	3750–2500
A. robustus	2100–1500
A. boisei	2500–1200

Considering its antiquity, the species *Australopithecus afarensis* is surprisingly well represented in the fossil record. The oldest well-dated specimens are from Laetoli in Tanzania (securely dated at 3.5 mya). Even older hominid fragments are known from Kanapoi (about 4.0 mya) and Lothagam (about 5.5 mya) in Kenya, which, on the basis of their limited anatomy, may possibly be early members of this species. Apart from the Kanapoi specimen (a distal humerus fragment), all these remains are of jaws and teeth; the Laetoli collection numbers 14 specimens. Much more striking, however, are the extraordinary and somewhat later finds from Hadar, which include the most complete *Aus-*

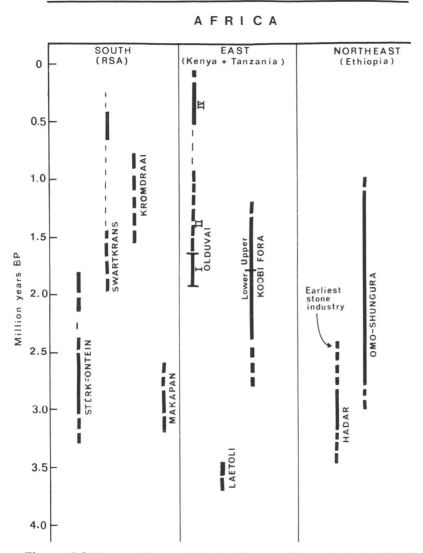

Figure 4.3 Ranges of uncertainty in the age of important *Australopithecus* fossils and the deposits in which they were found.

tralopithecus specimen we have, known colloquially as *Lucy* (AL-288). This skeleton is probably female and stood about 3 ft 6 in. tall (Fig. 4.5). It is 40% complete, but because one side of the skeleton is a mirror image of the other and we have different bones from each side, we can effectively reconstruct over 70% of it. It dates from about 3.0 mya and, together with a mass of other fossil material of both skulls and other

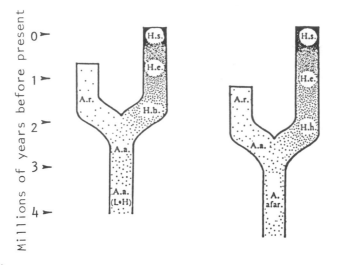

Figure 4.4 Diagrammatic representation of the last phases of human evolution showing two different interpretations of the evidence. *Left* (according to Tobias, 1980): here the ancestral species *Australopithecus africanus* includes the early forms called *A. afarensis* (L & H). *Right* (according to Johanson and White, 1979): here the early *A. afarensis* is the ancestral species and *A. africanus* is ancestral only to the robust branch (from Johanson and White, 1979).

skeletal elements from this period (3.3–2.8 mya), we can say a good deal about this ancient species. The size of the samples will become apparent as we review the anatomical evidence in later chapters.

Australopithecus afarensis, represented by the Hadar and Laetoli specimens, is characterized by having a skull somewhat reminiscent of that of a chimpanzee, but with a relatively large brain (380–500 cc), large masticatory apparatus (jaws and supporting structures), and extensive pneumatization (air pockets and sinuses within the thicker bones of the skull). See Fig. 4.6 (Kimbel *et al.*, 1984). The dentition, however, shows characteristics that clearly distinguish it from that of the apes, indicating that evolution toward the human condition was already under way. It is clear from the structure of the base of the skull that it was balanced on the vertebral column as in humans, and is fundamentally different in this respect from that of the apes. Evidence of the limbs shows that the creatures were bipedal—a bipedalism characteristic of all known hominids, but in this species accompanied by many apelike features of the hand and foot.

The best preserved skeleton belonging to *A. africanus* comes from the Sterkfontein site near Johannesburg. There are fragments of the scapula and humerus, some rib fragments, nine vertebrae from the thoracic and lumbar regions, the sacrum, pelvis, and a poorly preserved femur. The

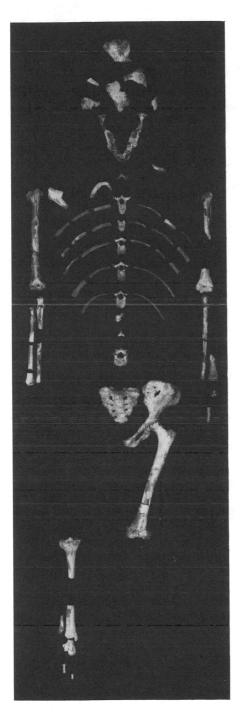

Figure 4.5. The fragmentary skeleton of "Lucy," discovered by D. C. Johanson in 1975. The skeleton is about 40% complete, and is the most complete specimen of *Australopithecus* known (courtesy Institute of Human Origins).

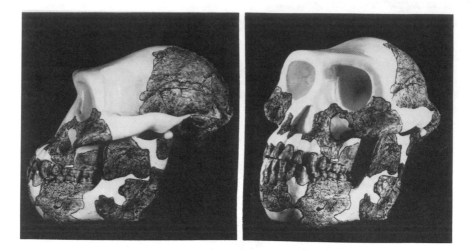

(A)

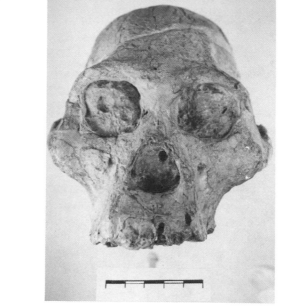

(B)

Figure 4.6. A composite reconstruction of the skull of *Australopithecus afarensis* (A); a well-preserved cranium of *Australopithecus africanus* from Sterkfontein (B); the skull of *Homo habilis* from Koobi Fora (C); and a reconstruction of the skull of *Homo erectus* from Peking (D). [A, by courtesy of Institute of Human Origins; B, the Transvaal Museum; C, R. E. Leakey; and D, by courtesy Trustees of the British Museum (Natural History).] Not reproduced to scale, but all about 20% of natural size.

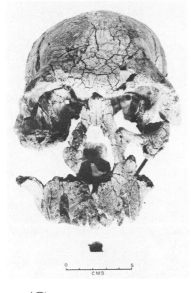
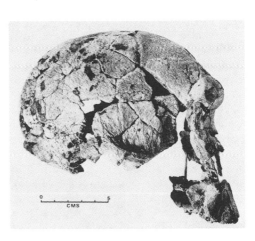

(C)

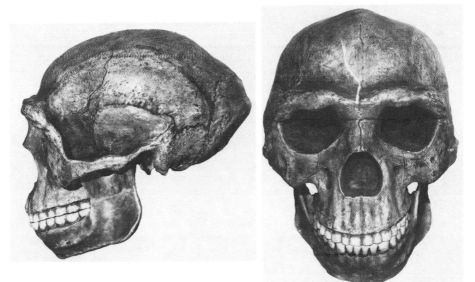

(D)

stature of this skeleton was possibly about 4 ft 3 in. The skull and other parts of the skeleton of this species are known from fragments of other individuals from this and other sites. Some idea of the quantity of material can be seen from the reconstructed skeleton shown in Fig. 4.7. The bones are named in the modern human skeleton in Fig. 4.8

The skull of *A. africanus* is again superficially apelike in appearance (though much more lightly built than that of the large African apes). Cranial capacity ranges from 435 to 530 cc. As a result, the braincase is slightly larger and more rounded than *A. afarensis*, especially in the larger-brained individuals (Figs. 4.6 and 4.9). The face is again relatively massive; the jaws are large and the supraorbital torus (brow-ridges) well developed. The big jaws have no chin, but a dental arcade and dentition close to that of modern humans is present. The canine teeth are distinct from those of the apes and similar to our own, being nearly spatulate like the incisors, and all the front teeth are much reduced. The molar and premolar teeth are relatively large but humanlike in general appearance.

The anatomy of the postcranial skeleton is more obviously humanlike than that of the skull, though there are minor differences in most features, which we shall describe in later chapters. It is quite different in almost every feature from that of living monkeys and apes. The implications of a detailed analysis of the skeletal material are that both *Australopithecus afarensis* and *A. africanus* were efficiently adapted bipedal animals. The brain, however, was still small relative to body size, compared with a modern human's, though relatively larger than that of an ape, and the large jaws gave the head an apelike appearance. In summary, these species carry an extraordinary combination of human and apelike characteristics.

The features of *A. afarensis* are broadly similar to those of *A. africanus* (though *A. afarensis* is almost certainly the older taxon). Because of this, some authors (e.g., Tobias, 1980) consider *A. afarensis* no more than a northern subspecies of *A. africanus*. (The latter name has priority, as it dates from 1925; *A. afarensis* dates from 1978.) White, Johanson, and Kimbel (1981), however, have carefully reviewed the characteristics of both groups and have identified distinctive features that they believe justify recognition of the groups as two distinct, but successive, species. Clearly, if they are successive, their features will blend in a continuous series, even though their mean characteristics are distinct. For descriptive purposes herein it is simpler to use the terminology of species than subspecies, and the evidence seems to weigh in favor of this approach. What is important is the anatomy and age of these fossils, and the part they have played in human evolution.

The species *A. robustus* comes from sites in the Transvaal that are broadly similar to but later than those that have yielded *A. africanus*. Again, there has been controversy as to whether these fossils are really specifically distinct from the older *A. africanus*. They could prove to represent a descendant population that may be either subspecifically

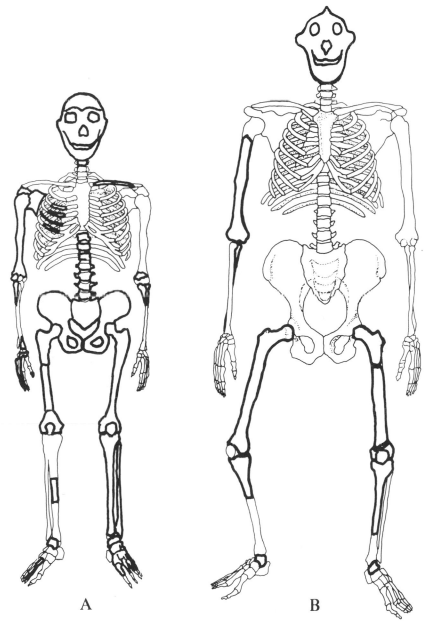

A   B

Figure 4.7 Reconstructed skeletons of *Australopithecus africanus* (*A*) and *Australopithecus boisei* (*B*) with those fossil bones that have been recovered (drawn with heavy line). The skeletons are reconstructed from fragments of many different individuals. In the case of (*B*), the reconstruction is highly imaginative, since much fossil material is missing.

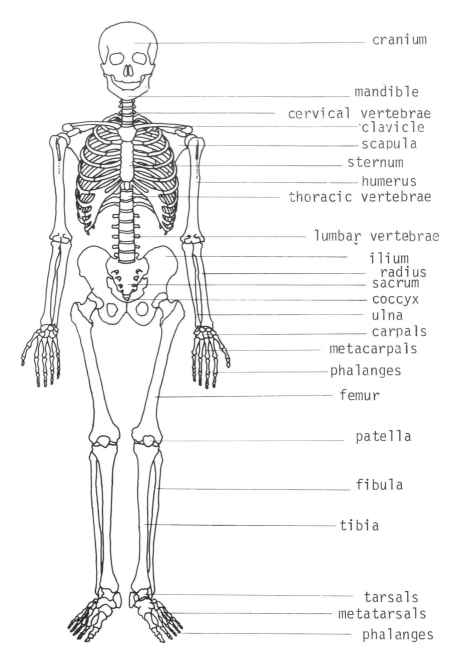

cranium

mandible

cervical vertebrae

clavicle

scapula

sternum

humerus

thoracic vertebrae

lumbar vertebrae

ilium

radius

sacrum

coccyx

ulna

carpals

metacarpals

phalanges

femur

patella

fibula

tibia

tarsals

metatarsals

phalanges

Figure 4.8 The human skeleton with the most important bones labeled.

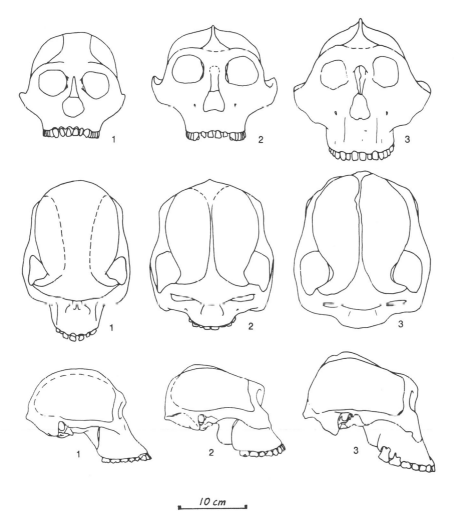

Figure 4.9. The skulls of *Australopithecus africanus* (1), *A. robustus* (2), and *A. boisei* (3), in frontal (*above*), superior (*middle*), and lateral views (*below*). Note the cranial characteristics that change with increasing cheek tooth size (*from left to right*), especially the different development of the sagittal crest and the size of the face and upper jaw (after Tobias, 1967).

distinct or specifically distinct from its ancestors. For our purposes it seems most convenient to recognize the group as a distinct species, even though the matter is not finally resolved.

Australopithecus robustus differs from *A. africanus* in having a much more robust skull (Fig. 4.9) and possibly a slightly larger and more robust skeleton. The differences in the skull are related to the dentition, in the molar and premolar teeth, and the heavier bone and muscle apparatus needed to carry them. This means that the important distinctions lie in the teeth and skull. Only one fairly complete skull with a cranial capacity of about 500 cc is known. Little is known of the postcranial skeleton, but it appears to indicate the same kind of bipedalism as that found in *A. africanus*.

The species *Australopithecus boisei* stands out clearly from the other three species of the genus. This group of early hominids, of which fossils have been found only in eastern Africa, responded to the grassland environment not as the gracile species, which was destined to become an adventurous tool-using predator, but more like the other higher primates with terrestrial adaptations, as rather orthodox herbivores. The teeth of *A. boisei* underwent exactly the kind of evolutionary changes that we associate with herbivores; that is, a vast increase in the grinding surface of the molars and a reduction of the front teeth—features also characteristic of but less pronounced in *A. robustus*. That the species is an *Australopithecus* is obvious at a glance (Fig. 4.9), but it is also clear that the species' adaptation was somewhat different and involved a commitment to the mastication and digestion of tough vegetable food. The molars are vast, bigger than those of any other primate, and the jaws are the most efficient known in terms of the power brought to bear on these grinding teeth. Like many other herbivores, this is a big animal, far more heavily built than the other species of the genus.

Beyond this we know very little about *A. boisei*. The few bones of its skeleton that we possess do not tell us for sure if the species was as fully bipedal as other species of the genus. But *A. boisei* had a brain of about 530 cc and brought all the advantages of advanced primate nature to its new environment. The fossil record of the species begins just over two million years ago, and the last remains date from a little over one million years ago. Theorists have suggested that the species succumbed in tooclose competition with its cousin, that slender but witty creature that was to become human.

A detailed discussion and description of the anatomy and adaptations of *Australopithecus* follow in later chapters, and our purpose here has been merely to introduce the genus. It spans over three million years and shows much change in form between the earliest populations that carry more apelike characteristics and the later and better known populations that can be sorted quite easily into two lineages, one of which leads to *Homo*.

A formal definition of *Australopithecus* follows (modified from Le Gros Clark, 1978):

Australopithecus: A genus of the Hominidae distinguished by the following characters:
Cranial capacity ranging from about 380–530 cc.
Relatively thin-walled cranium with strongly built ectocranial superstructures and robust supraorbital ridges.

A low sagittal crest in species and individuals with heavy masticatory apparatus.

Occipital condyles well behind the midpoint of the cranial length but on a transverse level with the auditory apertures.

Nuchal area of occipital bone restricted and downward facing, as in *Homo*.

Some development of pyramidal mastoid process but less striking than in *Homo*.

Jaws variable in size but often massive.

Chin absent but inner symphyseal surface strengthened by heavy mandibular torus.

Dental arcade generally parabolic with no diastema, except in *A. afarensis*.

Moderate-sized canines wearing down from the tip only; spatulate in later species.

Relatively large premolars and molars.

Anterior lower premolar bicuspid with subequal cusps, except in *A. afarensis*, where it is unicuspid.

Pronounced molarization of first deciduous molar.

Progressive increase in size of permanent lower molars from first to third.

Limb skeleton conforming in its main features to that of modern humans but differing in a number of details, such as relatively smaller sacroiliac articulation and acetabulum.

Blades of the ilia splayed widely and birth canal relatively narrower than in *Homo*.

The successful adaptation of *Australopithecus* to woodland and grassland savanna evidently allowed this genus to expand rapidly over a wide geographical area. Their primate heritage of evolved social behavior, intelligence, and vision, together with efficient bipedalism, enabled these creatures to adapt to a wide range of local conditions. Whether they managed to spread beyond the confines of the African continent is not yet clear, but their successors certainly did, as did many others of their versatile primate relatives.

This history of *Australopithecus* takes us through the Pliocene epoch. Throughout this time, the Old World land masses continued experiencing upheaval and increasing aridity. The retreating Tethys Sea finally disappeared; only relict seas remained: the Mediterranean, Black, and Caspian seas. At first the climate was generally warm, and the grasslands at this time extended from Europe to China, reaching their greatest known extent. These huge areas supported vast herds of grazing animals ancestral to our horses, cattle, antelopes, and elephants. Savanna primates appeared alongside *Australopithecus*: the macaques, baboons, geladas, and the great terrestrial ape *Gigantopithecus*, which is now extinct. Ancestral hyenas, cats, and dogs came to replace more primitive predators.

Toward the end of the Pliocene a succession of cool periods began to change the northern latitudes drastically. These now became colder, and in the ensuing Pleistocene the oscillating temperatures brought into existence the well-known ice ages (Figs. 4.10 and 4.11). Meanwhile, even

during the warmer interglacial periods the equable seasonal temperatures of the early Pliocene were lost, and cold winters became a factor of
great significance for the evolving flora and fauna. As we shall see, they
also had a profound effect on the successor of *Australopithecus*, the genus
Homo.

II. *Homo*
habilis

By the later Pliocene, *Australopithecus* must have
come to occupy the savanna grasslands and wood-
lands stretching from southern to northern Africa
and even perhaps into Asia, and it seems possible
(though evidence is at present lacking) that they were beginning to
supplement a primarily vegetarian diet with meat of many kinds. We
can guess that with the help of a very primitive technology, this creature's efficiency at extracting vegetable and perhaps some animal resources from the environment was continually increasing. Its successor
developed these traits and is recognized as the earliest form of the genus
Homo, called *Homo habilis*.

This species is poorly represented in the fossil record, but a small
number of specimens have been found in Africa that appear to stand
both morphologically and chronologically between the most advanced
gracile *Australopithecus* and the later *Homo erectus* (see below). The first of
these to be recognized was a group of fragmentary bones found by the
Leakey family at Olduvai Gorge in Tanzania. More recent discoveries
were made at Omo in southern Ethiopia, Koobi Fora in Kenya, and at
two of the Transvaal sites that gave us *Australopithecus*—Sterkfontein
and Swartkrans.

The diagnostic characteristics of *Homo habilis* are found mainly in the
skull and teeth (few long bones are known). The cranial capacity (based
on five specimens) ranges between 600 and 750 cc, and yet the stature
remains small, so that we are witnessing the beginning of a rapid increase in relative brain size. This increase is one of the most remarkable
trends in the evolution of the genus *Homo*. The skull, however, is lightly
built; the dentition is not unlike that of *Australopithecus africanus*, though
the premolars are longer in mesiodistal diameter (see Chapter 9, VI, C).
On the basis of the few bones preserved, the limbs appear to be modern
in form, which is to say that hand, leg, and foot bones show many
modern manipulative and bipedal characteristics.

The Olduvai remains constitute the first hominid fossils associated
with the remains of a stone tool industry. This is not to say that *Homo
habilis* was the first tool-using primate, because the tradition is almost
certainly older than that (see Chapter 9, IX). It is, however, quite clear
on the basis of the presently available evidence that *H. habilis* was the
first creature on earth to make regularly patterned stone tools in a reasonable quantity. About two million years ago is the first point in geological time that any stone industry can be confidently attributed to
any organism, though evidence of stone working has been found from
deposits believed to date from 2.5 mya.

The Olduvai evidence also makes it clear that the stone industry was associated with the butchery of animals of all kinds, though in the lower levels the animals are small. Meat-eating became a feature of the adaptation of these early members of the genus *Homo*.

The other important feature of the African *Homo habilis* fossils is that, morphologically and chronologically, they form a link between the ancestral African hominids, the later African specimens and the first non-African fossil hominids—those found in China and on the island of Java. The earliest of these fossils (from Java) may be as old as 1.7 million, but an age of 0.8 mya seems more probable. The Javanese fossils have been shown to have clear morphological continuities with the much earlier *Homo habilis* fossils from Olduvai (Tobias and von Koenigswald, 1964).

III. *Homo erectus*

If *Homo habilis* did not spread into Eurasia from the African continent, its successor, *Homo erectus*, certainly did. Adapting first to the tropical regions along the southern parts of Asia, this group slowly ranged northward into the center of the landmass. In doing so, they came face to face with the successive ice ages that spread southward throughout North Temperate Eurasia and began adapting to the changing climate. They built shelters and lived in caves to protect themselves from the cold; eventually, they collected and even domesticated fire. As they learned to cope with this new seasonal environment, their style of life changed and their anatomy and behavior evolved accordingly. Their success as hunters implies a well-developed social structure and extensive cooperation. During the warm interglacial periods they expanded northward and eventually learned to survive the cold winters of the North Temperate Zone.

Fossils from this period are placed in the genus *Homo* but designated *Homo erectus* to distinguish them from modern humans. The earliest are found in Africa (Koobi Fora, Ileret, and Olduvai Bed II) and date from about 1.5 mya (Fig. 4.11). The next group come from China and Java (as we mentioned earlier) and are dated about 0.8 mya. Soon after this we have skulls and jaws of slightly more modern appearance from many sites in the Old World. These date from between 600,000 and 300,000 BP; the most valuable come from Algeria (Ternifine), Germany (Heidelberg), Hungary (Vertesszollos), and China (Choukoutien). They reach latitude 41 degrees in Germany. From ca. 500,000 years BP, hearths are found in caves and other occupation sites, together with stone tools and bones.

As we might expect, the fossil remains of *Homo erectus* show much geographical variation in their general appearance and size. In view of their wide temporal and geographic range, however, they appear remarkably homogeneous. Unfortunately we do not have as big a collection of fossils of this species as of *A. africanus*; nevertheless, we can assess their anatomy fairly precisely. Although the skull still looks primitive, even somewhat apelike, the postcranial skeleton was becom-

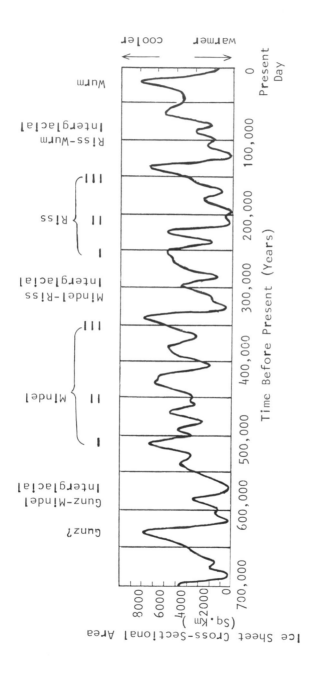

Figure 4.10. The Pleistocene ice ages were most probably caused by periodic variations in the geometry of the earth's orbit, and such variations as are believed to have occurred concur with evidence from other sources (oxygen isotope analysis of deep-sea cores). From these data it has proved possible to calculate the cross-sectional area of ice sheets at any time during the last 700,000 years. As we see in this chart, the main peaks of cold weather occur approximately every 100,000 years, but many other cold oscillations come between these peaks. Although correlation with the named northern European ice ages is not secure, those indicated may be correct.

ing very close to that of modern humans—so close in fact that individual bones are often indistinguishable from those of *Homo sapiens*. Compared with its ancestor *Homo habilis*, *H. erectus* seems to have been taller and much more heavily built. In comparison to modern humans, it was a powerful, muscular, stocky, thickset creature. It was almost certainly as efficient at walking and as well balanced in posture as we are.

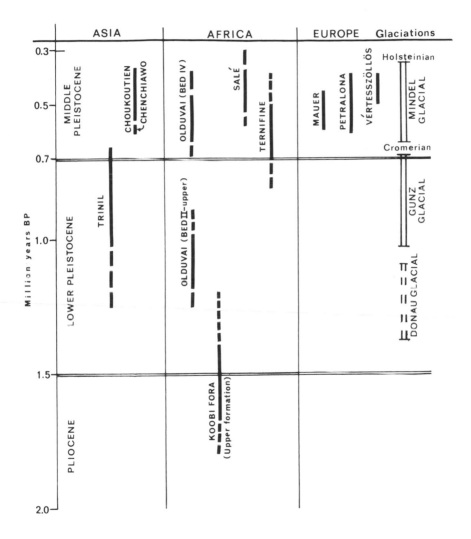

Figure 4.11. Ranges of uncertainty in the ages of important *Homo erectus* fossils and the deposits in which they were found. Some of the later specimens may be classified as *Homo sapiens* by some writers.

While that thickset body might have been taken for an unusually robust modern, the head of *H. erectus* was quite different. It still carried clear traces of the powerful jaws of *Australopithecus*, and although the brain was now much larger (close in size to our own), the skull was heavily built, thick, and carried a powerful musculature to support itself and the jaws (Fig. 4.6). The total effect was of a head still clearly distinct from that of modern humans.

The first finds of this species date from 1891, when the Dutchman Eugene Dubois unearthed the first really ancient fossil skull ever found, at Trinil in Java, amid much controversy. He called it *Pithecanthropus erectus* ("erect apeman"), and that is the origin of the name of this species. Dubois did not find any cultural remains with his discovery and seriously underrated the level of cultural achievement of these people. Today we have extensive remains from sites further north, especially at Choukoutien, near Peking in China, where there is a vast cave that was occupied for over 100,000 years by a band of hunters who left us their hearths, their tools, their food traces, and their own bones in profusion. Somewhat similar sites are known from Europe (Torralba, Ambrona, Terra Amata, and Vertésszöllös) and from Africa (Ternifine and Olduvai Bed II). We now have a fairly clear picture of the adaptations and life-style of *Homo erectus*. The achievements of this species were considerable: it mastered fire, and survived the northern winters by its superior technology and remarkable skill as a hunter.

As we have seen, one of the problems of studying a fossil lineage such as this is to determine at what points we should draw a line to separate successive species. Fossils that lie near these boundaries are classified in different ways by different scholars. Without becoming involved in this difficult question, we shall for convenience assign the species name *Homo erectus* to our ancestral hominid lineage during the period 1.5 to 0.3 million years BP. After that time our ancestors become almost indistinguishable from ourselves.

IV. *Homo sapiens*   By about 300,000 years ago, the people living in Eurasia and Africa were practically modern in appearance if not in culture. Their brain size was well within the range of that of modern humans, and their bodies were indistinguishable from ours, though some of the early populations were considerably more robust. Only their heads still looked strange, some with thick-boned skulls, some with long skulls and heavily built faces and jaws. Through the evolution of *Homo sapiens* during the last 300,000 years, we see the thinning of the skull bones, the final reduction of the jaws, and the appearance of the chin. As the jaws became smaller, the whole face shrank and receded under the braincase, so that it became surmounted by a vertical forehead. This changed the balance of the head, and the rather long narrow skull of *Homo erectus* became the rounded skull of modern populations.

During this final period, which brings us to the present day, the succession of ice ages continued, with two more major advances separated by warm interglacial periods (Fig. 4.10). Humankind's response to the ice was to become its greatest cultural achievement. By about 70,000 years ago, humans had made the final climatic conquest and developed the ability to survive the year-long cold of the arctic biome. The first evidence of this extraordinary feat comes from France, where the Neandertal people survived for more than 10,000 years in a climate as rigorous as that many Eskimos enjoy today. We do not yet know the whole story of their adaptation, but survive they did, to demonstrate that *Homo sapiens* was indeed the most adaptable creature on earth.

From the boundary zone between *H. erectus* and *H. sapiens* we have a number of very important fossils whose taxonomic status is controversial. They come from Petralona (Greece), Bilzingsleben (East Germany), Vertésszöllös (Hungary), and Mauer (Germany) in Europe, from Bodo and Omo (Ethiopia), and Ndutu (Tanzania) in Africa, and from Ngandong (Java). See Fig. 4.9 and 4.12. Many of them are not well dated; some are closer to modern *H. sapiens*, and some seem more closely related to the Chinese or Javanese *H. erectus*. It is to be expected that such intermediate forms should be found and that they should present problems to the taxonomist. More research will be needed to determine their final taxonomic and phylogenetic position.

From the next period the fossil record is extremely sparse. We have only four substantial finds dated between 300,000 and 200,000 BP. They all come from Europe: Montmaurin and Arago (France), Swanscombe (England), and Steinheim (Germany), reaching north to a latitude of 52 degrees (Fig. 4.13). The last three sites have yielded us skulls of great value; here we find considerable variation in the extent of jaw reduction, as well as the usual variability in size, robusticity, and cranial capacity.

Four later sites that belong to the next 100,000-year period have yielded fragmentary remains. It is not until the final 100,000 years of our history that the fossil record supplies us with good evidence of the most recent stages of our evolution, and even in this last period we find that the story is unclear. For the period's first half (100,000–50,000 BP), the known sites carry fossils that still have relatively large jaws, though they are modern in most other respects. These are the Neandertal, Rhodesian (Fig. 4.14), and Shanidar people. Then, between 50,000 and 30,000 years BP, we see these big-jawed people disappearing and being replaced by small-jawed people like ourselves. The earliest examples of this type come from South Africa (Border Cave, Natal, ca. 60,000 BP; Klasies River, ca. 40,000 BP), China (Tali, ca. 100,000 BP; Luijian, ca. 40,000 BP), and Southeast Asia (Niah, Sarawak, ca. 40,000 BP), and somewhat later from Europe (Dolni Vestonice, Czechoslovakia, ca. 30,000 BP). The only fossils that have been claimed to be intermediate are those from Es-Skhul and Djebel Kafzeh in Israel, dating from about 50,000 years BP, but many scholars consider them to be anatomically modern.

This transformation from a large- to a small-jawed modern type has

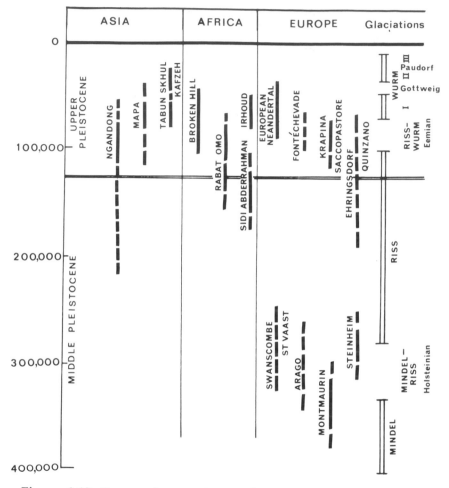

Figure 4.12. Ranges of uncertainty in the ages of important early *Homo sapiens* fossils and the deposits from which they were derived. The classical names of the European ice ages are shown on the right.

always been something of a mystery. In many areas the replacement appears to be so sudden that an invasion of modern forms from elsewhere has been postulated (e.g., in France), while in other places the two forms seem to have overlapped in time (e.g., southern Africa). Only in western Asia are found what appear to be intermediate specimens, but it may be pure chance that they have not been uncovered elsewhere. The simplest hypothesis on the basis of the present evidence is that modern forms first appeared in western Asia or northern Africa and soon spread south, moving later to eastern Asia and finally to Europe. We might expect the big-jawed forms to have survived longer in outly-

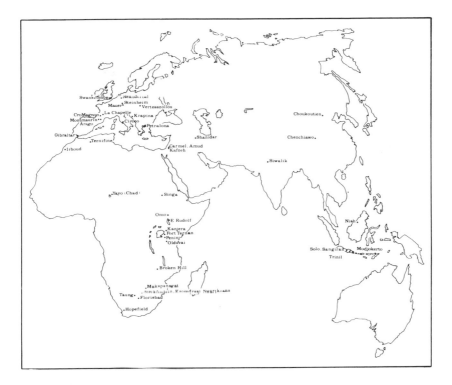

Figure 4.13. Map of the Old World showing the locality of some important sites mentioned in the text and listed in the Appendix, p. 431.

ing areas. We do not at present have enough evidence to determine whether the big-jawed populations were genetically swamped by the invaders, were exterminated by them, or evolved into their successors. Just one thing is certain: the genes of the big-jawed people survived longer in Southeast Asia than elsewhere, and have made a substantial contribution to the aboriginal people of Australia, who have the largest jaws and teeth among living peoples.

The search for the origin of modern people is today being pursued on many continents. In a few years we may have better data with which to test our hypotheses.

A formal definition of the genus *Homo* (based on Le Gros Clark, 1978) runs as follows:

> *Homo:* a genus of the Hominidae distinguished by the following characteristics:
> A large cranial capacity with a mean value of more than 1100 cc but with a considerable range of variation, from about 600 to over 2000 cc.

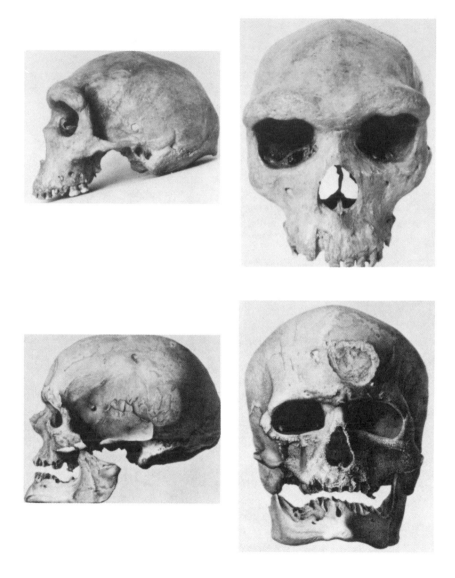

Figure 4.14. A well-preserved cranium of an early large-jawed *Homo sapiens* from Broken Hill, Zambia (*above*); a 19th century drawing of the first fossilized small-jawed skull of *H. sapiens* found, from Cro-Magnon, France (*below*). [Courtesy Trustees of the British Museum (Natural History).] Not reproduced to scale.

Masticatory apparatus variable in size but relatively smaller than the neurocranium when compared with *Australopithecus* and much reduced in modern *H. sapiens*.

These two primary features involve the following associated changes:

Flattening of the face.

Forward movement of occipital condyles in relation to skull as a whole.

Reduced masticatory musculature, with temporal ridges never reaching the midline of the cranial vault, their height being variable.

Nuchal area relatively smaller than in *Australopithecus*, variable, and much reduced in *Homo sapiens*.

Skull more rounded, and muscle attachment areas in general variable in size but less well marked than in *Australopithecus*, such areas almost imperceptible in modern humans.

Supraorbital torus variably developed, being enlarged in *Homo erectus*, though with reduced postorbital constriction compared with *Australopithecus*; torus absent in some populations of *Homo sapiens*.

Dental arcade evenly rounded, usually with no diastemata.

Mandible less deep than in *Australopithecus* but variable in its horizontal stressing, from an internal torus in *Homo erectus* to an external chin in living subspecies of *Homo sapiens*.

Canines relatively small, with no overlapping after the initial stages of wear.

Molar teeth variable in size, with a relative reduction of M3 in later populations.

In the postcranial skeleton, we may note:

Pollex well developed and fully opposable, so the hand is capable of a precision grip.

Hindlimb skeleton not very variable and fully adapted to bipedal locomotion, as in living populations.

V. Summary   The human family includes those living and fossil forms that are more like living humans than other living primates. The assessment of similarities and differences is the basis upon which the classification of fossil remains must be made. By convention, taxonomic groups at the level of the family should be differentiated not by slight anatomical differences but by a distinct complex of adaptations. The adaptations that distinguish the hominids, apes, and monkeys are primarily locomotor, and these have been discussed above; the Hominidae are quite strikingly distinct from all other primate taxa as a result of their terrestrial bipedalism. Other major differences include their dentition and cranial capacity.

They have accordingly been defined by Le Gros Clark on the basis of their total morphological pattern, and the following has been modified from his summary definition (1978). The meaning of these statements and the terminology will become clear in later chapters.

Family Hominidae: a subsidiary radiation of the Hominoidea distinguished from other Hominoidea by the following evolutionary trends:

Progressive skeletal modifications in response to erect bipedalism, shown particularly in a relative lengthening of the lower extremity and changes in the proportions and morphological details of the pelvis, femur, and foot skeleton related to the mechanical requirements of erect posture and gait and to the muscular development associated therewith.

Preservation of well-developed opposable pollex, ultimate loss of opposability of hallux.

Relative displacement forward of the occipital condyles and restriction of nuchal area, associated with the flexion of the basicranial axis.

Progressive diminution of canines to a spatulate form interlocking slightly or not at all and showing no pronounced sexual dimorphism, with disappearance of diastemata.

Replacement of sectorial first lower premolars by bicuspid teeth (with later secondary reduction of lingual cusp).

Alteration in occlusal relationships, with the result that in later forms all the teeth tend to become worn down to a relatively flat, even surface at an early stage of attrition.

Development of an evenly rounded dental arcade.

Marked tendency in later stages of evolution to a reduction in relative size of the molar teeth and in subnasal prognathism.

Progressive acceleration in the replacement of deciduous teeth in relation to the eruption of permanent molars.

Marked expansion (in the later products of hominid evolution) of the cranial capacity, associated with reduction in size of jaws and area of attachment of masticatory muscles and the development of a chin.

As we have proposed, the story of our evolution can best be told by studying in turn the evolution of each functional complex of characteristics according to its changing function. In treating each functional complex, we shall compare modern humans with the African apes (and occasionally Asian apes and monkeys) in order to investigate our functional anatomical relationship to them. Where fossils are available, we shall also evaluate the evidence from that source in order to establish in more detail the functional morphology of our hominid ancestors.

In that framework we shall study the special adaptations that distinguish us from other primates—adaptations to a terrestrial lifestyle. Besides the anatomical adaptations, we shall consider some behavioral changes and above all cultural advances, for hominids are indeed unique as culture-bearing animals. It was ultimately the hominid culture that accelerated the evolutionary process and generated that extraordinary animal called humankind.

The movement from the forest was a remarkable and rather mysterious event. Some authors see it as the direct result of a lowered rainfall, which restricted the extent of the tropical forests and caused increased population pressures. Others see it simply as resulting from the normal competitive pressures of organic life, which favor the exploitation of all exploitable environments. The point to note, however, is that if a population is to exploit an environment different from that of its ancestors, it must to some extent be adapted to the new environment already, or, to use a common term, it must be *preadapted*. Of course, many animals are preadapted in varying degrees to environments not their own, and this is what is meant by a generalized species. All quadrupedal arboreal monkeys are in their locomotor character preadapted to terrestrial life, and vice versa. The extent of preadaptation and the kinds of selection pressure involved will be discussed in Chapter 12. Our purpose in the next and later chapters is to examine human adaptations in some detail in a historical setting, so that it may be possible at a later stage to piece together the broad outline of the story of human evolution.

Suggestions for Further Reading

The most compact text on fossil primates is F. S. Szalay and E. Delson, *Evolutionary history of the primates* (New York: Academic Press, 1979). Some of the most important fossil Hominidae are briefly described in M. Day, *Guide to fossil man* (New York: World, 1965). For a complete catalog of Hominid fossils refer to K. P. Oakley, B. G. Campbell, and T. I. Mollison, *Catalogue of fossil hominids* [3 vols.; London: British Museum (Nat. Hist.), 1967–1977]. For a detailed review of some aspects of human paleontology, see S. L. Washburn and P. Dolhinow (Eds.), *Perspectives on human evolution*, Vol. 2 (New York: Holt, Rinehart and Winston, 1972). For an account of the nature and the history of paleoanthropology, see B. G. Campbell, *Humankind emerging*, 4th ed. (Boston: Little, Brown, 1985). The most recent publication on the evolution of the Hominoidea is by R. L. Ciochon and R. S. Corrucini, *New interpretations of ape and human ancestry* (New York and London: Plenum, 1983).

For a review of *Australopithecus*, see C. Jolly (Ed.), *Early hominids of Africa* (London: Duckworth, 1978); on *Homo erectus*, see K. W. Butzer and G. L. Isaac (Eds.), *After the australopithecines* (The Hague: Mouton, 1975), and on the Neandertal people, see E. Trinkhaus, *The Shanidar neandertals* (New York and London: Academic Press, 1983). For a review of the origin of anatomically modern humans, see F. Smith and F. Spencer (eds.), *The origins of modern humans: A world survey of the fossil evidence* (New York: Liss, 1984).

5

Body
Structure
and
Posture

**I. Body
Structure and
Locomotion**

A vast majority of the visible and gross features of the vertebrate body are related to the function of locomotion. Primitive locomotion in fish was achieved by a wavelike movement of the vertebral column in a lateral plane. The trout moves its tail from side to side, and the waves of muscle contraction pass forward up the animal's body. In primitive fish the fins are used mainly as stabilizers and for steering, but in later forms they may be used for propulsion, especially the paired fins: the *pectoral* fins behind the head, the *pelvic* fins nearer the tail.

The first fish to climb from the water used their pectoral and pelvic fins as props for their bodies, and thus came to initiate the process whereby the locomotor function was eventually transferred from the vertebral column to what became the limbs. The paired pectoral and pelvic fins were attached to the backbone or vertebral column by three paired bones of the same name, later to became the shoulder and pelvic girdles (Fig. 5.1). The lateral muscles at the sides of the vertebral column (the *myotomes*, which effected the wave motion of the body) were then greatly reduced, and the limb musculature developed in their place. An intermediate stage can be seen in the newt (Fig. 5.2), which uses both its myotome muscles and its limbs for propulsion.

As the limbs took over the locomotor function in terrestrial vertebrates, the vertebral column assumed a new function, that of supporting the body above the limbs. The flexibility so necessary for swimming was replaced by a stiffness necessary to carry the head and body. The soft parts already *ventral* to (on the underside of) the vertebral column in fish became suspended from it in vertebrates, and the head became supported by a cantilever. The resulting structure was compared by

119

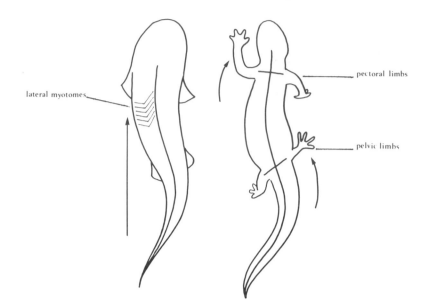

Figure 5.1. Locomotion in the fish and newt. Note that the same wave motion is used by both animals, but in the newt it is reinforced by the action of the pectoral and pelvic limbs, which evolved from the paired fins in the fish.

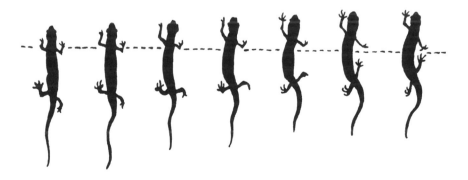

Figure 5.2. The locomotion of the newt on land combines features of fish and quadruped locomotion. Note that the limbs are splayed sideways; they cannot support the body, which is dragged along. The musculature is still lateral, not dorsal and ventral, as in evolved quadrupeds.

D'Arcy Thompson (1942) to that of a cantilevered bridge (Fig. 5.3) and can be understood as a weight-bearing compound girder of webbed construction. The important muscles are now ventral and *dorsal* to (that is, below and above) the vertebral column—no longer lateral. This important change in the evolution of vertebrates is clearly demonstrated in the secondarily aquatic whales, which evolved from terrestrial quadrupeds and which propel themselves in the sea by moving their tails up and down rather than laterally.

In the terrestrial quadruped, therefore, the vertebral column is under compression—the *body* of each vertebra (and the *intervertebral discs*) sustains the load, and the *supraspinous* ligaments are under tension, while the *nuchal muscles* of the neck support the head (see Fig 5.4). The bending moment (that is, the force tending to cause bending) in the vertebral column because of the weight of the body and head is reflected in the length of the *spinous processes*, which, together with the oblique *interspinous ligaments* and muscles, constitute the web of the compound girder. The spinous processes are rarely so greatly developed over the pelvic girdle as over the pectoral because in most vertebrates the weight of the tail is less than that of the head, though exceptions to this are known (dinosaur, kangaroo—see Fig. 5.5). Further stressing over the center section of the girder between the limbs is achieved by an inverted parabolic girder between the limbs of which the *sternum* and abdominal muscles form the ventral tension member (see Fig. 5.4).

Bridges, however, are designed for rigidity, and the anatomy of bridges is only comparable to that of animals insofar as animals require rigidity in their skeletons. Flexible structures are of a different order of complexity and do not so clearly follow the simple mechanical principles of the bridge-builder. That is why the cantilever structure is seen more clearly in the large slow-moving animals than in the small, active forms. Flexibility means that the bones must be jointed to allow movement, and the spinous processes of the vertebrae must be short enough not to touch each other when the body is flexed. As a result, the girder is supported by ligaments under very great stress; it has gained flexibility at the cost of strength.

The second factor to be recalled when studying skeletal architecture is that when the loads due to gravity are very slight, as in small animals, the structures associated with locomotion will mask those designed to withstand the force of gravity. In large animals, like the bison, the structural postural elements will be more strikingly developed than the locomotor elements.

In this chapter we shall trace the evolution of posture from the simple quadrupedal plan with horizontal trunk to the species with erect trunks that, as we have seen, are found among prosimians as well as many other higher primates, including humans. The vertebral column, which began in fish as a flexible rod under intermittent compression (when the myotomes contract) is, in quadrupeds, the compression member in a

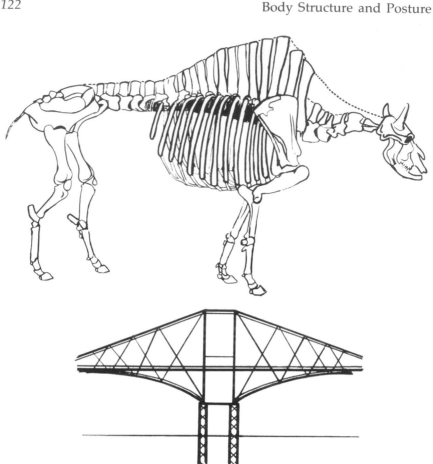

Figure 5.3. The bison and the cantilever bridge are constructed according to the same mechanical principles. The nuchal muscles (dotted line) in the bison and the upper horizontal member of the steel bridge absorb the tension due to gravity; the vertebral bodies and the lower horizontal member of the bridge withstand the compression. The vertical and diagonal struts separate the horizontal members and form the webbing. The nerve cord and road are suspended within this webbed compound girder (after D'Arcy Thompson, 1942).

weight-bearing compound girder. In primates it eventually becomes a vertical weight-bearing flexible rod, held erect by muscular "rigging," like the mast of a yacht. We shall see that this change of function is accompanied by an appropriate change in form.

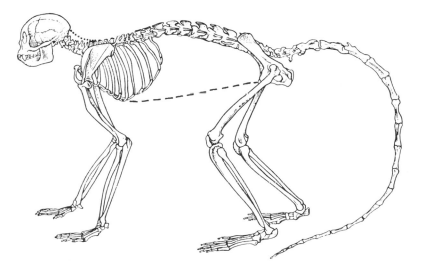

Figure 5.4. Architecture of a quadrupedal monkey (*Cercopithecus*). Note that the nuchal muscles (dotted) are still important but that the head is much lighter than in the bison so that the cervical spines are much shorter. The trunk is supported by an inverted parabolic structure under tension—the sternum and abdominal muscles (interrupted line) (after Le Gros Clark, 1971).

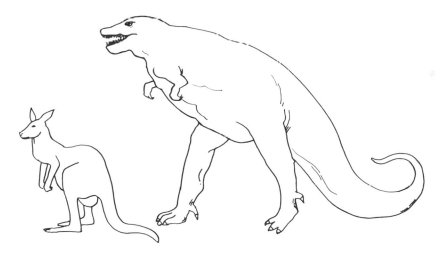

Figure 5.5. The marsupial kangaroo and the 20-foot reptile *Tyrannosaurus* from the Upper Cretaceous are both bipedal but gain support from a heavy and powerful tail. In each, the spinous processes of the vertebrae reach their greatest development over the hindlimbs. Note the greatly reduced forelimbs. (Not drawn to scale.)

II. Center of Gravity

Differences in the form of the vertebral column among quadrupedal animals are due to differences in posture, in locomotion, and so in the center of weight of the animal, better known as the center of gravity. This center will move forward or backward along the backbone depending on whether there is more body weight in the head and shoulder region or in the hindquarters and tail region. In the bison (see the skeleton shown in Fig. 5.3), the majority of weight, as in most terrestrial quadrupeds, lies over the forelimbs, and that is where the center of gravity lies. They have been described as forelimb dominant. In contrast, the dinosaur and kangaroo, for example, both have far more weight over the hindlimbs, and that is the feature which allows them semierect posture with tail support, so that they can gather vegetation with their forelimbs (Fig. 5.5). The hindlimbs are dominant.

A somewhat analogous adaptation can be seen in all the primates regardless of their limb proportions (Kimura et al., 1979). Here the center of gravity is well back for a number of reasons. The first is that the musculature and length of the hindlimbs has increased, especially in the leaping and grasping prosimians, to give extra locomotive power: the hindlimbs are dominant. The second is that the tail has enlarged in many groups into an organ for maintaining balance when jumping. It is also necessary that, when an animal is propelled forward, the force of propulsion should pass through the center of gravity; otherwise the animal would revolve in the air. It therefore follows that a center of gravity near the organ of propulsion, in this case the back legs, will add to the stability of the jumping primate. It is noteworthy that in this feature all the primates contrast with almost all other quadrupedal terrestrial mammals (except the leaping cats), which have forelimb dominance.

The energy required for locomotion is also a factor related to the position of the center of gravity. It appears that at small body sizes a posterior center of gravity is energetically advantageous, while at larger body sizes an anterior position is more efficient (Kimura et al., 1979). Most primates are relatively small animals, so this factor again supports the evolution of hindlimb dominance. The fact that the only fully terrestrial large primate (Homo sapiens) is bipedal is also significant, because this adaptation has been shown to be more efficient than that of a (theoretical) large hindlimb dominant quadruped.

It is for these reasons that in the early stages of primate evolution we see the center of gravity well back over the hindlimbs. This development makes practical the sitting position (Fig. 5.6) and frees the forelimbs for manipulation, just as we see in the squirrel and kangaroo. It also makes possible the key adaptations of human evolution, when the earliest hominids became able to stand and run on their hindlimbs. The forest adaptation of a low center of gravity was a necessary prerequisite for this final evolutionary development. In human evolution, while the tail was

Figure 5.6. The ring tailed lemur (*Lemur catta*) achieves a comfortable sitting position, but the forearms help to maintain the balance. Note the immense tail with heavy fur (from Le Gros Clark, 1971).

lost, the low position of the center of gravity was enhanced by still further development of the hindlimbs, somewhat reduced musculature in the forelimbs, and some changes in the form and position of the trunk and abdomen.

In living humans the whole skeleton is modified so that the body's weight can be carried with minimum effort and maximum stability. This means that most of the weight is transmitted by bones with the shortest possible cantilevers. The center of gravity passes from the occipital condyles down just in front of the vertebral column to the upper sacral vertebrae, then through the pelvic bones and down the hindlimbs to the tripods of the feet (see Chapter 6, IV). As we shall see, the bioengineering of the human body is unique in the animal kingdom.

III. Evolution of the Vertebral Column and Thorax

Adapted as many of them are for arboreal locomotion, the higher primates have evolved very flexible vertebral columns compared with most quadrupedal plains mammals. Flexion of the backbone has been added to that of the legs to give extra propulsive power in jumping, as well as to absorb the shock of impact. Flexibility was achieved by some small modifications of the vertebral column, which are most clearly seen in the quadrupedal forms:

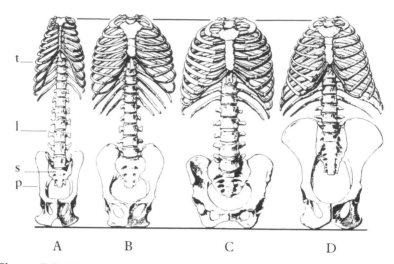

Figure 5.7. Trunk skeletons of the macaque monkey (*Macaca*) (*A*), the gibbon (*B*), a modern human (*C*), and the chimpanzee (*D*) drawn the same size. Note the different forms of the thorax (*t*), pelvis (*p*), and sacrum (*s*) in these species and the different relative lengths of the lumbar region (*l*) (from Schultz, 1950).

1. There are well-developed intervertebral disks that act as elastic buffers and allow tipping movement of the vertebrae.

2. The spinous processes are relatively short, with the result that extensibility is not restricted by their proximity.

3. The *thorax* (the rib cage and its musculature), which is deep and which gives the kind of support to the trunk that is achieved by a parabolic girder, stiffens the vertebral column along the thoracic region. In contrast the *cervical* (neck) and *lumbar* (waist) regions of the vertebral column in quadrupedal monkeys are relatively long, and these regions in particular allow bending movements of the body (see Fig. 5.7A).

In climbing and brachiating forms, however, this flexibility is somewhat lost, since the vertebral column plays a far smaller part in locomotion. We find, for example, that three of the nine lumbar vertebrae found in quadrupedal monkeys are absent in gibbons (Washburn, 1963b, and see Fig. 5.7B), and that the lumbar region is reduced in all the great apes. The numbers of such vertebrae are shown in Table 5.1, where it can be seen that in this respect the human is stiff-backed and morphologically approaches the arboreal climbers. It is, however, interesting that the stiffening has proceeded further among the great apes than it has in humans. As a whole, these data suggest strongly that we evolved from a primate that engaged in arboreal climbing rather than one with a history of quadrupedalism. But other fundamental changes

TABLE 5.1. Average Numbers of Thoracic and Lumbar Vertebrae

Primate	Thoracic vertebrae		Lumbar vertebrae	
	Mean	Range	Mean	Range
Papio	10	9–11	9	9
Hylobates	13	12–14	6	5–7
Great apes	13		4	
Humans	12	10–12	6	5–7

have clearly occurred in the spine and thorax during human evolution that were adaptations both to a history of erect posture and climbing and to the later evolution of bipedal locomotion:

1. The size of the vertebral body increases at the pelvic end of the vertebral column (Fig. 5.8) because the forces of compression are no longer constant along the trunk as they are in a quadruped with a horizontal backbone. As the force of compression increases from head to pelvis because of the weight of the animal, so the cross-section of the vertebral body, which bears it, must also increase (Schultz, 1953). This increase is present in the great apes but is greatest among humans. The proportions of the vertebral body also change (Schultz, 1960), and the weight distribution on the vertebrae themselves changes in order to carry the vertical forces of compression.

2. The spinous processes are more or less equally developed along the vertebral column because the bending moment—the tendency to bend—is no longer restricted to the points of support, the limb girdles, but is now more evenly distributed. For example, the gorilla has well-developed cervical spines that anchor the nuchal muscles, but they are reduced in modern humans. As a result of our fully erect stance and bipedal locomotion, the head, although heavy, is more or less balanced on top of the vertebral column (Fig. 5.9).

3. Among all the climbers, the inverted parabolic girder effect of the sternum and abdominal muscles, which stiffens the backbone (see Fig. 5.4), becomes less important. The sternum tends to lie nearer to the vertebral column, its segments become fused (Schultz, 1930) and broadened, and the form of the thorax is wider and shallower (Fig. 5.10), bringing the center of gravity nearer the vertebral column. Schultz (1960) has studied the angle between the "neck" and "body" of the ribs and finds that among the monkeys the angle varied from 89 to 100 degrees, while among the more erect apes and humans it varied from 50 to 65 degrees (see Fig. 5.10).

A wider but flatter thorax also gives greater power to the lateral abdominal muscles (Keith, 1923). Ventral displacement of the vertebral

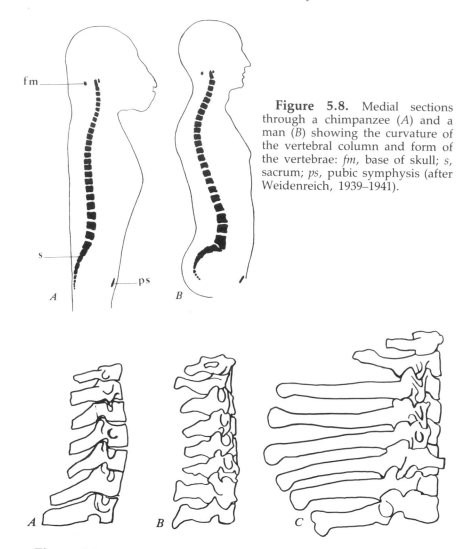

Figure 5.8. Medial sections through a chimpanzee (*A*) and a man (*B*) showing the curvature of the vertebral column and form of the vertebrae: *fm*, base of skull; *s*, sacrum; *ps*, pubic symphysis (after Weidenreich, 1939–1941).

Figure 5.9. Cervical vertebrae of a modern human (*A*), gibbon (*B*), and gorilla (*C*). The length of the spinous processes of the vertebrae is porportional, not to the weight transmitted down the vertebral column, but to that carried by the nuchal muscles. (Not drawn to scale.)

column may similarly improve the leverage of some of the back muscles (e.g., iliocostalis). Finally, it has the effect of allowing increased shoulder mobility (Chapter 7, II), since it alters the position of the shoulder girdle, as well as moving the shoulders apart (Cartmill and Milton, 1977). It is clear that a wide shallow thorax is advantageous to an arboreal climber.

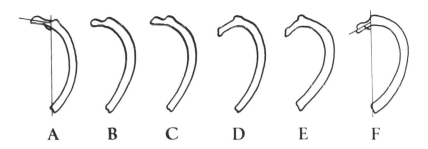

Figure 5.10. Drawings of the second rib of three Old World monkeys [macaque (*A*), baboon (*B*), *Presbytis* (*C*)] and two apes [gibbon (*D*) and chimpanzee (*E*, together with a human rib (*F*), reduced to the same length. Note that the different form of the ribs reflects the different form of the thorax in these groups. The angle between the "neck" and "body" of the rib is shown for *Macaca* and *Homo* and is quoted in the text (redrawn after Schultz, 1960).

4. Increased weight transmission through the pelvis and legs has resulted in an enlargement of the *sacrum*, the zone of fused vertebrae that carries the pelvic girdle. This implies the "sacralization" of more vertebrae at the base of the spine. Among the lower primates two to four vertebrae are sacralized, while among the great apes and in humans the numbers average 5.5 and 5.2, respectively (Schultz, 1930). The sacrum is also relatively wider in erect species and more convex on the internal surface (see Figs. 5.7 and 5.8). It forms a fixed base for the action of the powerful muscles (the erector spinae) that support the back (Keith, 1923).

We see from these data that we have a relatively stiff backbone, which suggests that our origin lies among climbers, and we no doubt benefited by this preadaptation in the evolution of a fully erect bipedal posture. In achieving the posture of a climbing species and later a bipedal one, the vertebral column evolved to form a vertical weight-bearing towerlike structure in place of a compound girder, which means (as we have seen) that the vertebral bodies, which transmit the weight to the pelvic girdle, show an increase in size as we pass down the vertebral column, and the spinous processes, which supported the webbing of the girder, have become much reduced. The thorax has flattened to allow the center of gravity to move back toward the vertebral column, which now effectively lies within the thoracic area (Fig. 5.10 and 5.11). The weight of the body is transmitted to the legs via the pelvic girdle (the changes here will be discussed in Chapter 6).

The evidence of human ancestry derived from the vertebral column is important and is supported by the evidence of the fossilized vertebrae of *Australopithecus* (Fig. 5.12), which show most of the humanlike features associated with an upright posture, though (as might be predicted) the

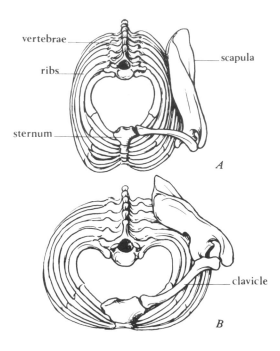

vertebrae

ribs

sternum

scapula

A

B

clavicle

Figure 5.11. View from above, of the thorax and shoulder girdle of a macaque (*A*) and a modern human (*B*) drawn to the same anterior-posterior depth. Note the very different shapes of the thorax and the different positions of the shoulder girdle (from Schultz, 1950).

bodies of the lumbar vertebrae are still relatively small (Robinson, 1972). Known fossils of early *Homo sapiens* populations in general have vertebrae similar to modern humans (Straus and Cave, 1957; Trinkhaus, 1983), and, as we shall see, there is much further evidence that the erect climbing posture was the first necessary preadaptation for the evolution of hominid bipedalism.

IV. The Head and Neck   We shall consider the functions of the head itself in Chapters 8 and 9. Our purpose here is to consider the head in its relation to the rest of the body and the cervical vertebrae in particular. Just as semierect and erect postures have affected the form of the *cervical vertebrae,* so have they also affected the position of the head in relation to the spine and the connecting musculature.

We have seen, in such quadrupedal forms as the bison or monkey, how the head is supported by a webbed girder consisting of the cervical vertebrae under compression and the nuchal muscles and ligaments under tension (Figs. 5.3 and 5.4). It is clear that as a more erect posture evolved, less strength was required in the nuchal muscles as well as in the spinous processes of the cervical vertebrae, since more of the weight was carried directly by the bodies of the vertebrae.

At the same time, we have already mentioned that in primate evolu-

Figure 5.12. View from the right side of the lumbar region of the vertebral column together with the sacrum and left pelvic bone of *Australopithecus africanus* from Sterkfontein (from Robinson, 1972).

tion generally the muzzle was reduced and the brain enlarged. This development resulted in the center of gravity of the head itself moving more nearly over the point of pivot of the head upon the vertebral column. This evolutionary trend is shown in Fig. 5.13, where different skulls are aligned by their *occipital condyles,* the convex bone surfaces with which the skull pivots upon the first cervical vertebra or *atlas.* Here we clearly see the center of gravity moving back until, in humans, where the masticatory apparatus is also reduced, it is almost exactly over the occipital condyles. It has not yet achieved a perfect balance and probably will not do so, because there are no powerful ventral muscles present to support the skull from the front and stop it from tipping backward. Those that are ventral to the occipital condyles (the *longus capitis* and *rectus capitis anterior*) arise very close to the occipital condyles and so exert only slight leverage on the head. The condition of modern humans is seen in Fig. 5.14, where the weight of the head, ventral to the occipital condyles, is balanced by the nuchal muscles, dorsal to this pivot.

Le Gros Clark (1950) has proposed a useful means of assessing the extent to which this movement of the center of gravity has occurred in different primates: by measuring the position of the occipital condyles in

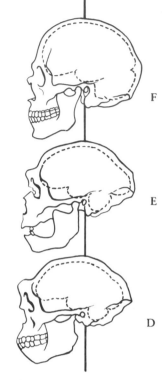

Figure 5.13. Skulls of a tree shrew and various primates aligned upon the point on which they pivot on the spine, the occipital condyles. Note that in passing from the lower to the higher primates, the center of gravity moves back, the brain expands, and the jaws recede. Tree shrew (*A*), *Cercopithecus* (*B*), gibbon (*C*), *Homo erectus* (*D*), *Homo sapiens neanderthalensis* (*E*), modern human (*F*). (Not drawn to scale.) (After DuBrul and Sicher, 1954.)

relation to the rest of the skull. The *condylar position index* is calculated according to the simple geometrical basis described in Fig. 8.19, and some figures for this index are recorded in Table 5.2. Note that the increase of the index has just begun in *Australopithecus*, but has not at this stage progressed very far in the human direction. The index numbers may be compared with the drawings in Fig. 5.13. This double change in posture and in the head's center of gravity has had the effect of removing work from the nuchal muscles, which have in turn been reduced. Schultz (1942) has shown that taking size into account, six times as much power is needed in the nuchal muscles of monkeys and apes than is needed to support the human head. This reduction in musculature has been reflected in the area of insertion of the nuchal muscles upon the back of the skull (termed the *nuchal area*; Fig. 5.15).

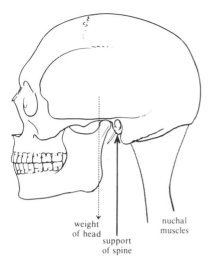

Figure 5.14. In the course of human evolution the center of gravity of the head has moved nearer to the point of pivot upon the spine—the occipital condyles. In modern humans it approaches this point but is still slightly anterior to it. The weight is still balanced by the now rather reduced nuchal neck muscles.

TABLE 5.2. The Position of the Occipital Condyles[a]

Primate	Mean condylar position index
Proconsul africanus	30
Pan gorilla	24
Pan troglodytes	23
Pan paniscus	35
Australopithecus africanus	39
Homo sapiens (modern)	77–81

[a]Condylar position index measured in Frankfurt plane; CD/CE (see Fig. 8.19). Data from Le Gros Clark, 1950; Ashton and Zuckerman, 1951; and Davis and Napier, 1963.

When we compare, for example, the gorilla and modern humans, the difference is striking. The fossil skull of *Homo erectus* is intermediate in this respect. Like the Neandertal people that followed somewhat later, *Homo erectus* also has a well developed masticatory apparatus, a long head, clearly defined nuchal area, and well-developed nuchal musculature. The balance of the head was less perfect that it is in modern humans.

The balance of the head, and the power required in the nuchal musculature, is also reflected in the development of the spines of the cervical vertebrae. We have atlas and axis fragments, and a sixth or seventh cervical vertebra from Hadar; the latter carries a very long spinous process. We have no other fossil hominid cervical vertebrae until we come to the Neandertal people; here again, the spines are significantly longer

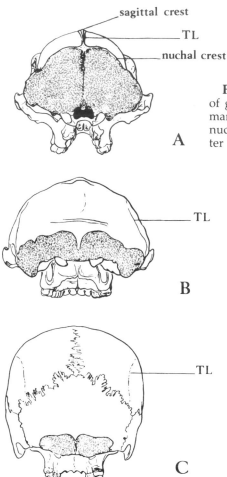

sagittal crest

TL

nuchal crest

A

Figure 5.15. Posterior view of skulls of gorilla (*A*), *Homo erectus* (*B*), and human (*C*). Note the decrease in size of the nuchal area (*shaded*); *TL*, temporal line (after Weidenreich, 1939–1941).

TL

B

TL

C

that those of modern *Homo sapiens* (Trinkhaus, 1983). The final reduction of the spinous processes appears to have been a very late development.

The reduction of the nuchal musculature has affected the form of the skull, for the area of attachment in humans in not only much reduced, but it is no longer bordered by a clearly defined nuchal crest. The back of the skull is relatively smooth and follows more closely the outline of the brain, for in humans the outer layer of bone of the skull (the outer "table") responds very little to the much weaker nuchal muscles during growth (see Figs. 5.13 and 5.15.).

A characteristic of the base of the skull that is related to its balance is the orientation of the foramen magnum—the opening through which the spinal cord passes from the brain. In quadrupeds, the foramen faces

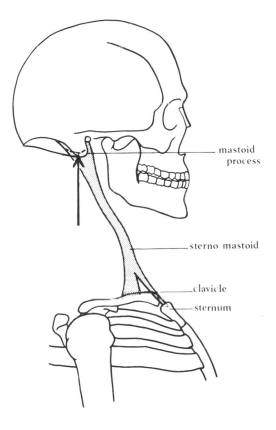

mastoid process

sterno mastoid

clavicle

sternum

Figure 5.16. Because of the way the head is carried by quadrupeds, there is sufficient support for it in the nuchal muscles, and there are no powerful muscles that attach the front of the skull to the thorax or neck. To pull the head forward after it has been thrown back, humans rely on two rotator muscles, the sternomastoids, which are contracted together. Sufficient leverage is obtained to pull the head forward only by a slight elongation in the human mastoid porcess, which comes to project forward of the point of pivot, shown here (dotted) behind the mastoid process.

more or less backward, but in modern humans it has come to face more or less vertically downward. This feature was noted by Raymond Dart in 1925 when he described the first *Australopithecus* skull which suggested to him that the skull had belonged to a bipedal animal.

It has been mentioned that there are no powerful ventral muscles in the neck to keep the head from tipping back. This presents a problem probably unique to humans, because the center of gravity of the head is well forward in all other primates. We can tip our head back beyond the point of balance, and we must be able to pull it forward again. The forward movement of the head is made possible not only by the *longus capitis* and *rectus capitis anterior* mentioned above, but also by the *sterno-mastoid muscles*, which, lying on each side of the neck and inserted upon the *mastoid process* of the skull and the sternum and clavicle at the top of the thorax, are individually able to rotate the head upon the vertebral column (Fig. 5.16). Their importance to the primates is great, since this group is dependent on rotary head movement to maintain effective all-around vision. When in the very last phase of human evolution the center of gravity of the head moved back near the occipital condyles (the pivot), the sternomastoid muscles could just counteract the backward forces upon the head by contracting together. In Fig. 5.16 we see that, viewed from the side, their insertion is more or less in line with the occipital condyles. They are really only effective in this role when the head is tipped back.

Krantz (1963) has shown that the sternomastoid muscles obtain effective leverage to pull the head forward only because an enlargement and lengthening of the mastoid processes has occurred in human evolution. This lengthening alone makes it possible for the sternomastoids to raise the head after it has tipped back; only by lengthening is the insertion point of the muscles brought forward beyond the point of pivot when the head is thrown back (Fig. 5.17). As it is, the leverage is very poor, a fact that can be verified by throwing your head well back over the back of a chair and slowly raising it again, with your hands on your sterno-mastoids. The enlargement of the mastoids in human evolution is thus a direct result of our erect posture and of the forward movement of the occipital condyles in relation to the skull as a whole.

In this section we have seen how we come to have a reduced nuchal musculature, a reduced nuchal area, and elongated mastoid processes. All these developments are facets of a double adaptive complex that (a) involved the adoption of a fully erect posture, and (b) involved changes in the olfactory region, the masticatory apparatus, and the brain, which together moved the head's center of gravity and the relative position of the occipital condyles. All these changes moved the center of gravity of the head to a position only slightly ventral to the position of the occipital condyles themselves.

V. The Tail

As we have seen, the tail is used by many primates for balance in jumping, and has evolved in such a way as to be one factor in bringing the center of gravity back toward the hindlimbs. It is also used in many groups as a rudder and elevator (such as the tail of a bird or airplane), to adjust the position of the body as it passes through the air so that the animal will arrive with its limbs in a position to grasp on impact. This function accounts for the thick hair on the tail of most jumping primates, which increases its aerodynamic effect (Fig. 5.6). We have also noted that in the New World monkeys the tail has evolved as a grasping organ—a fifth limb—adding to mobility as well as to stability.

The tail is also used in other ways by terrestrial monkeys, among which it tends to be reduced. It acts as a support for the baby riding on its mother's back like a jockey, as a device for signaling (as in dogs), as a fly-whisk for the anogenital region (as in cattle), or it may be almost completely lost, as in the stump-tailed macaque. Among macaques alone it varies from 27 to 3 vertebrae.

As we have seen, the apes and humans—the Hominoidea—have lost their tails. Reduction has gone furthest in the gibbons: as compared with 20 to 30 caudal vertebrae found among most monkeys, the gibbon has

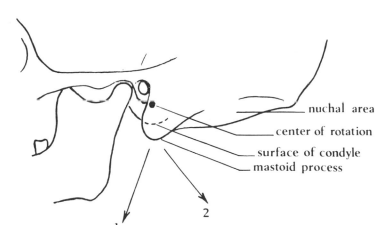

Figure 5.17. The mechanical advantage bestowed by the mastoid process is apparent only when the head is moved from its normal position [when the sternomastoid muscle lies in direction (1)] and is tilted back, when the sternomastoid pulls in the direction indicated as (2). Under these circumstances the mastoid can be seen to give some leverage about the center of rotation of the skull upon the spine. A smaller mastoid would reduce to nil the leverage available (after Krantz, 1963).

only one or two. The apes as a group are tailless, and this seems to be a product of their climbing mode of locomotion, even if the gibbons also brachiate. In a climbing species the tail serves no aerodynamic or other purpose and has been lost. The only apelike creature with any evidence of a tail is the very early primitive genus *Aegyptopithecus* from the Oligocene of Egypt, which was much smaller and probably much more active than modern apes. It is also of interest that the slow-climbing (SQC) prosimians, such as the lorises, are also tailless.

In humans the tail is also reduced, but we retain from three to five small caudal vertebrae known as the *coccyx*. As this reduction is shared with all the Hominoidea, we can attribute it to an ancestral condition which we believe was that of an arboreal climber. The adoption of bipedal locomotion did not require any major reversal of the condition, though some modification seems to have occurred.

It is of interest that reduction of the tail in modern humans has not progressed as far as in the gibbon. This may be because the coccyx has assumed a totally new function arising from our fully erect posture. The tail vertebrae, the coccyx, are now curved ventrally and help form the basin-shaped structure that carries the viscera. Ligaments run from the forward-pointing coccyx to the *ischium*—the lower part of the pelvis.

Figure 5.18 shows how the viscera in humans are carried in a sack-shaped cavity: the vertebral column at the back and the *rectus abdominis* muscles in front. At the base, the sack is rounded off by the curved

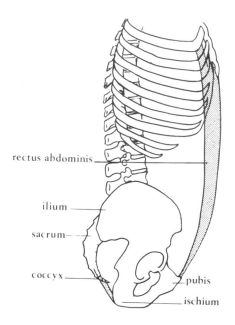

rectus abdominis

ilium

sacrum

coccyx

pubis

ischium

Figure 5.18. Diagrammatic side view of the pelvis and associated musculature, which form a basin and support the viscera in humans. Note that the coccyx, the remains of the tail, has an important function as a supporting structure—a totally new use for a tail.

bones of the pelvis, the sacrum, and the coccyx, as well as the *sacrotuberous* and *sacrospinous* ligaments. This viscera-carrying function of the pelvis is again something new and is associated with bipedal locomotion. We shall consider the complex evolutionary changes in the pelvic bones in the next chapter.

VI. Summary   In this chapter we have seen how the arboreal life of the primates has influenced the evolution of the vertebral column in a number of ways.

1. The center of gravity has moved toward the hindlimbs.
2. The tail has assumed special functions, as a balancing rudder and elevator in many prosimians and Old World monkeys, and as a fifth limb in New World monkeys.
3. Erect sitting and climbing postures have developed in some groups.
4. The vertebral column maintains considerable flexibility in quadrupedal forms but loses flexibility in climbing forms.

This process leads to the form of many modern terrestrial and arboreal monkeys without important modification, but a number of other trends have appeared in the Hominoidea.

1. The evolution of climbing locomotion: the tail is lost, the flexibility of the spine is reduced, the thorax is flattened, and the posture is erect (*Pongo* and *Pan*).

2. The evolution of smaller brachiating climbers (*Hylobates*).

3. The evolution of terrestrial bipedalism: the posture is fully erect, the function of the vertebral column is finally modified as a vertical weight-bearing structure, the thorax is flattened to bring the center of gravity over the pelvis, the tail is modified to form a floor to the pelvis, and the relation of the head to the neck approaches balance (*Australopithecus*, *Homo*).

We have seen in this chapter how the different locomotor adaptations discussed in Chapter 3, Sections IV and V have fundamentally affected the form of the vertebral column, the ribs, the neck, and the skull. We have also seen how humankind's unusually large brain and small muzzle have brought the human head nearly into balance, and we have noted the modifications in the nuchal muscles and mastoid process that have resulted. All these changes, however, are relatively slight compared with those that have affected the locomotor apparatus.

Suggestions for Further Reading

Most of this chapter is based on the extensive investigations of Professor Adolf Schultz. A lifetime of research in primate morphology is briefly summarized in A. H. Schultz, *The life of primates* (New York: Universe Books, 1969); more detailed information may be obtained from the references therein. Two other publications summarize work relevant to this chapter: A. H. Schultz (1950), "The specializations of man and his place among the catarrhine primates," *Cold Spring Harbor Symp. Quant. Biol.* **15**. 37–53; "Age changes, sex differences, and variability as factors in the classification of primates," in S. L. Washburn (Ed.), *Classification and human evolution*, pp. 85–115 (Viking Fund Publs. Anthrop., No. 37)(Chicago: Aldine, 1963.)

A recent paper on primate evolution has proved of great interest and value: P. Andrews and L. Aiello, "An evolutionary model for feeding and positional behavior," in D. Chivers, *Food acquisition and processing* (New York: Plenum, 1984).

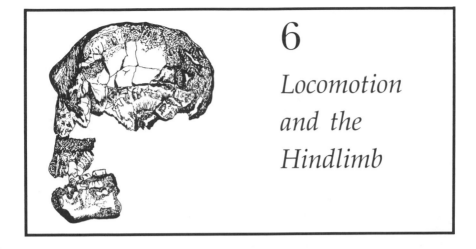

6

Locomotion and the Hindlimb

I. Allometry and the Generalized Primate Limb

It is the limbs rather than any other part of the primate body that can be described as "generalized." The limbs having varied functions can generate a broad range of potential behavior, and are thus capable of evolution in a number of different directions. Primate limbs are not used like props, merely to support the body on a horizontal surface as in most quadrupedal vertebrates. Among the primates, limbs have evolved to grasp tree branches and to support the body at various angles in relation to the branch. They have evolved to support the weight of the animal not only standing (under compression), but also clinging and hanging (under tension), as food-gathering and sense organs, and for cleaning the body. The primates are characterized, therefore, by an efficient "universal" ball and socket joint at the point of the attachment of the limbs at the shoulder and hip, which allows movement in any direction. The musculature has evolved accordingly, allowing the proximal limb bones (those nearest the trunk) to be moved in almost any direction and to support the body under compression or tension at a wide range of angles.

Flexibly mounted limbs are not so much a specialization of the primates as they are a primitive mammalian characteristic that has been lost in other orders of mammals. Most terrestrial mammals, such as the ungulates, have undergone limb modifications in evolution that provide greater stability at the expense of flexibility. Among horses, for example, not only is the direction of movement of the proximal limb bones (those nearest the trunk) limited to one plane, but the power of rotation of the distal bones (those farthest from the trunk) is also lost, and the digits are reduced in number, no longer having any grasping or clawing function.

In the primates, however, we find that the mammal limb evolved without losing its generalized form, in response to the three-dimensional forest environment.

With a very flexible universal proximal joint to the limbs, the elbow and knee joints evolved not so much to give more flexibility as to provide stability. Stability was achieved by retaining the primitive hinge joint, with movement limited to one plane and more limited musculature, an arrangement that has the advantage of retaining the main mass of muscles near the trunk and keeping the limbs slender—an essential mechanical arrangement for a stable center of gravity in a fast-moving animal.

When we compare the skeletons of different species, we clearly need to take into account body size. It is obvious that the gorilla has longer arms than the gibbon, for the gorilla is a much bigger animal. What is more interesting and of much more significance for our study is the fact that *in relation to its body size*, the gibbon has longer arms than the gorilla. This statement involves scaling the arm length according to body size—a process called *allometry*.

When we discuss limb length, cranial capacity, or any other metrical feature of the primates, therefore, we shall always be considering them in relation to body size. This is what is meant when we say that the forelimbs are relatively long and the brain relatively big, even if we do not explain on each occasion relative to what. (If we are investigating the morphology of a fossil species and do not know its overall body size, we may need to estimate it or use another skeletal dimension that we believe will reflect it.)

The question then arises as to how we can determine body size: is it a dimension or a matter of weight? In the past limb proportions have been related to body length or trunk length, but as we saw in the last chapter, both these dimensions vary among living primates in relation to body size, and they do not reflect it exactly, only approximately. In living populations it has proved best to use live body weight as an indication of overall size, and this variable gives the best results in allometric analyses (Aiello, 1981b).

An alternative approach to the study of relative limb length (or other dimensions of the body) has been to compare them to each other. For example, Schultz (1937) has compared fore- and hindlimb dimensions directly as an index, the *Intermembral Index*, which has the formula $100 \times$ humerus + radius/ femur + tibia (Fig. 6.1). This is a very suggestive way of illustrating differences in limb proportions, but when limb proportions differ, the index does not tell us which limb has altered in length or if they both have changed. Thus, the formula does not display individual limb differences. It is much better to compare the dimensions of each limb in turn to body size.

This is best achieved by preparing a logarithmic plot of each variable under study against body weight. The resulting bivariate display can be used to distinguish those specimens that show a constant linear relationship with body weight of the variable from those specimens that deviate from this relationship. Thus, we can clearly distinguish the changes in size and shape of a bone that have occurred in evolution merely as a result of change in the size of an animal from those that have occurred as a result of a particular specialization in a species, which is independent of body size. An example of such a plot can be seen in Fig. 6.2, taken from the detailed studies of Aiello (1981b).

When we look at Schultz's Intermembral Index (Fig. 6.1), we can see that humans and all Old World quadrupedal monkeys (of which the macaque is typical) have hindlimbs longer than forelimbs, while the other Hominoidea have forelimbs longer than hindlimbs. These data emphasize the great length of human legs and the shortness of human arms. The allometric data give us a more revealing picture: Fig. 6.2 shows clearly that in relative leg length, humans are typical higher primates, and it is the great apes that have somewhat shorter hind legs than the usual pattern (and this raises their intermembral index). We shall see in the next chapter that human arms are also of typical length for our body size: it is the gibbon alone of the Old World species that is exceptional in this characteristic (and this raises *their* intermembral index).

While the allometric plot is a far more valuable tool in the study of limb proportions and skeletal evolution than Schultz's index, we have referred to the use of the intermembral index because (together with a range of other similar indices) it appears frequently in the literature and is often used in the attempt to understand fossil locomotor adaptations.

Only one fossil discovery of *Australopithecus* supplies even some of the necessary documentation to calculate limb proportions—the *A. afarensis* skeleton called Lucy. The bones, though not all complete, broadly indicate an intermembral index of 82 to 85. The arms are longer in relation to the legs than in modern humans, and Lucy's index lies closer to the climbing (ape) pattern than does the human index. To examine the limb proportions of Lucy allometrically we have to estimate body size; this has been done by Wolpoff (1983) using the thickness of her limb bones and the height of the body of her third lumbar vertebra. Using this rather indirect route and the limited data, Wolpoff has concluded that her femur length is normal for her size compared with humans, but her humerus is at the upper end of the range of sizes for humans of similar body size. A second, poorly preserved femur belonging to *A. africanus* from Sterkfontein also falls into the human range. However, it must be remembered that the technique has introduced many possible errors and that the data are based on a damaged sample of only two individuals. Though highly suggestive, the data are inadequate to allow us to come to any firm conclusions, and we must, therefore, await the discovery of further sets of fossil limb bones in a good state of preservation so that the limb proportions of a series of ancestral groups may eventually be calculated.

The proportions of the limbs of other early hominids are as yet un-

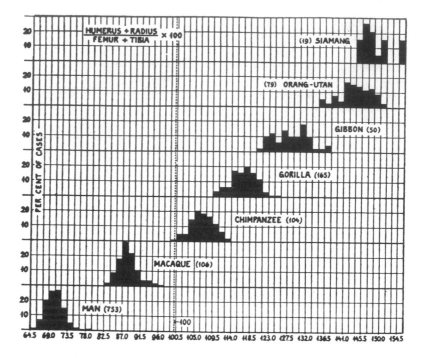

Figure 6.1. Frequency polygons showing the distribution of variations in the intermembral index of different primates. Figures at the bottom indicate the intermembral index, while those in brackets the sample size. The longer the hindlimbs in relation to the forelimbs, the lower the index; an index of 100 indicates limbs of equal length (from Schultz, 1937).

known, as the fossil record is so sparse. However, remains of Neandertal people are sufficiently complete for this purpose, and we find their intermembral index and limb proportions to be similar to those of modern humans. However, the late Neandertal people of Europe and the Near East appear to have short distal limb segments in relation to their proximal limb bones. This is possibly an adaptation to the cold climates of the last Ice Age because it is also found today among cold-adapted races of modern humans (Trinkhaus, 1981).

When a detailed, feature by feature comparison is made between the locomotor adaptations of humans and other higher primates, it is a curious fact that humans prove to lie anatomically closer to the New World howler monkeys (genus *Alouatta*) than to any of the great apes, to which they are certainly much more closely related (Stern, 1971). The howler monkey is a climbing genus that still sometimes travels quadrupedally, and so shows an intermediate spectrum of locomotor be-

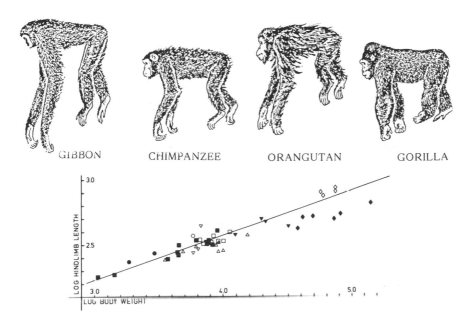

GIBBON   CHIMPANZEE   ORANGUTAN   GORILLA

Figure 6.2. *Above,* profile views of the four anthropoid apes, drawn with similar trunk length to indicate proportions of limbs (from Erickson, 1963). *Below,* bivariate plot showing the relationship between log body weight and log length of the hindlimb (from Aiello, 1981b). The solid line represents the principle axis. Note that the great apes alone have exceptionally short legs for primates of their size. Other specimens cluster near the principle axis. Males and females of each species are included separately in the analysis. □, *Colobus* and *Presbytis;* ■, *Cercocebus* and *Cercopithecus;* △, *Macaca;* ▲, *Papio;* ∨, *Alouatta, Lagothrix* and *Ateles;* ●, *Cebus;* ◇, *Homo sapiens* (Caucasian and Negro). ◆, Apes.

havior. (The possession of a prehensile tail does not appear to alter significantly the forces operating on its four limbs.) These anatomical similarities suggest a close functional link between *Homo* and *Alouatta* due, we must suppose, to both genera having a climbing ancestry. If we overlook its tail, the howler monkey provides a model of the locomotor pattern that could give rise to either the African great apes or humans. The superficial similarities between the Miocene *Proconsul africanus* and *Alouatta* further support this suggestion. The African great apes differ from this basic pattern as a result of their large size and their commitment to quadrupedal knuckle-walking.

In comparison with our close cousins the great apes, humans show only minor structural similarities in the pelvic region. As we shall see, human evolution has modified our locomotor anatomy drastically, which goes some way to account for the curious parallel mentioned in the preceding paragraph.

**II. The
Evolution
of the Pelvis**

The pelvis links the hindlimbs to the vertebral column. It is a structure of complex shape, curved in all three planes of space, and together with the base of the vertebral column (the sacrum) forms a rigid hollow bony structure. The pelvises of three primates are illustrated in Fig. 6.3; in Fig 6.4 the relationship between vertebral column, pelvis, and legs is shown. The pelvis consists of four parts, the left and right *hip* bones, which meet in the midline ventrally and are fixed to the *sacrum* dorsally (as is shown in the upper row of Fig. 6.3) and the *coccyx*. Each hip bone consists of three components, the *ilium, ischium,* and *pubis,* and from about the time of sexual maturity these bones become fused together (though there is never any movement between them). The ventral joint of the pubic bones (the *pubic symphysis)* is closed in humans between the ages of 17 and 25, and in other primates at an equivalent stage of development. (The human hip bone is shown in Fig. 6.5; the three bones and their most important landmarks are labeled.)

The ilium, which is firmly jointed to the sacrum, is a broad flattened bone with a bladelike extension spreading in a curve from each side of the sacrum. At its lower end it fuses with both the ischium and the pubis, and at this point of fusion is formed the cup-shaped socket into which fits the head of the thighbone (the femur). This socket is called the *acetabulum,* and through the pair of acetabula is transmitted the whole weight of the body in bipedal primates. From this central point the ischium extends backward (dorsally) and downward (to the base of the buttock in humans). In some primates it is expanded where it approaches the skin in the form of two *ischial tuberosities,* which give rigid support in the sitting position. Each *pubis* extends forward (ventrally) to meet its fellow in the midline, where it holds the two sides of the pelvis together at the pubic symphysis (see Fig. 6.3).

When we examine nonhuman primate pelvises, however, we note some striking differences from our own. Although the general function and relation of the bones is similar, a number of obvious differences are apparent in the monkey and ape pelvises drawn in Fig. 6.3, differences that are even more striking in the lower primates. In the pelvis of the little tree shrew, a primitive mammal, it is possible to see the arrangement of the pelvic bones typical of the early mammals and some prosimians (Fig. 6.6). The most strikingly different characteristic here is the long narrow blade of the ilium, which extends forward on each side of the sacrum. The most notable trend in the evolution of the primate pelvis has been the widening and shortening of this blade, from the condition seen in Fig. 6.6 through Fig. 6.3A to that in Fig. 6.3C. The widening of the ilium appears to be an allometric effect of increase in body size, while the shortening is a feature peculiar to hominids.

If we compare the larger Old World monkeys with the small prosimians, we find that some widening of the blade of the ilium has occurred, upon which the important groups of muscles that effect move-

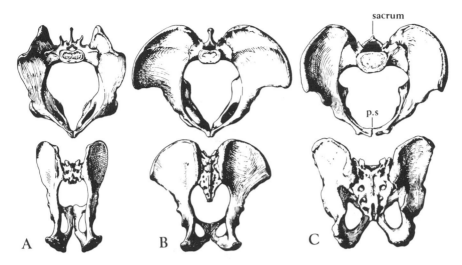

Figure 6.3. Views from above and behind of pelvis of macaque (*A*), gorilla (*B*), and modern human (*C*). *p.s.*, Pubic symphysis. All drawn the same size (from Schultz, 1963b).

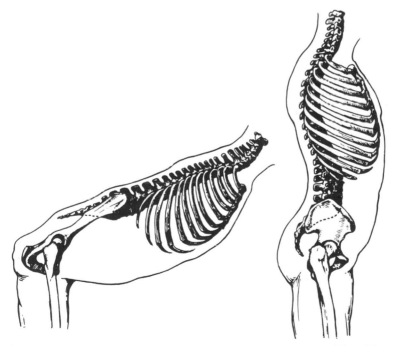

Figure 6.4. The pelvis in chimpanzee (*left*) and human (*right*). Note the difference in shape and size and the relationship between the pelvis, vertebral column, the femur (after Schultz, 1963b).

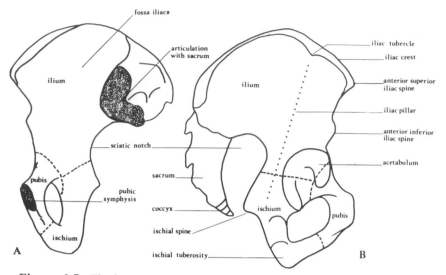

Figure 6.5. The human right hip bone. An internal view (*A*) and an external view from the right (*B*) showing the sacrum in position. The dashes indicate the boundaries of the fused bones; the dots indicate the thickening of the ilium called the "iliac pillar." The shaded areas are articular surfaces.

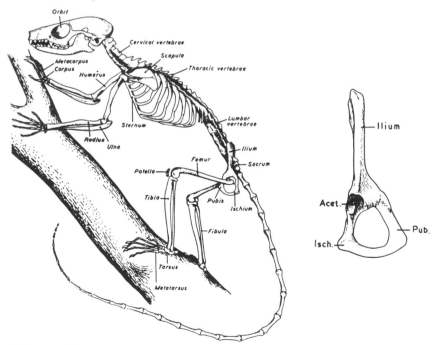

Figure 6.6. Skeleton of *Ptilocercus*, a tree shrew, drawn about two-thirds natural size, and an enlarged drawing of the right hip bone (from Le Gros Clark, 1971).

ment of the femur originate (Figs. 6.3A and 6.6). This enlargement of the areas of origin coincides with the enlargement of the muscles as a whole in the larger animals caused by the increased size and power developed in the hindlimbs. It is allometrically related to body size (Aiello, 1981a). The two ischial tuberosities are also farther apart to give a more stable sitting position, and form a foundation for the *ischial callosities*—the especially hardened areas of skin characteristic of the Old World monkeys (Fig. 10.2), which commonly not only sit, but sit while sleeping, often with their feet drawn up.

Among the apes we see that the larger animals have still heavier musculature to support a larger body, and the blade of the ilium is still wider (see Fig. 6.3). The crest (the top ridge of the ilium) assumes more importance; the lateral abdominal muscles inserted upon it act as stress members of the parabolic girder supporting the trunk. But the generally elongated form of the pelvis survives.

The human pelvis shows remarkable differences compared with the pelvises of other primates; the Latin word *pelvis* means basin, and the basinlike shape that it assumes in humans gives it that name. In the human family, the attainment of bipedal locomotion has fundamentally changed the stresses set up in the pelvis that are correlated with a change in shape (see Fig. 6.3). This change in shape has been analyzed into a number of components (Schultz, 1963b; Le Gros Clark, 1971; Washburn, 1950).

1. The muscles of the thigh move the thigh forward and backward and provide the power for both quadrupedal and bipedal locomotion. The muscles that move the leg forward are called the *flexors* because they bend the thigh at the hip; those that move the leg backward are called the *extensors* because they straighten the leg at the hip joint. The development of these muscles is relatively greater in humans than in any other primate because humans alone depend entirely upon them for locomotion, and the leg is proportionally heavier. At the same time, there are changes in bone structure that tend to increase the leverage of the muscles. These changes involve as a whole the movement of the areas of origin of the muscles outward from the point of pivot: a small horizontal extension here greatly increases effective muscle power (Fig. 6.7).

Probably the most important flexor muscles are the *ilio-psoas* and *rectus femoris;* the areas of origin of the latter are indicated in Fig. 6.8. The human pelvis has a much more pronounced *anterior inferior iliac spine* (where a portion of the muscle and the iliofemoral ligament originates) that has the nonhuman pelvis.

The most important extensor muscles among higher primates are the so-called "hamstring" muscles, the *biceps femoris*, the *semitendinosus* and the *semimembranosus*, the former two having their origin on the ischial tuberosity and their insertion on the tibia and fibula, respectively. (The origin and position of the biceps femoris is shown in Fig. 6.8.) However, in humans the extensor action of these muscles, so important in locomo-

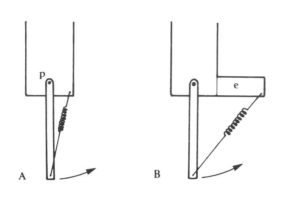

Figure 6.7. In this diagram the effective power of the spring to move the lever to the right (*A*) is increased by moving the fixed end of the spring farther away from the pivot (*p*) as in (*B*). The structural extension (*e*) is a model of the increased horizontal dimension that is typical of the human pelvis as compared with the nonhuman pelvis and that increases the effective power of all muscles used in locomotion as well as of those required to maintain the balance of the trunk above the legs.

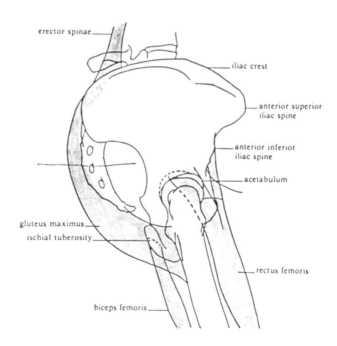

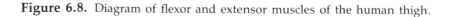

Figure 6.8. Diagram of flexor and extensor muscles of the human thigh.

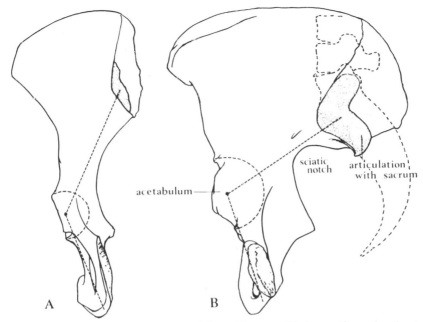

Figure 6.9. Pelvis of chimpanzee (*A*) and human (*B*) drawn the same size to show that as the articulation with the sacrum (shaded) moves down toward the feet, it must also move away from the acetabulum so that the size of the pelvic canal will be maintained. The dotted line shows the change in structure of the pelvis that results in its "bent" appearance.

tion, is reinforced by the action of another muscle, the *gluteus maximus*, the upper part of which is an *abductor* muscle in other primates. This change in function has been brought about by a complex change in the form of the ilium and, in particular, in the proportion and curvature of its blade, which has brought the muscle to lie behind the acetabulum rather than to one side.

The most obvious difference between the nonhuman and the human ilium is that in humans, the bone has shortened as well as widened so that the articular surface, where the bone is connected to the base of the vertebral column (the sacrum), has been brought nearer to the acetabulum, the socket of the femur. This change gives greater stability, since the weight of the trunk is transmitted more directly to the leg. As Schultz (1963b) has shown, the ilium occupies only 24% of trunk length in humans, whereas it occupies 36–38% in the great apes.

In human evolution, as the ilium becomes shorter, its axis—from sacrum to acetabulum—must have a greater angle with the ischium in order to keep the same diameter in the pelvic birth canal (Fig. 6.9). Shortening of the bone must therefore go with outward displacement, and this "bending" causes the change in the position of the origin of the gluteus maximus (which arises on the posterior extremity of the iliac crest) in relation to the acetabulum and femur. As Fig. 6.8 shows, the gluteus maximus in humans may act as a very powerful extensor of the thigh, but it has been shown not to be used during relaxed bipedal walking on a level surface (Basmajian, 1978); it functions only to stop the forward movement of the limb as the heel touches the ground. That is, in relaxed walking it does not propel the body forward, but merely controls limb movement by stabilizing each leg as it begins to carry the weight of the body. However, the muscle has great power and comes into play in running or climbing or walking upstairs. It also raises the trunk from a bent position and when standing up after sitting or squatting.

This discussion highlights an important point made by Tobias (1982): in standing still or walking in a relaxed manner on the level, humans use only a small part of their available locomotor power and retain enormous muscle reserves. This means not only that the reserves are available for demanding situations, such as having to run uphill, but that the energy consumption in standing or relaxed walking is minimal, because of the mechanical and energetic efficiency of the human body.

From Fig. 6.8 it can be seen that the greater the extension of the pelvic bones in a ventral or dorsal direction, the greater the leverage a given muscle will exert on the leg for locomotion. Although Fig. 6.8 is a great oversimplification of the musculature, the mechanics of the muscle action are straightforward. The leverage of the muscles is increased by the changes that have occurred in the pelvis during human evolution. The ischial tuberosity has moved outward and upward to give more leverage to the biceps muscle about the acetabulum—the pivot of the femur. Another advantage of the increasing extension of the pelvis dorsally is the greater leverage given to the *erector spinae* muscles, which are inserted on the ilium and which maintain the upright posture of the vertebral column.

2. The "bending" of the ilium (mentioned above) has resulted in the formation of a relatively deep sciatic notch, which is associated with an accentuation of the ischial spine, the point upon which the sacrospinous ligament and the important muscles forming the floor of the pelvic basin are inserted.

Since we have seen that the "bending" of the ilium affects the diameter of the pelvic birth canal, we find here a sexual characteristic. The "bending" has proceeded further in males than in females; the canal is smaller in males, and this difference in size is correlated with a more acute angle of the *sciatic notch*. The size of the canal in women is functionally related to the size of a baby's head, which must pass through it at birth; the size of the canal in men is controlled mainly by locomotor, not reproductive, factors. These differences in the pelvis contribute to the difference in locomotion and posture between men and women. The canal in the human female only just accommodates the head at birth, which is relatively larger than in other primates. The cross-sectional area

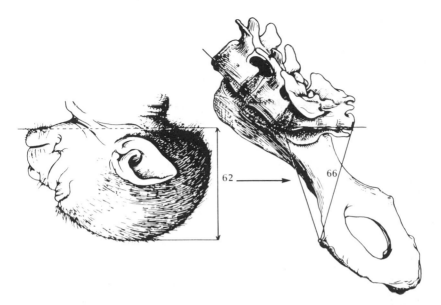

Figure 6.10. The pelvis of an adult baboon and the head of its full-term fetus, drawn to the same scale. Mother and young died immediately after the difficult birth (from Schultz, 1963b).

of the canal may well be one limiting factor in the length of the period of gestation in humans, although the cranial bones of a human baby exhibit remarkable flexibility at this stage of development. Figure 6.10 demonstrates that this limiting factor is not unique to *Homo*. The figure is taken from Schultz (1963b), who states that the canal is proportionally larger in adult females than in males of the same species in all primates except the great apes. This is, therefore, a recognizable secondary sexual characteristic.

3. The "bending" of the ilium and the human erect posture bring about a reorientation of the sacrum in relation to the ilium, with the result that the axis of the pelvic canal lies almost at a right angle to the vertebral column; the pubic symphysis and sacrum have become more nearly the floor and roof of the pelvic cavity rather than the ventral and dorsal walls (Fig. 6.4; see also Fig. 5.8). At the same time, because greater weight is transmitted through the area of contact between sacrum and ilium, this area has increased relative to that of the ilium as a whole (Fig. 6.9). For the same reason, the acetabulum and the head of the femur have also increased in relative size during the course of human evolution.

4. The *adductor* and *abductor* muscles, which move the limb from side to side in an arboreal primate according to the position of a branch or the demands of the terrain, are essential in a bipedal creature for maintaining the lateral balance of the trunk upon the legs.

During bipedal walking all the weight of the body is carried by one leg at a time, and it is the abductor muscles (together with the lateral abdominal muscles that extend from thorax to ilium) that raise the trunk at each step and pull it vertically over the thigh, so that the center of gravity lies over the triangle of the foot and the weight is transmitted directly down the leg through the knee. The adductors, on the other side of the thigh, help to hold the balance (Fig. 6.11).

In apes, the *gluteus medius* and *minimus* lie dorsal to the hip joint and act mainly as extensors of the thigh at the hip joint, while gluteus maximus acts as an abductor (Tuttle *et al.*, 1975, 1979). In humans, the two smaller gluteus muscles (the medius and minimus) are the major abductors with only some help from gluteus maximus which is primarily an extensor. It has been shown (Mednick, 1955) that the *iliac pillar* and *iliac tubercle*, a thickening of the ilium unique to humans among living primates (see Fig. 6.5), take the compression exerted by the gluteus medius when it lifts the trunk to keep the unsupported side of the pelvis from tilting downward, during bipedalism. Clearly the leverage of these muscles will be enhanced by lateral extension of the ilium and the lengthening of the neck of the femur. Since they support the weight of the whole body, they are muscles of prime importance in maintaining erect posture. The whole abductor mechanism is distinctive of humans.

Associated as most of these changes are with erect posture and bipedal locomotion, they constitute one important set of adaptations that human ancestors underwent, and by general consent they are accepted as among the most striking diagnostic complex of characteristics of the Hominidae. These characteristics are already present in *Australopithecus* in an evolved condition. One of the fossil discoveries that has helped to elucidate the course of human evolution is that of the pelvis of *Australopithecus*. Pelvic bones from Hadar include a complete left hip bone (of Lucy) and a left ischium. They conform closely to the more extensive South African finds. There are also some fragments of a right and left ilium from Koobi Fora. Finally, remains of seven fragmentary hip bones from South Africa have been described by Robinson (1972):

Sterkfontein	Mature pelvis without coccyx right ilium and pubis
Kromdraai	Left ilium
Swartkrans	Imperfect right hip bone
Makapansgat	Juvenile left ilium

All these specimens are remarkably similar to those of modern humans, and nearly all the characteristics of the human pelvis listed above are present. The differences between the *Australopithecus africanus* pelvis and that of modern humans are slight and include the following features (Le Gros Clark, 1955; Napier, 1964; Robinson, 1972; Lovejoy, 1973; Day, 1973):

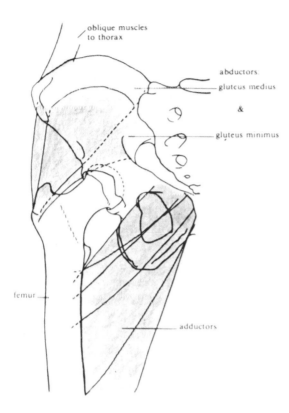

Figure 6.11. Diagram of abductor and adductor muscles of the human thigh.

1. The bones are smaller and lighter, being more comparable to those of pygmies than to the large races of modern humans. The muscle attachment areas on the blade of the ilium are less marked than in *Homo sapiens;* the acetabulum is relatively small; the iliac pillar and iliac tubercle are present but not as well developed as in *Homo sapiens.* (These differences can be seen in Fig. 6.12.)

2. The blade of the ilium is more flared to each side (see Figs. 6.12 and 6.13) than in humans: the ilia are oriented frontally, not anteromedially.

As a whole these differences indicate that *Australopithecus* was smaller and more lightly built than *Homo.* Based on the pelvis alone, *Australopithecus* was bipedal. Because the abductors were positioned to provide greater leverage than in modern humans, the pressures generated between the acetabulum and the head of the femur are calculated to have been half those developed in modern humans (Lovejoy, 1973)—(see Fig. 6.13). In this sense, *Australopithecus* was a mechanically more

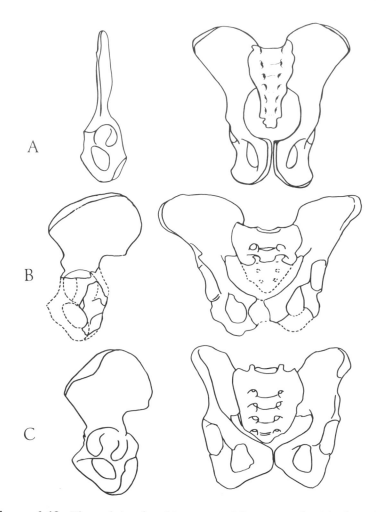

Figure 6.12. The pelvis of a chimpanzee (*A*) compared with that of *Australopithecus africanus* (*B*) and a San Bushman (*C*): *left*, left lateral view; *right*, the entire pelvis from the front. Note the strong similarity between the human and *Australopithecus* pelvis. (Not drawn to scale.) (Redrawn from Le Gros Clark, 1971; and Dart, 1949.)

efficient biped than modern humans* can claim to be, although specimens of modern hip bones can occasionally be found with similar laterally flared ilia.

From the evidence of the *Australopithecus* pelvis we can conclude that the form of the ilium has changed slightly in human evolution since that

---

*Zuckerman *et al.* (1973) conclude through complex biometrical analysis that abduction was poorly developed.

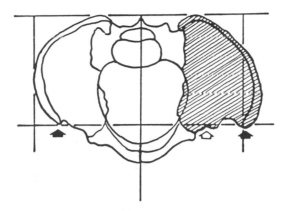

Figure 6.13 Superior view of the left ilium of *Australopithecus* (*right, shaded*) articulated with the pelvis of a San Bushman. The greater lateral flare and prominent ilio-psoas groove (open arrow) of *Australopithecus* are evident (from Lovejoy, 1973).

time, so that the blade does not flare out so widely. This is probably due to the fact that the birth canal has increased in size to accommodate an enlarged infant's head, while the maximum pelvic breadth between the hips was limited due to the balance and mechanics of walking. Thus, we have less leverage available for the abductors in modern humans than in *Australopithecus*, and this has resulted in greater forces being transmitted through the acetabulum and head of the femur. Both are significantly larger in *Homo*. This particular development is the reverse of that indicated in these pages (and shown in Fig. 6.7) for the pelvis as a whole. It is a secondary response to the special condition of large cranial size in the newborn that we associate with the genus *Homo* (Lovejoy, 1973).

Napier (1964) and Robinson (1972) have described some differences between the innominate fragments from Sterkfontein and Swartkrans. They state that bones from the latter site, besides being larger and more robust, are less human and more apelike in form. One Swartkrans specimen has a longer ischium, which suggests that the population it represents may have been active arboreal climbers, since this feature gives greater leverage to the biceps femoris. In view of the small size of the sample ($N = 1$) and the lack of relative measurements, they cannot be taken to signify the presence of two groups with different locomotor adaptations. On the contrary, measurements of all specimens fall into the range of variation of modern humans (Lovejoy, 1973).

Fragmentary innominate bones classified (Fig. 6.14) as *Homo erectus* have been found at Koobi Fora (ca. 1.6 mya) and Olduvai Gorge (ca. 0.6 mya), and there is a much more recent discovery from Arago in southern France (ca. 0.45 mya). All these bear many resemblances to the innominates of modern humans, but are distinct in their great robusticity and in a number of minor characteristics described by Day (1982). These include a large acetabulum with a well-developed supporting iliac

Figure 6.14 Three specimens of the pelvis of *Homo erectus* from Arago (*left*), Olduvai (*center*) and Koobi Fora (*right*). The outer or gluteal surface is uppermost. Note the large acetabulum and broad iliac blade (from Day, 1982).

pillar, a forwardly raked and well-developed anterior superior iliac spine, and a ventrally deflected anterior inferior iliac spine, together with a long iliac blade (Sigmon, 1982). It seems possible, however, that these differences are a product of the unusual robusticity of this species, rather than signifying a slightly different locomotor adaptation from that found in living humans.

A few specimens of early *Homo sapiens* innominates have been collected. Those attributed to the Neandertal people in Europe and the Near East are generally robust and suggest a very muscular physique. They are peculiar, however, in the form of the upper branch of the pubic bone, which is long and slender in both sexes. By moving the pubic symphysis outward (ventrally), this lengthening would increase the circumference of the birth canal, so it may be an adaptation to large-headed babies (Trinkhaus, 1983). This more delicate bone is not found in any *H. erectus* finds, so we do not know the history of this adaptation. The fossils from Skhul and Qafzeh in Israel are modern in form as are all the succeeding upper paleolithic European fossils.

III. The Hindlimb

As we have seen, the morphology of the limb bones reflects the mode of locomotion of the animal. The lengths of the different bones in relation to body size and the extent of movement possible are related to the function of the limb as a whole. The thickness of the bones reflects proportionately the size of the animal and the weight and stresses transmitted by each bone; it is to be expected on allometric grounds that the bigger animals would have longer limbs and heavier bones.

The *femora* (singular *femur*) or thigh bones of the lower primates and small monkeys are, on the whole, slender and straight, while the *great* and the *lesser trochanters*, upon which abductors and rotators respectively, are inserted, are prominent processes. The femora of the larger quadrupedal monkeys and the great apes are proportionately more robust and slightly bowed in the anteroposterior plane, but bones that are more often under tension than compression tend to have a straight shaft. Remains of fossil femora of *Proconsul* have been found (Le Gros Clark and Leakey, 1951) with a nearly straight shaft, and they have been compared and found to have much in common with femora of arboreal monkeys as well as with modern humans (Napier, 1964; Fig. 6.15). The human femur has a longer neck than that of apes, thus giving the abductor muscles more power. The larger femoral head transmits a much greater force (as we noted when discussing the acetabulum, and which is due to the extra weight that each leg must carry together with the muscle force generated) than would the head of a quadrupedal creature. The angle of the condyles at the knee, the "carrying angle," is also different from that found among the African apes because the transmission of weight down the leg is altered, as we shall see below.

Thirteen fragments of femora of *Australopithecus* are known from Hadar (Johanson *et al.*, 1982), twenty-one from Koobi Fora (Walker, 1973; Leakey and Leakey, 1978), one from Olduvai (Day, 1969), and three from the Transvaal (Robinson, 1972). Altogether thirty-eight specimens are known, some are fragmentary while others are almost complete. All these bones are fairly close to those of modern humans but show some distinctive characteristics: in particular, they carry a relatively longer and more anteroposteriorly flattened femoral neck that is set closer to a right angle to the shaft axis, together with less lateral flare to the greater trochanter, and a smaller femoral head. They also have a groove for the obturator externus tendon at the back of the femoral neck—a groove present in all human femora and absent in other primates (Day, 1973; Walker, 1973). Some belonged to small gracile individuals probably no more than 4 ft 3 in. in height (Sterkfontein), while others suggest a stature up to 5 ft 7 in. (Koobi Fora). The evidence indicates an increase in size with time, and nearly as much variability in robusticity as we find in modern humans.

The striking features of the small femoral head and long neck are fully concordant with the evidence from the pelvic bones that we have just considered. The small head fits the small acetabulum and indicates that a far smaller force (gravity and muscle force combined) is transmitted through this joint due to the greater leverage available for the abductor muscles. At the same time the longer neck moves the great trochanter out from the midline, but the trochanter shows less lateral

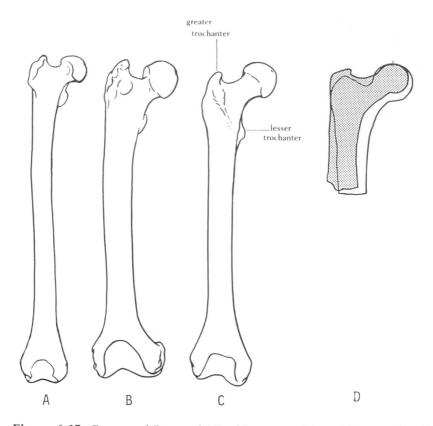

Figure 6.15. Femora of *Proconsul* (*A*), chimpanzee (*B*), and human (*C*), all drawn the same size. (*D*) Outline of the proximal femur of *Australopithecus* (shaded) overlain on the femur of a female human. In (*D*) the weightbearing points have been aligned. Note the neck length and angle, the size of head, the size and form of the condyles, and the degree of lateral projection of the greater trochanter (*D* after Lovejoy *et al.*, 1973).

projection from the shaft than in *Homo*. Nevertheless, this is all concordant with the theory of increased leverage suggested by the laterally flared blades of the ilia (Fig. 6.13.—Lovejoy, 1973).

Unusually well preserved are parts of five femora believed to belong to *Homo erectus* from Java. One was complete and measured 455 mm, giving a supposed stature of 5 ft. 6 in. (168 cm). They appear to be similar to those of modern humans, though they are very robust. A femur of *H. erectus* from Peking (Weidenreich, 1938) and a shaft fragment from Olduvai Gorge (Day, 1971), in contrast exhibit a combination of traits that do appear to be distinctive of this species (Day and Mollison, 1973). Although of late date (ca. 450,000 BP) the femur fragment from Arago also shows similarities to these *H. erectus* femora. As a group these femora are characterized by a large femoral head, shorter and less flattened neck (than in *Australopithecus*), a flattening of the shaft (platy-meria), and other minor features found separately but not usually in combination among modern humans (Day, 1976).

Femora of Neandertal people are also known and again appear to differ little from those of modern humans, though they are rather robust and the joint surfaces may be somewhat larger in area in relation to length. As in all these forms, the circumference of the human femur is second only to that of the very heavy gorilla due to the increase in weight borne by the leg bones (Schultz, 1953), and it is greater in relation to femur length than in any other primate (Aiello, 1981b).

Human legs are, on the average, the longest among all the primates relative to trunk length (Schultz, 1950), but when compared to body size they are not significantly longer than the legs of other higher primates, though ape legs are shorter (Fig. 6.2—Aiello, 1981b). As the size of the animal increases in evolution so does the size of each bone, and the bearing surfaces tend to increase proportionately in area. Both the ar-ticular head of the femur and the acetabulum into which it fits on the pelvis are, as we have seen, significantly larger in humans than in other primates. However, at the distal end the two condyles (at the knee) form a bearing surface relatively narrower than that found in the large apes. This would appear to be a result of improved weight transmission, which passes through the outer rather than the inner condyle and en-ables the foot to be placed beneath as well as outside the body's center of gravity at each step (Fig. 6.16).

Full extension at the knee does not occur among nonhuman primates. Humans can not only fully extend the knee, but by a slight medial rotation of the femur on the tibia can lock the knee joint in the fully extended position. This makes it possible to stand with almost no muscular exertion at this joint.

We have the distal ends of ten femora belonging to *Australopithecus*, which show that the human condition is evolving. The bicondylar angle (the carrying angle) is significantly larger among modern humans and *Australopithecus* than among the African apes, in which it averages zero (Fig. 6.16). Among the great apes, the orangutan most closely approaches *Homo* in this characteristic (Stern and Susman, 1983).

Other features of the *Australopithecus* femur fall among the range of variation found in modern humans. In particular, a lateral view of the condyles shows that the increased joint contact seen in humans (Fig. 6.17) is already present in *Australopithecus*, though absent in apes and monkeys.

Based on these features of the pelvis and femur, the case for a well-developed bipedalism in *Australopithecus* is overwhelming.

The distal bones have also been modified in the course of human evolution. In the primitive mammal, the existence of two distal limb bones rather than one made possible the rotary movement of the hand and foot (*manus* and *pes*). The *tibia* can be considered the fixed bone that

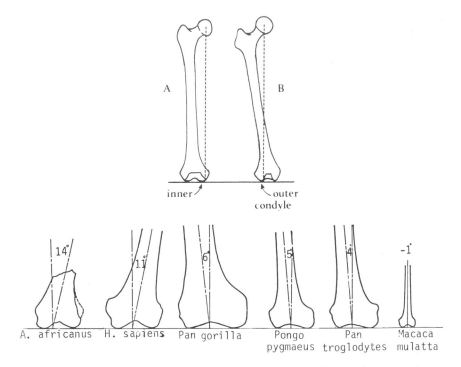

Figure 6.16. Femora of gorilla (*A*) and human (*B*) drawn the same size to show difference in the line of weight transmission down the leg and the bicondylar angle. *Below*, the bicondylar angle is shown for a number of higher primates. Note that this feature in *Australopithecus* is even more extreme than is found in the human species. (Lower figure from Lovejoy and Heiple, 1970.)

transmits the majority of the weight (the *ulna* in the forelimb), while the *fibula* (and *radius*) at its distal end can move in a rotary manner around the base of the tibia, and this movement in turn revolves the pes (or manus) (Figs. 4.7 and 6.6). Such rotation is essential to an arboreal creature for grasping the trunk and branches of trees at all angles. The retention of this generalized feature of primitive limb structure is characteristic of the primates, though modified in *Tarsius* and some lemurs. In the terrestrial forms the rotation is somewhat limited, and in humans the function of the fibula is limited and there is no rotation.

The size of the tibia is related to the weight transmitted through it, and in the circumference of the tibia in relation to body size humans exceed all other primates; as a corollary, no weight is transmitted through the fibula, and in humans it has become reduced, as has our ability to revolve the foot at the ankle joint.

Fossil tibiae and fibulae of *Proconsul* are known and the fibula is strong and suggests good climbing capability (Walker and Pickford, 1983). The earliest hominid specimens we have are ten tibial and six

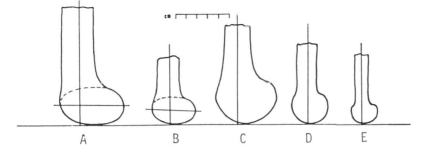

Figure 6.17 Tracings of the lateral view of the condylar surface of the femur of a human (*A*), *Australopithecus* (*B*), gorilla (*C*), chimpanzee (*D*), and baboon (*E*) (from Lovejoy, 1973).

fibial fragments from Hadar. Sixteen tibial and five fibial fragments are known from Koobi Fora. We also have an almost complete tibia and fibula from Olduvai Gorge. All these belong either to *Australopithecus* or *Homo habilis.* The bones belong to a creature habitually bipedal, but whose gait may well have differed slightly from that of modern humans (Davis, 1964). The bones appear long and slender and the fibula is reduced in mobility; it is indistinguishable from that of modern humans of similar size. Fragments of these bones are also known from Neandertal and Rhodesian people; they appear to fall within the range of variation of modern humans.

All these changes, which have clearly occurred in human evolution, are directly related to the development of erect posture, to the position of the femur in relation to the pelvis, to its position in relation to the tibia, and to the transmission of weight directly through the bone rather than through a system of stressed and tensioned components of bone and muscle. The reduction of the fibula in *Homo* is related to the reduced need to rotate the foot, an adaptation no longer necessary in a creature that walks and runs upon the ground. Rotation within the foot (inversion and eversion) is discussed in the next section.

IV. The Evolution of the Foot   The hind foot or *pes* has a number of fairly distinct functions in primates, and it is the changing balance between these functions that characterizes its evolution. The functions may be considered in turn.

The foot acts as a lever that adds to the propulsive force of the leg, a function clearly seen in all primates. As we saw in Chapter 3, the distal segment of the *calcaneus* and associated *tarsal bones* is elongated with respect to the heel segment: this increases the load arm in relation to the power arm and gives greater movement for the force applied. The smaller primates, especially the prosimians and those under 10 kg, have

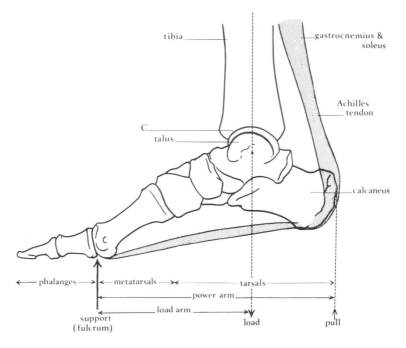

Figure 6.18. Diagram of the structure and mechanics of the human foot. C, crural joint. The drawing is of the inner border of the right foot.

a long load arm (Martin, 1979). In higher primates and humans the load arm is extended by the lengthening of the *metatarsals*. By contraction of the *gastrocnemius* and *soleus* muscles of the calf of the leg, the human body is raised upon the ball of the foot. The point of pivot is the *ankle joint* between the *talus* (or *astragalus*) and the tibia and fibula. The power arm of the lever is the calcaneus, tarsals, and metatarsals; the contraction of the gastrocnemius is transmitted through the *Achilles tendon* to the heel. The load arm is the tarsal section of the foot anterior to the pivot, together with the metatarsals (Fig. 6.18).

The relative lengths of the load arm and power arm have changed in primate evolution according to the mode of locomotion. Extensive movement (i.e., a long load arm) of the foot lever is characteristic of the VCL leaping prosimians, where power is less essential and maximum movement required. The load arm of *Tarsius* represents eight-ninths of the power arm. In the heavier Hominoidea, power is more important, and, in humans, the load arm is less than three-quarters of the power arm of the foot lever.

Within the load arm itself, the proportions of the tarsal and metatarsal segments have varied, but only in a few primates (e.g., *Tarsius*) has the tarsal segment assumed great importance. Among the Anthropoidea, the metatarsals are longer, the tarsal bones remaining short and compressed. Slight lengthening of the tarsal segment is, however, characteristic of humans (Fig. 6.19).

When we examine the mechanism for the transmission of weight from above, we see some significant variations in the load line. Among the monkeys that have retained a quadrupedal mode of locomotion, weight is transmitted mainly through the middle digit of the foot, which is the longest. These monkeys retain the primitive phalangeal projection index of decreasing length common to quadrupedal mammals, so from the longest to the shortest the order of digits is 3>4>2>5>1 (Fig. 6.20). Among prosimians and apes, as body weight increased and the grasping functions of the foot increased at the expense of the propulsive function, the hallux (great toe) became increasingly opposable to the other digits. In the feet of apes, we find the weight is transmitted to the branch not through digit 3 but through the webbing between digits 1 and 2 (Figs. 6.20 and 6.21). This shift in the load line has extended in the human foot almost to digit 1 (the hallux), and the existence and origin of the shift is of the utmost importance in human evolution. It suggests very strongly that the Hominidae evolved from an arboreal climber with a partially opposable hallux. In contrast, the terrestrial monkeys have retained the primitive weight line through the third digit.

The human foot has the formula of phalangeal projection 2>1>3>4>5, and Schultz (1950) has shown that this change has been brought about by a reduction of digits 2 to 5 rather than by an elongation of the hallux. A significant comparison can be made with two subspecies of gorilla. The mountain gorilla (*Pan gorilla berengei*) is almost entirely terrestrial, and its foot is much closer to that of humans, since the opposability of the hallux is partially lost. It should be compared with the lowland gorilla (*P. gorilla gorilla*), which is to a greater extent arboreal (Fig. 6.22). The pattern of robusticity of the metatarsals is also significant in this respect: in the fully terrestrial forms alone, weight is also transmitted through digit 5, which is more robust than digits 2 and 3 (Day and Napier, 1964) and which, by distributing the weight like a tripod, helps lateral balance (Fig. 6.23). This is particularly characteristic of humans in which the foot gives not two-point support, as in other primates, but the three-point support of a tripod.

The foot may also act as a means of gripping the branches and trunks of trees. As we have mentioned, claws are effective for a small animal that moves on the trunks of trees and upon the larger branches. In larger arboreal animals claws are less effective, and in the primates the phalanges (fingers and toes) have evolved sufficient length to enable the animal to grasp the branches among which it moves.

The effective grasp of the phalanges can be increased not only by lengthening but by the evolution of an opposable hallux (or *pollex* in the forelimb), a striking adaptation among the prosimians, the small size of which poses special problems in gripping the larger branches. Among the relatively large quadrupedal monkeys, opposability of the hallux has

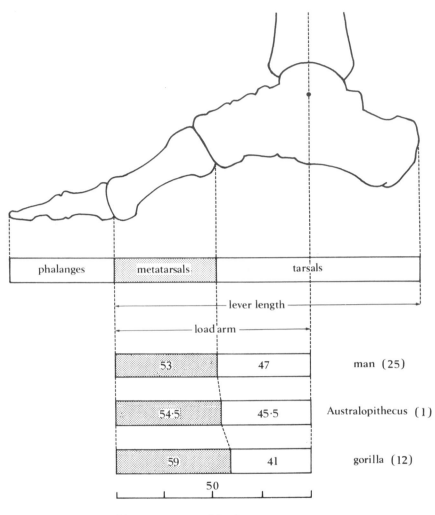

Mean percentage of load arm

Figure 6.19. Mean proportions of different parts of the feet of three primates suggest that the tarsal section of the load arm has lengthened in human evolution. The number following each name shows the sample size (*n*) used for this calculation. The proportion in humans is, of course, variable and overlaps the figures for *Australopithecus*. As always, the significance to be attached to a sample of *n* = 1 is questionable, but in the absence of further evidence there is no reason to discard these figures as a basis for discussion (data from Day and Napier, 1964).

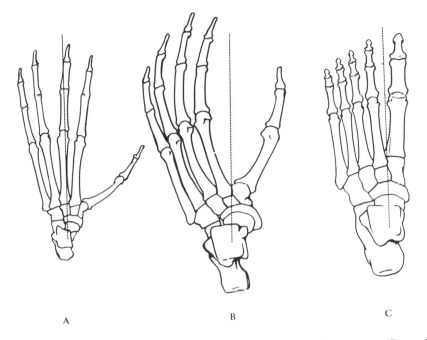

Figure 6.20. Foot skeleton of macaque monkey (*A*), chimpanzee (*B*), and human (*C*), showing the load line in each genus (after Morton, 1927).

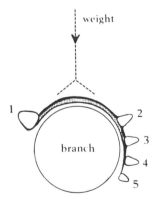

Figure 6.21. The transmission of weight to the branch in apes is not through a digit but through the musculature between digits 1 and 2. This figure shows a diagrammatic transverse section through foot and branch (after Morton, 1927).

not evolved to any great extent; instead, long phalanges are the rule. Only among the slower-moving great apes do we find full opposability of the hallux, and, as has been described, this opposability gives rise to a change in the load line, an important feature in human evolution. Among humans, opposability has been effectively lost, and although it is still possible to grip and hold objects between the hallux and second digit, the hallux cannot be widely abducted.

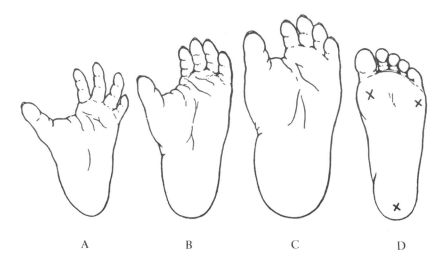

Figure 6.22. Feet of chimpanzee (*A*), lowland gorilla (*B*), mountain gorilla (*C*), and human (*D*). The three centers of weight in the human foot are marked with crosses (from Morton, 1964).

Although the foot cannot be rotated at the ankle joint, it can be in-verted or everted as a result of *intratarsal rotation*. Running or walking barefoot on rough uneven ground requires these movements, and some eversion is necessary at the final toe-off thrust in bipedal locomotion to maintain stability.

A second development of great importance has effected the grasping power of primate hands and feet (together called the *cheiridia*). The function of the hooked claws of the primitive mammals has been taken over in the primates not only by the long phalanges but also by the evolution of *volar pads* (friction pads) on the palms, which were enlarged and eventually fused. Volar pads are characterized by a special ridged friction skin with a microscopic folding of the outer layer. This skin lacks hair but is well supplied with sweat glands. The claws themselves were reduced in evolution and modified to flat plates, the nails, which gave support to the terminal pads (nail pads) that are so important in main-taining a firm grip on a branch.

The foot may also act as a tactile organ, and there was an immense evolutionary proliferation of the tactile receptors in the ridged friction skin. In this way, the cheiridia came to replace the tactile rhinarium and whiskers on the snout, serving as prime organs of touch in the pro-simians and most other mammals. The importance of this development, particularly as it applies to the hand, is immense in human evolution; it is discussed further in Chapters 7 and 8. In the human foot we find an organ perhaps unnecessarily well-endowed with tactile receptors, since

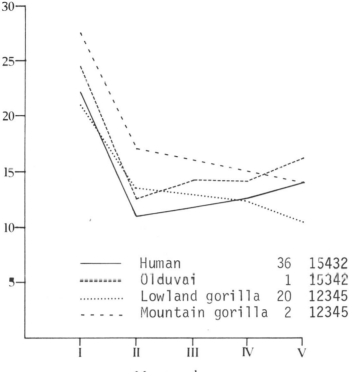

Figure 6.23. The robusticity of the metatarsals is an important indicator of weight distribution through the foot. The robusticity index is defined as mean diameter × 100 divided by mean length of the different metatarsals. In this diagram, this index is shown for the metatarsals of humans, the Olduvai foot, and the two gorillas. In both humans and the Olduvai foot the fifth metatarsal is the second most robust after the first. In the gorillas it is the least robust, though in the more terrestrial mountain gorilla the index of all metatarsals is greater, especially that of the fifth metatarsal. The five-figure numbers in the key show the order of robusticity (the robusticity formula) of the different samples; the smaller numbers indicate the sample size (*n*) (after Day and Napier, 1964).

during its evolution its functions are reduced to that of locomotion alone.

Fortunately, we have some knowledge of early hominid feet. There is a partial foot from Hadar and a nearly complete foot from Bed I, Olduvai Gorge. There are other foot bones from Hadar and Koobi Fora and one from Kromdraai. Most authorities agree that the Hadar tarsal bones are those of a plantigrade hominid, though a few minor apelike features can be detected. Tuttle (1981) and Stern and Susman (1983) show that the proximal phalanges from Hadar are both longer and much more curved than those of modern humans. The middle phalanx is also long relative to the proximal phalanx, and the Hadar hallux is somewhat divergent and can be slightly abducted to grasp—a function lost in the final stages of human evolution. Thus, in this foot we see some characteristics of a grasping organ, and in the curvature of the third and fourth proximal phalanges the foot is chimpanzeelike (Fig. 6.24) (Tuttle, 1981).

These specimens also suggest that the line of weight transmission through the foot was slightly different from that in modern humans. The presence of a facet between the first and second metatarsals of the Olduvai foot suggest that the hallux was less divergent than in apes and closer to that of modern humans. In other respects the foot is reminiscent of *Pan* and suggests a transitional stage, still with some arboreal capability, in the evolution of the modern human foot (Lewis, 1981). (See Fig. 6.19.)

We have, by a miraculous chance of fossilization, a whole series of footprints of an early *Australopithecus* dating from 3.5 million years BP, discovered at Laetoli in Tanzania. Extensive examinations have been made by a number of authors who claim that the prints signify a modern type of bipedalism (e.g., Day and Wickens, 1980), though some authors feel that something more transitional is implied (Stern and Susman, 1983). In general it seems likely that they are broadly compatible with the skeletal material known from Hadar, which is roughly contemporary. All the evidence points to the fact that bipedalism was well established by 3.5 million years BP, even if it was not exactly as we see it in modern humans. Indeed, such is hardly to be expected, and the minor apelike features of the foot such as curved phalanges are entirely appropriate to fossils from this early stage in human evolution. The behavioral implications of these discoveries will be discussed in a later chapter, but we can suppose that as well as their bipedalism, there was at least some degree of arboreal activity still shown by *Australopithecus* at this time.

No fossil foot bones are known of *Homo erectus*, but of Neandertal people we have both bones and a footprint. They indicate a broad, short foot that can be matched in modern populations and, although the proportions are unusual, they are not apelike, as was once claimed (Trinkhaus, 1983).

In summary, it is possible to distinguish six different but related trends in the evolution of the human foot:

1. There was probably a reduction of the load arm of the foot lever compared with the power arm in the evolution of the anthropoid foot. This reduction would have occurred as the monkeys became heavier animals. In the Hominoidea we see a shortening of the phalanges themselves.

2. The volar (friction pad) area of the foot has increased to cover the whole contact surface.

3. With the hairless volar skin, we find sweat glands and a great density of tactile sensory nerve endings; both these features were perhaps a little reduced in the most recent stages of human evolution.

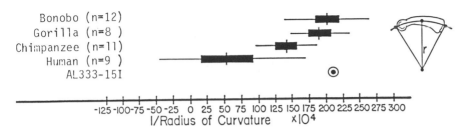

Bonobo (n=12)
Gorilla (n=8)
Chimpanzee (n=11)
Human (n=9)
AL333-15I

-125 -100 -75 -50 -25   0   25   50   75 100 125 150 175 200 225 250 275 300
1/Radius of Curvature $\times 10^4$

Figure 6.24. Curvature of the proximal phalanx of the fourth toe is shown in this diagram as the reciprocal of the radius of the bone. The bonobo is *Pan paniscus* and AL333–151 is a specimen from Hadar. Three other toe phalanges from this site show similar degrees of curvature (from Stern and Sussman, 1983).

4. There was a change of load line from metatarsal 3 to between metatarsals 1 and 2; this change was a correlate of the freely opposable hominoid hallux. In human evolution we have evidence that the load line moved toward the hallux, which became relatively large and lost the ability to be abducted into a divergent position.

5. The high proportion of load arm made up by the metatarsals (as compared with tarsals) is somewhat reduced in humans by a shortening of metatarsals 2 through 5. The phalanges lost their curvature.

6. The foot is narrowed, yet at the same time a new secondary load line evolves along metatarsal 5, which becomes more robust and this enables the weight of the bipedal animal to be distributed through three centers in each foot—the heel, the hallux, and the small toe—giving the human foot the stable character of a tripod, the pedal triangle.

These changes, which summarize the evolution of the human foot, involve not so much the evolution of a new function as a reduction in the original primate functions. We see the foot changing from a tactile grasping and locomotor organ to a simple locomotor lever. The phalanges are straightened and shortened, and their musculature is reduced. Although some nonlocomotor function is still possible in the foot (especially if the full use of the arms and hands is restricted or inhibited over an extended period of time), we see that it has clearly been evolving from a generalized to a specialized organ.

V. Summary   From a consideration of the lower limbs it is possible to conclude that the evolution of erect posture and bipedalism has involved fundamental changes in the function of two structures, the pelvis and the foot, together with a host of minor changes at the knee and elsewhere.

The pelvis, already structurally complex, has changed shape markedly in accordance with the different characteristic position of the femur in relation to the vertebral column, the different weight line, and the changed function of the associated musculature. In particular, the pelvis, in its role as anchor for the important muscles of forward propulsion and lateral balance, has broadened to give these muscles greater leverage around the hip joint. This broadening has been accompanied by a shortening of the ilium and a closer approximation of the joints through which the body weight is transmitted—the sacral articulation and acetabulum. The vertebral column has undergone realignment with the pelvic basin which provides essential support to the viscera.

The leg is straight, from the side, and in humans we see the whole body weight transmitted directly through the outer condyle of the knee joint from acetabulum to ankle. This gives humans considerable mechanical and energetic efficiency.

The feet themselves have undergone evolution from a generalized to a specialized condition. They have been modified to carry the whole weight of the body in walking, one foot at a time. The points of weight transmission have become precise and threefold rather than general, and the primary line of weight has moved close to the hallux. Digits 2 through 5 have shortened, and the two outer metatarsals have thickened.

The immense power of the human stride comes, in turn, from three muscles. The hamstrings bring the body forward over the knee; the gluteus maximus controls extension at the hip; the gastrocnemius gives a final lift by raising the foot lever on the distal end of the metatarsals. The human bipedal stride is unique in the animal world and, as we have seen, was already advanced in *Australopithecus*, although there are still features in the limb bones of this genus that are distinct from those of modern humans. *Australopithecus* was a biped at least 3.5 mya; by the time of *Homo erectus* the evolution of human bipedalism was probably complete.

It seems clear that bipedalism was the first character-complex to evolve in hominid evolution. Its origin is still obscure, but experimental work summarized by Fleagle *et al.* (1981) that has involved extensive electromyographic studies of the hindlimb muscles of ceboids and apes has shown that climbing and bipedalism entail similar patterns of limb excursion and muscle activity. Thus, in this very important matter of structure and function, arboreal climbing is preadaptive to bipedalism. If bipedalism was already established over 3 mya, it must have begun to evolve much earlier, at least a million years before. We shall return to this important matter in Chapter 12.

Suggestions for Further Reading

The evolution of the primate and in particular the hominid pelvis has been treated in some detail by W. E. le Gros Clark in two books: *The antecedents of man,* 3rd ed. (Chicago: Quadrangle Books, 1971) and *The fossil evidence for human evolution,* 3rd ed. (Chicago:Univ. of Chicago Press, 1978). The evolution of the human foot has been described by D. J.Morton in *The human foot* (New York: Haffner, 1964). The hindlimb of *Australopithecus* has been discussed by J. T. Robinson, *Early hominid posture and locomotion* (Chicago: Univ. of Chicago Press, 1972). Its mechanics have been reviewed by C. O. Lovejoy, G. Kingsbury, G. Heiple, and A. H.Burstein (1973), "The gait of *Australopithecus*," *Amer. J. Phys. Anthropol.* **38**, 757–780.

Other useful articles include J. T. Stern (1975), "Before bipedality," *Yearb. Phys. Anthropol.* 19, 59–68. L. C. Aiello and M. H. Day, "The evolution of locomotion in the early Hominidae," in R. J. Harrison and V. Navaratnam, *Progress in anatomy*, Vol. 2 (Cambridge: Cambridge Univ. Press, 1982) and the book by V. T. Inman (1980), *Human walking* (Baltimore and London, Williams & Wilkins).

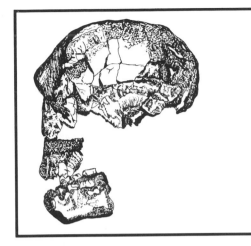

7

Manipulation and the Forelimb

I. The Mammalian Forelimb

In the locomotion of the terrestrial quadrupedal mammals, including the quadrupedal primates, the main driving force is derived from the hindlimbs. The body is propelled forward from the rear (not pulled), and mechanical efficiency requires rigid joints between the pelvic girdle and the vertebral column. The function of the forelegs, even in a generalized mammal, is different. They carry the majority of the weight, and in movements such as running and jumping, usually hit the ground first and take a great deal of shock, which would be transmitted throughout the body by a rigid bone-to-bone connection. Such a connection does not exist; the shock is absorbed by the mass of muscle that suspends the spine and thorax from the shoulder girdle (Fig. 7.1).

Although the shoulder or *pectoral* girdle consisted in its primitive condition of three bones, as does the pelvis, it is reduced in mammals to two—the *scapula* and *clavicle*. The scapula is the dorsal bone and is equivalent to the ilium; the clavicle is ventral and equivalent to the pubis. Unlike the innominate, however, they are not fused but are connected by a moveable joint (*acromioclavicular joint*). The clavicle makes contact with the rest of the skeleton through a second movable joint, the *sternoclavicular*, where it articulates with the *manubrium* or upper part of the sternum (Fig. 7.2). The clavicle maintains the distance of the scapula from the sternum and, acting as a strut, allows movement of the scapula at a constant radius from the manubrium. The weight of a quadrupedal animal is suspended on the *serratus anterior* muscle. The clavicle keeps the *glenoid cavity* (the pectoral equivalent of the acetabulum), into which the *humerus* fits, at a fixed distance from the midline of the body. Figure 7.1 shows, however, that when the body weight alone is transmitted by

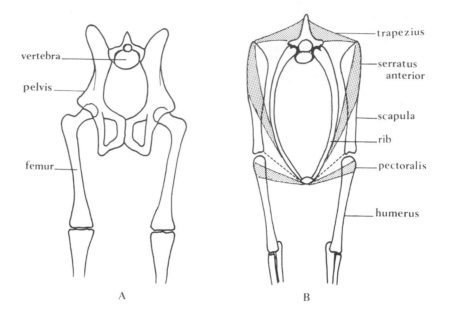

Figure 7.1. Diagrammatic representation of the limb "suspension" in a typical quadrupedal mammal (such as a horse). The pelvic girdle, supported by the hindlimbs, is fixed to the backbone by a rigid bone-to-bone connection (*A*). Upon the pectoral girdle (*B*), the body is suspended mainly by two muscles, the *trapezius* and *serratus anterior*, which transmit the animal's weight from the backbone and ribs. The lower end of the scapula is steadied by the *pectoralis*, which holds the head of the humerus near the trunk. The dotted line indicates the position of the clavicle if it is present. It is lost in ungulates, since it serves no function in a strictly quadrupedal animal.

the pectoral girdle (as among ungulates, which do not leap or climb), the clavicle is not an essential component because movement occurs only in the anterior-posterior plane, and there is no lateral swing of the forelimbs. The clavicle has been lost in these groups.

The muscle suspension of the forelimbs gives them greater flexibility than the hindlimbs, and this flexibility has been exploited, as we shall see, by the primates.

As noted in the last chapter, the intermembral index of the higher primates indicates that the apes have long forelimbs in relation to their hindlimbs (Fig. 6.1). However, an allometric analysis in which forelimb length is related to body size shows that only the gibbon and two New World genera (*Ateles* and *Brachyteles*) have forelimbs of exceptional length (Fig. 7.3). All these species depend to a large extent on suspensory locomotion, which is often rapid, and have specialized in the evolution of long forelimbs. It is notable that the pattern of elongation is

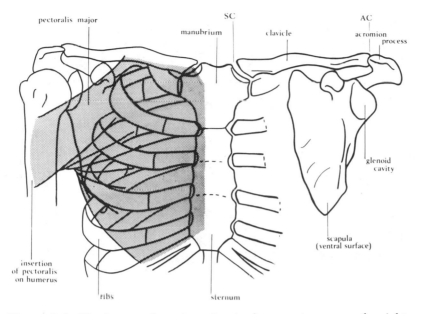

Figure 7.2. The human thoracic region is shown cut away on the right so that the scapula can be seen from the front. On the left the thorax is intact, and upon it is shown the *pectoralis major* muscle on its ventral side, which in quadrupeds propels the body forward and in climbing primates lifts the body. *SC*, sternoclavicular joint; *AC*, acromioclavicular joint.

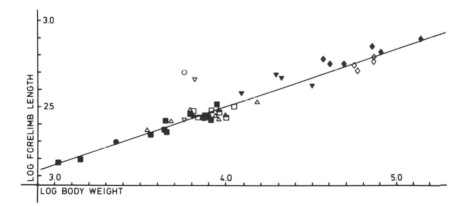

Figure 7.3. Bivariate plot showing the relationship between log body weight and log length of the forelimb (from Aiello, 1981b). The solid line represents the principle axis. Note that all the primates fall close to this axis except the gibbons and New World climbing and suspensory monkeys, which alone have exceptionally long forelimbs. Symbols are as in Fig. 6.2.

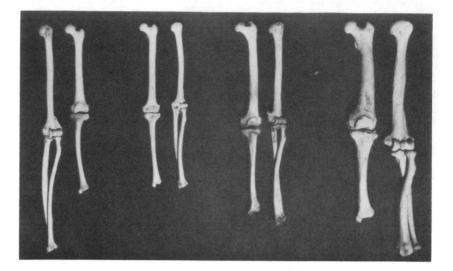

Figure 7.4. Forelimbs and hindlimbs of (*from left to right*) orangutan, pygmy chimpanzee, chimpanzee, and gorilla. The arms are relatively long in the orangutan, while the legs become relatively short in the knuckle-walking apes as body size increases (from Pilbeam, 1972). Note that the limbs of the orangutan are reversed in position compared with those of the other apes.

equally true of both components of the forelimb, the humerus and radius (Aiello, 1981b). The limb bones of the great apes are illustrated in Fig. 7.4.

II. The Pectoral Girdle

Unlike many quadrupedal mammals, the primates have retained the two-boned pectoral girdle, and both bones are flexibly mounted. They can move some 40 degrees around the sternoclavicular joint in each of two planes. At the same time, the scapula can move freely in relation to the clavicle around the acromioclavicular joint. These two movable joints allow the scapula to move over the surface of the thorax in both the vertical and the horizontal planes. The acromioclavicular joint allows the scapula to twist in relation to the clavicle and so remain flat on the surface of the thorax, in spite of the fact that the latter is not spherical. The scapula moves forward around the thorax in pushing or thrusting and to the back in pulling the body forward or upward in climbing. In addition, by the action of the *trapezius* and serratus anterior muscles, the scapula can be rotated (Fig. 7.5, right side) so that the glenoid cavity (equivalent to the acetabulum in the pelvis) can face in different directions (Tuttle and Basmajian, 1976). This rotation nearly doubles the range of movement of the limb in a vertical plane. It may also be noted that the glenoid cavity is more open than the acetabulum and therefore allows a greater arc of movement in the fore-limb than in the hindlimb. Here, stability has been sacrificed for flexibility.

From this short account of the pectoral girdle, it can be seen that, in accordance with its different function, its structure is very different from that of the pelvis. In fact, the structure is more generalized—that is, capable of more kinds of movement—and we might accordingly predict a greater variety of adaptations in the course of primate evolution. Such is indeed the case. While differences in the clavicles are slight, a glance at Fig. 7.6 will show the differences between the scapulae of the macaque, gibbon, gorilla, and human. More striking differences in the position of the shoulder girdle upon the thorax, differences related to the change of function of the forelimbs in the evolution of the primates, were shown in Fig. 5.10.

We can trace three distinct functions of the forelimbs and follow the changing proportions of each. These functions are: (1) to support the weight of the body in quadrupedal locomotion; (2) to suspend the weight of the body in arboreal climbing; and (3) to manipulate objects.

The change from quadrupedalism to climbing is reflected in the form of the clavicle and of the scapula and its musculature. We find that among primates with well-developed forelimb mobility the clavicle is longer; among quadrupedal species it is shorter. Its greater length alters the position of the glenoid fossa to facilitate suspensory arm movements.

The scapula and its musculature have a more interesting story to tell. The shoulder girdle can be rotated to allow the arm to rise vertically, a movement that is clearly important in locomotion by suspension. Oxnard (1963) has shown how the importance of the two rotatory muscles, the serratus anterior and the trapezius, varies among the main locomotor groups of primates (Fig. 7.7). The data indicate that, as might be expected, the muscles are better developed in apes than in quadrupeds and that in this characteristic humans are closer to the apes. At the same time, there is a relative increase in length of the bony lever to which these muscles are attached (as we saw in our consideration of leverage about the hip joint). That is, the insertion area of the muscles on the scapula tends to move away from the central point around which the scapula rotates. Therefore, in species using forelimb suspension we find that the scapula is lengthened; it is longer in the monkeys, especially the arboreal species, and longest in the Hominoidea. The *acromion process*, which, with the ventral bar of the scapula, carries the trapezius and *deltoid* muscles, is also well developed for the same reason (see Figs 7.5 and 7.6).

The actual plane of the glenoid cavity has similarly changed in relation to the rest of the scapula, reflecting the "characteristic" position of

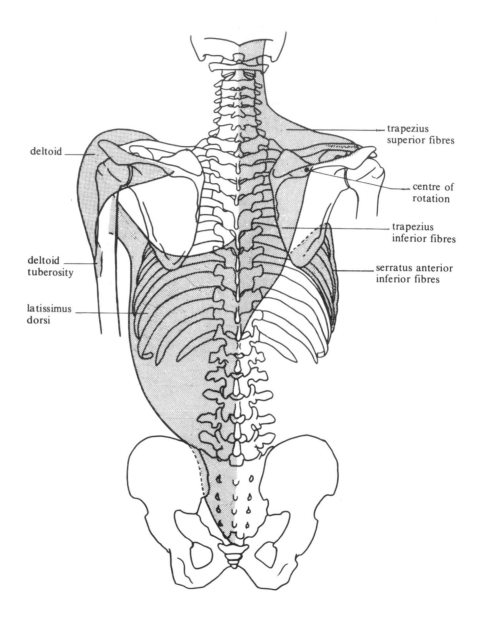

Figure 7.5. Dorsal view of the human trunk showing (*on the right*) the two muscles that can effect rotation of the scapula around the center of rotation and (*on the left*) muscles that raise and lower the humerus.

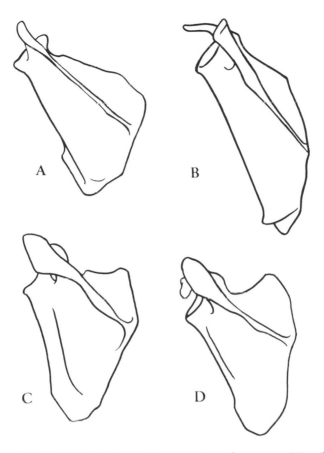

Figure 7.6. Dorsal (*outer*) view of the scapulae of macaque (*A*), gibbon (*B*), human (*C*), and gorilla (*D*), all drawn the same size.

the arms in relation to the body and the greater mobility of the shoulder joint in species that climb and swing by their arms. Quadrupeds, therefore, tend to have the glenoid cavity pointing more or less horizontally; climbers and brachiators have the cavity pointing cranially—upward. The way the angle is measured and some figures for it are shown in Fig. 7.8.

In this angle humans are aligned with the quadrupeds, since the plane of the human glenoid cavity is more nearly parallel to the lateral margin of the scapula than it is in climbing species due to the fact that humans do not raise their arms upward but, on the contrary, carry them hanging down from the shoulder. The human bipedal stance has clearly changed the shape of the scapula in this respect; we need not conclude that the larger angle indicates a history of quadrupedal locomotion!

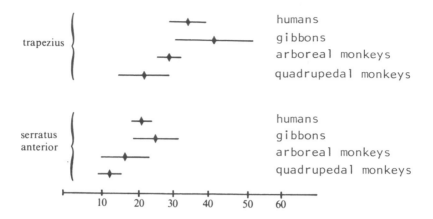

Figure 7.7. Diagram illustrating the relative mass of the trepezius and serratus anterior muscles in the three main locomotor groups of higher primates and in humans. Note that in these two features, humans lie near the arboreal primates. The numbers are units of relative mass (data from Oxnard, 1963).

On the other hand, when we turn to fossil hominid clavicles and scapulae we find a different situation. At present, fossil remains of pectoral girdles are rare. The lightness of the scapula and its position within the surrounding musculature probably render it liable to destruction by predators and scavengers. Clavicles are more common. There are two fragments belonging to the Miocene *Proconsul*, but, apart from their obviously hominoid character, they are too incomplete for further diagnosis (Le Gros Clark and Leakey, 1951). Of *Australopithecus* we have fragments of six clavicles, three from Hadar and one from Makapansgat, all fragmentary, and two from Olduvai. Of the latter, which are quite well preserved, one is described as only slightly different from that of modern humans but suggesting a history of suspensory locomotion (Oxnard, 1968). One clavicle is known from Choukoutien (*Homo erectus*), and it appears to be quite similar to a modern human clavicle (Weidenreich, 1938).

There is a fragmentary scapula of *Proconsul africanus* that is described as a diminutive version of a chimpanzee scapula. There is one fragmentary scapula from Hadar and one from Sterkfontein attributed to *Australopithecus*. (There is also a very small fragment from Koobi Fora, which may also belong here.) Fortunately, both consist of the anterior part, including the glenoid cavity and part of the ventral bar (Fig. 7.8A), and the Sterkfontein specimen includes part of the *coracoid process*. The coracoid is curved strongly and the bicipital tuberosity at the upper end of the glenoid fossa is well developed: these characteristics indicate suspensory use of the arms and especially climbing. At the same time, the small angle that the plane of the glenoid cavity makes with the ventral

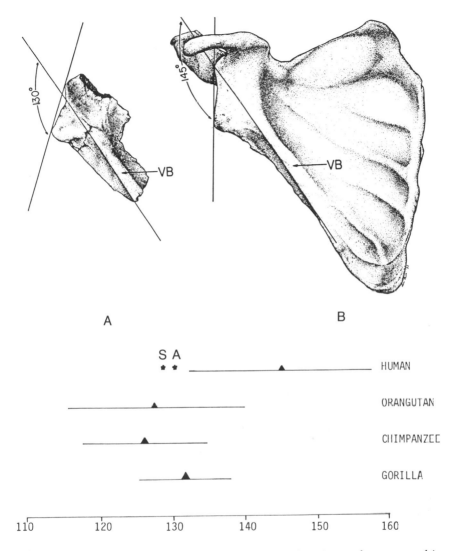

Figure 7.8. The angle of the plane of the glenoid cavity can be measured in relation to the ventral bar of the scapula, as shown in the human scapula (*B*). A typical value for this bar-glenoid angle is 145 degrees. *Left*, a fragment of the scapula of *Australopithecus* from Hadar (*A*). Here the angle is estimated at 130 degrees (without much error). This angle has been measured in the great apes and a closely related angle has been measured in quadrupeds. The data are shown in the diagram below, together with an estimate of the angle from a scapula fragment of *Australopithecus* from Sterkfontein (*S*). The glenoid fossa is directed significantly more cranially in *Australopithecus* and in apes than in humans and quadrupedal monkeys. (Drawings from Stern and Sussman 1983; data from these authors and from Oxnard, 1963 and Vrba, 1979.)

bar of the scapula (the *axilloglenoid angle*) again puts both bones into the suspensory class (Fig. 7.8—Stern and Susman, 1983). Broom summarized his description of the Sterkfontein scapula by stating that the bone falls between that of the orangutan and modern humans in its overall form; or, stated another way, it had a more suspensory function than the modern human scapula.

Although they are only small fragments, these scapulae from the opposite ends of Africa may provide an important clue to the locomotor history of *Australopithecus*. Two fragmentary bones constitute a small sample, but they do suggest either that *Australopithecus* used its arms in a suspensory mode for climbing, at least to some extent, or that its recent ancestors did so. Which of these hypotheses is more likely to be correct will be clarified further by the evidence of the skull and our discussion of the ecology and behavior of the group (Chapter 9).

We can distinguish, therefore, three types of scapulae associated with three locomotor patterns: the quadrupedal scapula, shorter and squarer in proportion, with a ventrally pointing glenoid cavity; the suspensory or climbing scapula, elongated, with a cranially pointing glenoid cavity; and the human scapula, still elongated, but with a ventrally pointing glenoid cavity. The anatomy of the *Australopithecus* scapula suggests some degree of climbing in the ancestry of hominids, and in the final evolution of *Homo* we see a change in the angle of the glenoid cavity associated with bipedalism.

III. The Evolution of the Upper Arm

The humerus is a powerful bone that carries a considerable proportion of the body weight in quadrupedal mammals; the forward muscles of propulsion, the *pectoralis* and *latissimus dorsi* muscles, are attached to it (Fig. 7.2, left). Among suspensory species, these muscles are used to lift and support the body; the latissimus dorsi muscles, which are dorsal to the thorax, also help to lift the body in climbing by retracting the arm (Tuttle and Basmajian, 1976—Fig. 7.5, left). Oxnard (1963) has shown that they are relatively larger in suspensory and climbing species than in quadrupeds, for only in the former do they lift the total body weight. The arm itself, on the other hand, is lifted by the deltoid muscle (Fig. 7.5), which again is larger in suspensory and climbing species. At the same time, the point of insertion of the muscle on the humerus moves down the bone to give greater leverage (Fig. 7.9).

Less striking differences include the plane of the articular head of the humerus in relation to the elbow joint. This relationship has changed in evolution according to the position of the scapula on the thorax (see Fig. 5.10). Thus, in quadrupedal forms the center line of the ball joint of the humerus, which fits into the glenoid cavity, is just about at right angles to the elbow (Fig. 7.10C) facing directly backward, while in the chimpan-

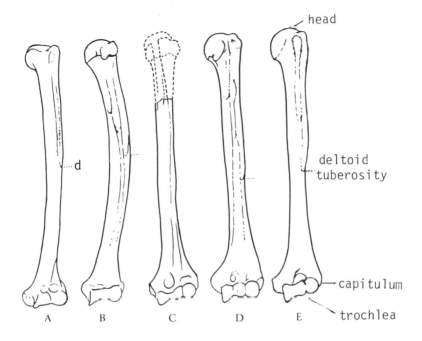

Figure 7.9. Humeri of langur (*A*), *Cercopithecus* (*B*), *Proconsul* (*C*), chimpanzee (*D*), and human (*E*), all drawn the same size. *d*, The lowest extent of the insertion of the deltoid muscle (after Napier and Davis, 1959).

zee the humerus has twisted so that the angle between elbow and head is nearer 30 degrees, and the head points inward (Fig. 7.10A). The arboreal langur monkey *Presbytis* is intermediate in this characteristic (Fig. 7.10D). Humans share this feature with the apes; in both climbing species and bipeds the arm is able to move laterally as well as backward and forward.

Finally, it is characteristic of climbing species as well as humans that the hinge joint at the elbow, the *trochlea* (Fig. 7.9), allows full extension of the forearm and gives considerable lateral stability at that point. In the characteristics of the trochlea, the great apes and humans are very close.

In the upper arm, therefore, we find that humans share two characteristics with climbing species: (1) relatively large arm-raising muscles and a low point of insertion of the deltoid, and (2) a humerus with head pointing inward rather than straight backward. However, our upright stance has led to a reduction in musculature in the climbing muscles, the pectoralis and latissimus dorsi, compared with all other primates and especially the suspensory species.

Let us now turn to the fossil evidence for the evolution of the hominoid forelimb. We have fragments of two humeri of *Proconsul*, over ten of *Australopithecus*, and one of *Homo erectus*. One *Proconsul* humerus is

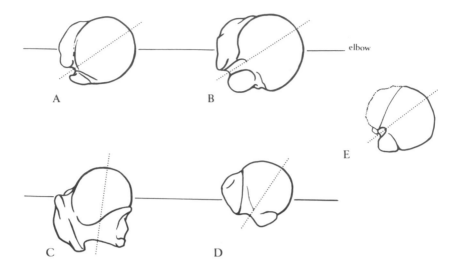

Figure 7.10. View from above of the heads of the right humeri of chimpanzee (*A*), human (*B*), macaque (*C*), and langur (*D*). They are aligned according to the plane of the elbow joint. (*E*) Humerus head of *Australopithecus*; it is not correctly aligned since the elbow joint is missing. It is added, however, for comparison of the form of the head and groove.

associated with a radius and ulna, and these important fossil bones, which belong to the small species *Proconsul africanus*, have been described in fifteen different studies, including Napier and Davis (1959), Morbeck (1975), and Feldesman (1982) (Fig. 7.11B). The humerus has been well reconstructed and in form and dimensions appears somewhat similar to that of an arboreal monkey. The point of insertion of the deltoid muscle is well down, and the trochlea closely approaches the condition seen in *Pan*, allowing great extension at the elbow joint. It has been suggested that suspensory locomotion was combined with intermittent quadrupedalism (Conroy and Fleagle, 1972; Morbeck, 1975). On the basis of this limited evidence we can conclude that *P. africanus* is a small ancestral ape, possibly specialized for below-branch feeding (Aiello, 1981a). The overall impression is of a powerful, climbing animal that was not much constrained by joint and ligament morphology to a limited range of postures (Walker and Pickford, 1983).

Small portions of two femora and a humeral shaft have also been reported from the chimpanzee-sized *Proconsul nyanzae* (Le Gros Clark and Leakey, 1951). These fragments do not bear clear suspensory characteristics, but their body size and limb proportions suggest they were below-branch feeders (Aiello, 1981a).

It is not possible to draw more than very limited conclusions from these few fragments of *Proconsul*, but it is fair to admit evidence of small climbing proto-apes in the Miocene—*Proconsul africanus*. While we may never know for certain what place this Miocene species had in the evolution of the Hominoidea, at least the limbs do not preclude its close association with a hypothetical Miocene ancestor for both apes and humans.

Useful humeri of *Australopithecus* were found at the following sites:

Kanapoi	Distal extremity (Patterson and Howells, 1967)
Sterkfontein	Proximal extremity (Broom *et al.*, 1950)
Kromdraai	Distal shaft and trochlea with ulna fragment (Broom and Schepers, 1946)
Olduvai	Distal shaft and trochlea
Koobi Fora	Parts of eight humeri (McHenry, 1973) (2 well preserved)
Hadar	Parts of nine humeri (Johanson *et al.*, 1982) (4 well preserved)

The proximal end of the humerus bears a few characteristics that may enable us to understand its function. It is impossible to know the angle of the head (Fig. 7.10) unless the bone is more or less complete. The development of the tubercles will indicate the importance of the muscles they carry: the lesser tubercle carries the *subscapularis*, a medial rotator and the greater tubercles the *supraspinatus*, the *infraspinatus*, and the *teres minor*, which raise the arm upward and rotate it laterally. The tubercles are separated by the bicipital groove, which lodges the long tendon of the biceps muscle. The Sterkfontein humerus consists of the head and greater part of the shaft and has been described by Broom (Broom *et al.*, 1950). In general, he finds the head of the humerus almost typically human, but the lesser tubercle is very well developed and suggests a powerful *subscapularis* muscle—an important muscle in climbing. The Koobi Fora specimens stand out as peculiar and distinct from existing Hominoidea (McHenry, 1973); their status is in doubt, but they probably belong to the heavily built *A. boisei*.

The trochlea and distal end of the humerus is easier to interpret but far less varied among hominids than the proximal end and cannot be easily used to distinguish human from ape. The earliest hominid find of such a fossil is the fragment from Kanapoi, interpreted as morphologically close to *Homo* by McHenry and Corruccini (1975), but close to the Hadar finds and, surprisingly, the robust fossils from Koobi Fora, as interpreted by Feldesman (1982). It seems to be an early *Australopithecus* of rather robust build. It is notable that, like modern humans, it lacks the trochlear ridge that stabilized the elbow joint of the African apes (McHenry and Corruccini, 1975). The Hadar specimens are slightly closer to modern humans but not much. The distal humerus fragments from Koobi Fora are of later date, are very robust, and are not closely related to the other specimens. They possibly belong to the little known

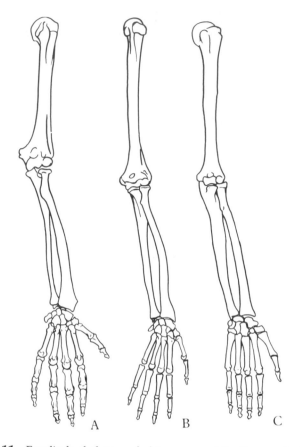

Figure 7.11. Forelimb skeleton of chimpanzee (A), *Proconsul africanus* (B), and human (C) (reduced to the same length) (after Le Gros Clark, 1971).

robust *A. boisei*. All the fossil specimens are typically hominid, however, and quite distinct from all other primates including *Hylobates* and the monkeys. The trochlea unites humans and their ancestors closely to the African apes.

A humerus shaft from Choukoutien associated with the fossils of the Peking population has been described. Though unusually thick-walled, this bone appears to fall within the human range. The specimens of femora and humeri are few, yet it seems clear that *Homo erectus* had limbs not dissimilar to our own. If it were not for the evidence of the skull, we might be tempted to assess these limb bones as belonging to a robust race of *Homo sapiens*.

From the data presented here it seems clear not only that the human arm carries characteristics associated with climbing primates, but also that our fossil relatives, *Australopithecus*, show strongly the characteris-

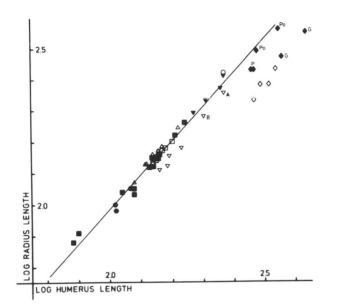

Figure 7.12. Bivariate plot showing the relationship between log length of the humerus and log length of the radius. Symbols as in Fig. 6.2 (from Aiello, 1981b). Note that only the gorilla and human have an exceptionally short radius compared with the humerus.

tics of an ape that frequently elevated its arms in climbing. With the minor exceptions noted, the total morphological pattern conforms closely to that seen in the African apes. The evidence for climbing locomotion in human ancestry accumulates.

IV. The Forearm

The brachial index was devised as an indicator of the relative lengths of the humerus and radius and is still sometimes used. It is better, however, to use a logarithmic display of the proportions of the forelimb. Figure 7.12 shows that, in general, the proportions are constant, though in humans and in the gorilla some shortening of the forearm relative to the upper arm has occurred (Aiello, 1981b), The reasons for this shortening are not understood, though in the case of the gorilla it is possibly associated with that heavy animal's more quadrupedal mode of locomotion: enormous weight is transmitted by the forelimbs. Keith (1923) has suggested that shorter human forelimbs would have resulted in more precise movements of the hand around the elbow.

The bones of the forearm, the radius and ulna, are flexed and extended about the hingelike elbow joint by muscles that originate on the humerus and lie along it. The *triceps brachii* is the extensor of the fore-

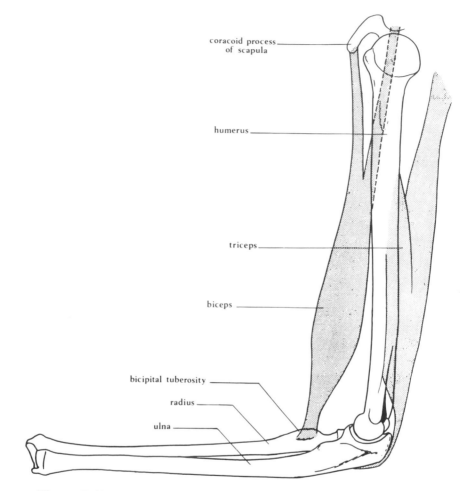

coracoid process
of scapula

humerus

triceps

biceps

bicipital tuberosity

radius

ulna

Figure 7.13. Diagram showing the flexor and extensor muscles of the human forearm.

arm; the *brachialis* and *biceps* the flexors (Fig. 7.13). As might be expected, the extensor is the important muscle in a quadrupedal mammal because it supports the weight of the animal at the elbow, and the flexors are important in climbing forms because they flex the elbow and lift the animal. Oxnard (1963) has again shown that the relative mass of the triceps is, as we might expect, diminished in climbing species. The length of the power arm of the forearm lever (that is, the distance of the *bicipital tuberosity* from the elbow joint) also varies with locomotor function (see Fig. 7.13). In this characteristic, humans lie between the quadrupedal and the suspensory condition.

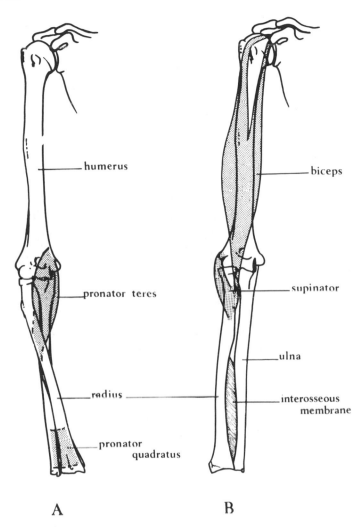

Figure 7.14. Pronation (*A*) and supination (*B*) in humans, shown from the front. The muscles used are shown in each drawing.

The retention of the two bones in the forelimb is due to the need for effective rotation of the hand (compare tibia and fibula), called *pronation* and *supination*. In pronation the thumb lies nearer the body; in supination digit 5 does so (Fig. 7.14). The articulation of the elbow joint is primarily between humerus and ulna, and at the wrist between the radius and the carpal bones (see Fig. 7.11). Forces of compression and tension are transmitted between the two forearm bones by the *interosseous membrane*. The extent of rotation is limited by the extent to which the radius and ulna are bowed. While the supination of quadrupedal monkeys is about 90 degrees, that of apes and humans is nearer 180 degrees. The power of rotation is dependent on the mass of the *pronator* and *supinator* muscles (see Fig. 7.14). There is, again, a correlation between extensive rotation of the forearm and arboreal climbing, as might be expected.

When the arm is fully extended, the *olecranon process* of the ulna fits snugly into the *olecranon fossa* of the humerus, so that the forearm cannot extend beyond 180 degrees. This is an important characteristic in quadrupedal species that take weight on their arms. However, it is interesting that among the African apes, especially the gorilla, the olecranon process is reduced. This appears to allow overextension of the forearm, which would give greater than usual stability in a heavy animal that has taken up quadrupedal knuckle-walking. In contrast, the human olecranon process is more protuberant than in apes. This has been interpreted by Tuttle and Basmajian (1974) to be a result of the use of tools and weapons, which required powerful and rapid extension of the elbow joint.

Fossil bones of the forearm are extremely rare. Some from *Proconsul africanus* suggested a relatively short forearm, but new discoveries suggest otherwise. In most characteristics of the forearm, the *Proconsul* fragments show an intermediate condition between quadruped and climber, as we have seen in earlier sections.

There are a number of fragments of radii and ulnae from Hadar on which no comparative studies have yet been made. A small part of the proximal end of a radius from Koobi Fora has been described as carrying a well-developed bicipital tuberosity (Day *et al.*, 1976). A proximal fragment of a small right ulna is also known from the same skeletal collection. Comparative studies are awaited. A very well preserved large ulna from Omo has been attributed to *Australopithecus boisei*. It is difficult to interpret but has characteristics which are present in both *Pan* and *Homo* (McHenry and Corruccini, 1976).

We have three fragments of radii of *Australopithecus* from Makapansgat and one from Swartkrans. There is a proximal ulna fragment from Kromdraai. Only the ulna has been studied in detail, and all are claimed to be similar to those of modern humans (Robinson, 1972). Fragments of the forearm bones belonging to Neandertalers though robust, are within the variation of form found among living *Homo sapiens* (Trinkhaus, 1983).

V. The Hands of Primates

The manus of the primitive tree shrew (Fig. 7.15A) is a simple, five-fingered (*pentadactyl*) organ, the rather short phalanges making it appear like a paw. It is not prehensile, or are the phalanges opposable. Each terminal *phalanx* carries a claw, and there are six volar pads of friction skin on the palm and one terminal pad on each digit

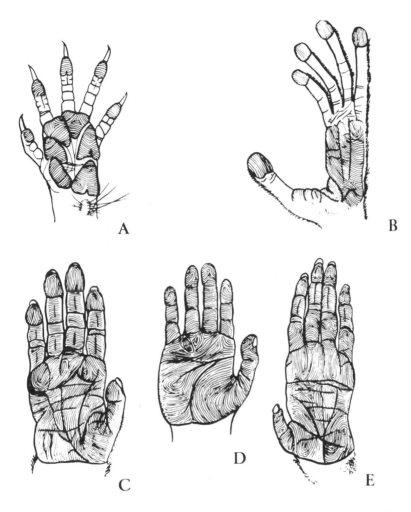

Figure 7.15. The hand of a tree shrew (*A*) and various primates: *Indri* (*B*), chimpanzee (*C*), human (*D*), and orangutan (*E*). Note the carpal papilla on the wrist of the tree shrew, and the increase in the area of the friction skin in the higher primates. Note also the elongated pollex of *Indri* and the short pollex of the orangutan (after Biegert, 1963).

(finger). Just above the wrist lies a small skin projection (*papilla*) from which long hairs grow, the *carpal vibrissae*. This papilla is a tactile sense organ common to many mammals, and similar to those that bear whiskers on the face in dogs and cats. The vibrissae are implanted in richly innervated skin and form an extremely sensitive but undiscriminating tactile sense organ.

Studies of comparative anatomy make it clear that, as the volar pads increase in area and sensitivity in primate evolution, the importance of the carpal vibrissae decreases. They are poorly developed in the New World and absent in the Old World anthropoidea. In the human fetus, a small transient cutaneous papilla can sometimes be seen in the carpal region (Schultz, 1924). During evolution the innervation of the volar pads and particularly of the terminal digital pads increases in density so that they become tactile sense organs of great importance.

While it happens that in humans the tactile function of the hand completely replaces the locomotor function, the manus remains an important locomotor organ in all other primates. The forelimb may not be as important as the hindlimb in the propulsion of the quadrupedal species, yet as a grasping organ it is always important. This grasping function is enhanced among the climbers.

All primates, including humans, possess prehensile hands and convergent-divergent fingers. The coming together of the fingertips (convergence) and their spread (divergence) result from the arched form of the carpus (wrist) and metacarpus (palm), for the fingers are set in a curve at the wrist. Prehensility is a very important characteristic of primates, since it permits one-handed feeding. The squirrel, which lacks prehensility in its manus, has to use both forepaws to feed.

There are two means of achieving a hold on a branch. The first, which is typical of the prosimians, involves a (*pollex*) thumb, which diverges from the phalanges and can be opposed to them around the branch, as we saw in Chapter 6, IV (Fig. 7.15B). The second, which is typical of suspensory species, especially brachiators, involves the lengthening of the phalanges and, in many cases, a reduction of the thumb (Fig. 7.15E). In the quadrupedal Old World monkeys, however, the opposable thumb is retained as a result of the evolutionary development of the hand for use in other functions. These include the procurement of food, grooming, hygiene, defense, the care and carrying of offspring, improved three-dimensional tactile sense, and so forth (Biegert, 1963). A baboon has a precision grip between thumb and fingers sufficiently well evolved to extract the sting from a scorpion (Schultz, 1969), and feeds by plucking individual grass stems by hand: manual grazing.

In certain arboreal monkeys and some Hominoidea we find the thumb much reduced, since the more the hand is involved in suspension, the less it is available for manipulation. This specialization reaches its most extreme form not in the gibbon, where the thumb is retained for limited manipulation, but in *Ateles* and the African *Colobus* monkey, where it is vestigial or absent.

A second characteristic of climbers and brachiators is the hooklike hand with its long phalanges. The proportions of the hands of apes have been investigated by Schultz (1956), and some figures are listed in Table 7.1. When hand length was compared to body trunk length, it was found that while in monkeys hand length is usually less than one-third of trunk length, it rises to an average of 47% in New World monkeys and as much at 59% in the gibbon.

TABLE 7.1. Mean Proportions of the Hand and Thumb in Primates[a]

Primate	Hand breadth × 100 / Hand length	Pollex × 100 / Hand length
Presbytis	29	40
Macaca	35	56
Pan troglodytes	34	47
Pongo	34	43
Pan gorilla	51	54
Hylobates	20	52
Homo	43	67

[a]Data from Schultz, 1956.

The evolution of an opposable thumb is a feature of great importance in human evolution. The thumb reaches its greatest relative size among prosimians, such at *Indri* (Fig. 7.15B), and in some prosimian species the second digit is much reduced to permit an even more effective grip on a branch. Although in these forms the thumb is completely opposable to the digits, Napier and Napier (1967) have described this condition as *pseudo-opposability* to distinguish it from the somewhat different human condition. In the latter, we share with the Old World monkeys and apes a modified kind of carpo-metacarpal joint, which gives the thumb a different range of movement and results in *true opposability*.

Among the suspensory species of monkey and great ape with true opposability, the thumb is relatively reduced from a longer primitive condition. The great apes *Pan* and *Pongo* have such short thumbs that they cannot achieve the powerful "pulp-to-pulp" contact of thumb and other digits (Fig. 7.15) that forms the anatomical basis of humankind's precision grip. While the chimpanzee thumb is somewhat longer than that of the orangutan, it is still short of the human condition (see Fig. 7.15C, E and Table 7.1). Apart from the prosimians, therefore, a fairly long thumb can be found only among the quadrupedal Old World monkeys, lesser apes and humans, and, as we have seen, it is widely used in many of these groups for the manipulation of objects.

A fairly complete fossil hand belonging to the Miocene *Proconsul africanus* has been described. It appears to be basically similar to the hand of a quadrupedal Old World monkey with certain signs of suspensory adaptations. The hand is unfortunately incomplete, and in particular it is not known how long the thumb was; nevertheless, its assessment as showing similarity to a suspensory or climbing species seems justified. This conclusion accords with that drawn on the basis of the forelimb bones. It seems clear that in *P. africanus* we have an incipient climbing ape, and one that, on the basis of the limb bones, could conceivably be ancestral to the living Hominidae.

**VI. The
Evolution
of the
Human Hand**

We have seen that monkeys use their hands for a wide variety of purposes, and one of the most important, which we must now consider, is their contribution to the satisfaction of the exploratory drive: the hands make possible a detailed examination of parts of the environment that can be manipulated. Monkeys, apes, and humans are almost the only animals that fiddle about with things, that turn them over and examine their form and texture. This manipulation of objects is not necessarily directly related to the procurement of food, but is simply a process of investigation equivalent to the olfactory investigation of the environment so characteristic of a dog. But in manipulation a monkey is investigating only one particular part of the environment and frequently a part that can be separated physically from the rest. This ability of higher primates to extract an object from its setting and examine it visually and three-dimensionally from all sides is a development of the utmost importance in human evolution. A carnivore will examine the olfactory nature of objects but cannot simultaneously see them as part of the visual pattern of the environment, because during examination by the nose the objects are more or less out of sight. Admittedly, a ball or bone can be manipulated with the paws and tossed with the mouth, but the range of objects examined is limited, and, compared with the primate hand, the paws and mouth give only a rough indication of shape and texture.

As will become apparent in Chapter 11, the detachment of objects from the environment appears to be a most important prerequisite for the evolution of primate perception. The examination of things as objects we owe to the evolution of the primate hand and the opposable thumb. The recognition of different kinds of objects we owe to primate visual and tactile examination. Only a primate can, as it were, extract an object from the environment, examine it by smell, touch, and sight, and then return it to its place. In this way the higher primates came to see the environment not as a continuum of events in a world of pattern but as an encounter with objects that proved to make up these events and this pattern. We shall return to this development in considering perception and conceptual thought (Chapter 11, II).

This all-important development may have come quite early in primate evolution, with the appearance of the opposable thumb and the use of the hands in feeding. In higher primates, as we have seen, a further advance in opposability has perfected this remarkable organ of manipulation. Napier (1961) writes that true opposability involves "a compound movement of abduction, flexion, and medial rotation at the carpo-metacarpal articulation of the pollex." A saddle joint evolves at the base of the thumb that allows a 45-degree rotation as well as movement in two planes (Fig. 7.16).

Though monkeys and apes have such an opposable thumb, it is only in humans that the thumb is long enough and divergent enough to carry a heavy musculature. This length and strength make possible a precision grip—strong yet delicate (Fig. 7.17). At the same time, with the

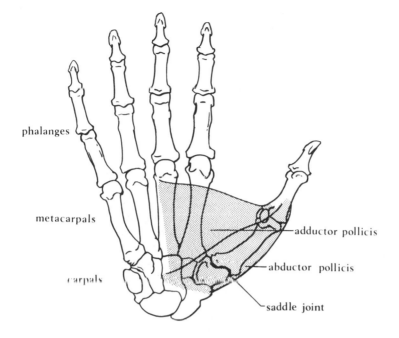

Figure 7.16. The musculature of the thumb is well developed on the palm of the human hand. The two powerful muscles shown move the thumb toward and away from the palm (after Napier, 1962).

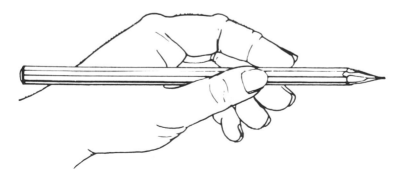

Figure 7.17. Our precision grip, because of the full opposition of thumb and finger, is precise and delicate, yet strong.

hand no longer functioning to grasp branches, the *metacarpals* and phalanges straighten out in the course of human evolution, and the terminal segments broaden (Fig. 7.18).

Fragments of the hands of *Australopithecus* individuals are known from Sterkfontein, Olduvai, and Hadar, and they are reported to be somewhat less than fully human (Fig. 7.18). The Hadar hand bones constitute the largest collection and have been described by Stern and Susman (1983), who stress their morphological similarity to apes, though Tuttle (1981) and Johanson *et al.* (1982) see both ape and human characteristics in them. Starting from the wrist, the two *capitate* bones from Hadar retain a generalized hominoid morphology. Both are waisted, as is the specimen from Sterkfontein (Lewis, 1973), and in this feature resemble *Pan* and some arboreal monkeys. The *pisiform* from Hadar, which is rod-shaped, is also different from the human bone (pea-shaped) and resembles those of apes and monkeys. The elongate pisiform is associated with the powerful *flexor carpi ulnaris* muscle in apes and gives increased leverage for flexion of the wrist (carpal) bones that must be important in climbing locomotion (Tuttle, 1981).

The Hadar trapezium—the wrist bone that carries the thumb—is an important bone since it carries the carpo-metacarpal articulation of the pollex; it closely resembles the joint in chimpanzees and is significantly more apelike than the specimen from Olduvai, though even that is less than fully human (Napier, 1962). The bones of the pollex form a thumb that lies in its relative length halfway between that of apes and that of humans. The first metacarpal of the thumb has a joint that matches the trapezium in its apelike structure, and the bone itself is not so strongly developed as it is in modern humans; the proximal phalanx has the same apelike characteristics.

The remainder of the metacarpals resemble recent finds from Sterkfontein of approximately the same age. They are somewhat curved but contrast with *Pan* in lacking a transverse dorsal ridge and expanded articular surface on the metacarpal head—features associated with knuckle-walking (Tuttle, 1967, 1970, 1981).

The proximal phalanges are slender and curved as markedly as in a chimpanzee, much more curved than anything seen in modern humans (Fig. 7.19). The muscle markings and joints in all phalanges show that the fingers could flex powerfully, and this implies suspensory and grasping function. They are all generally apelike.

There are no features in the *Australopithecus* hand suggesting a knuckle-walking adaptation or even a knuckle-walking ancestry (Tuttle, 1981; Stern and Susman, 1983). In contrast, the hand bones in our possession indicate clearly that the hand functioned as a grasping organ. The thumb shows little evidence of the opposability and strength characteristic of the modern human thumb.

Nothing is known of the hand of *Homo erectus*, but it is of interest that although Neandertal had a brain as large as our own, their hands were

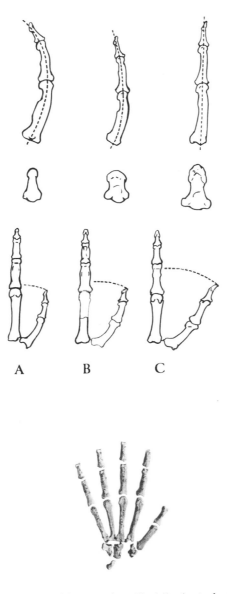

A   B   C

Figure 7.18. Above, hand bones of gorilla (*A*), *Australopithecus* (*B*), and human (*C*). Note the straightening of the fingers in this series, the broadening and lengthening of the last phalanx of the thumb, and the lengthening of the thumb as a whole. Some of the *Australopithecus* thumb bones (from Sterkfontein) are reconstructions on the basis of other evidence (redrawn from Napier, 1962). *Below*, composite hand based on hand bones of *A. afarensis* found at Hadar. Note that only the terminal phalanges are missing (courtesy D. C. Johanson).

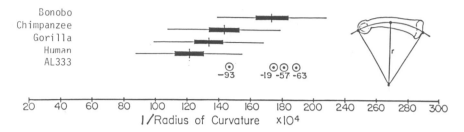

Figure 7.19. Curvature of the proximal phalanx of the fourth finger of various species shown in this diagram as the reciprocal of the radius of the bone. AL-333 includes four specimens from Hadar. Note that the Hadar bones of early *Australopithecus* are curved as much as those of apes (from Stern and Sussman, 1983).

slightly different and possibly a little less dextrous (Musgrave, 1971; Trinkhaus, 1983).

When apes and monkeys carry objects about, their locomotor efficiency is impaired. Tool carrying, food carrying, and food sharing (which is possible only if the food is first carried) all have been reported to a limited extent among chimpanzees (Goodall, 1968), but the animals cannot move quickly with a handful of objects and do so only for very limited distances. In early humans the survival value of being able to carry objects without a disadvantageous effect on locomotion is of the greatest importance. Not only is it most advantageous to be able to carry food when surprised by a competitor or a predator, but this ability can lead eventually to food gathering and so to food sharing. The survival value of carrying also serves in turn to promote a physical structure adapted to bipedal locomotion.

Tool using is not as rare as might be supposed in the animal kingdom. As we shall see in Chapter 9, IX, tool-using species exist among birds as well as mammals. The manufacture of stone tools, however, is confined to hominids. At Hadar in northwest Ethiopia, stone tools have been found that may be as old as 2.5 million years, but their date has still to be confirmed. At Sterkfontein in South Africa, tools associated with *Homo habilis* are estimated to be about 2 million years old. For the first well-dated industries we have to go to East Africa. At Koobi Fora, stone tools have been found dated at over 1.8 mya; they consist of battered cobbles, choppers, and flakes, and show an assured technique in their production (M. D. Leakey, 1970). At another contemporary site nearby, stone tools are found in association with the bones of a hippopotamus (Isaac *et al.*, 1971). From Olduvai Gorge, Bed I, we have a succession of increasingly varied tool kits just under 1.8 million years old, often in association with *Homo habilis*, which have been described in great detail (M. D. Leakey, 1971—Fig. 7.20). Here we see pebbles, probably first used for pounding roots and smashing bones, preserved after flaking as chopping tools for the preparation of meat. The history of tool use will

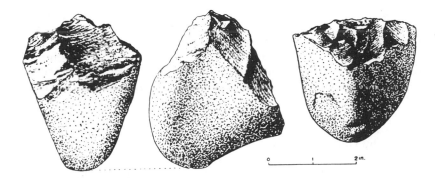

Figure 7.20. Primitive pebble tools of the kind associated with *Homo habilis* at Olduvai, Tanzania. [By permission of the trustees of the British Museum (Natural History).]

be discussed in Chapter 9, but for the present it is worth stressing that a skillfully made and skillfully used cutting tool was an asset of revolutionary importance. It was probably an essential element in the evolution of a carnivorous diet among hominids.

The tools we have found associated with and presumably manufactured by *Homo habilis* are relatively simple and crudely made. Finely flaked stone tools are found (to date) only in association with the bones of hominids that had larger brains. Fine Acheulian hand-axes are known from Ternifine as well as Olduvai, where they are associated with remains of *Homo erectus*. Clearly, the final perfection of the hand consisted not only in anatomy but in sensory perception and motor control associated with a highly developed brain. The most clear-cut neural correlate of the evolution of the human hand is the increase in representation that this organ carries on the motor and somatic sensory cortices of the brain (see Fig. 3.9). We have also noted significant improvement in the direct nervous pathways to the hand of the ascending and descending fibers.

We can summarize the evolution of the human hand by marking three probable stages in its development. First, the primate arboreal environment selected the pseudo-opposable thumb for grasping branches, and this thumb survived among the prosimians and some New World monkeys. Second, either from it or independently, the truly opposable thumb appeared among the Old World monkeys and apes. Finally, the evolution of bipedalism freed the hand from all locomotor function and allowed the perfection of the precision grip by a lengthening of the thumb metacarpal and phalanx, a slight further modification of the wrist, and some development of the musculature of the thumb. By this time the fingers had straightened, and the familiar human hand evolved.

The hand has contributed as much as the eye to the making of *Homo sapiens;* together they gave us a new perception of our environment and, with the development of technology, an increasing control over it.

VII. Our Ancestors: Arboreal Climbers

From the evidence presented in this chapter as well as in Chapter 6, it seems incontrovertible that our ancestors were climbing arboreal apes. Not only do *we* share many characteristics with the living climbing species, but *Australopithecus* carries an even greater variety of characteristics suggesting climbing. Oxnard has reviewed the evidence relating to the evolution of the human shoulder (1969) and concluded that the most likely origin for it lies in a fully arboreal ape genetically related to *Pan* but functionally similar to *Pongo*. Stern (1975) agrees, but adds that the ancestor was smaller than an orangutan, though it employed its long forelimb much as the living orangutan does. It used its hindlimb in a pronograde manner but was somewhat better adapted to this than the orangutan itself. Tuttle and Basmajian (1974) conclude that humans and *Pan* share a common heritage of arboreal adaptations including vertical climbing, hauling, hoisting, and suspensory behavior, while knuckle-walking played an inconsequential role in protohominid behavior. Fleagle and co-workers, in a summary paper (Fleagle *et al.*, 1981) have concluded that, on the basis of experimental work involving electromyographic studies of muscle activity and limb excursion in ceboids and apes, climbing and more suspensory modes such as brachiation entail similar patterns of muscle use and bone strain in the forelimb. It follows, then, that there is no longer any reason to attribute suspensory modes of locomotion such as brachiation to the ancestral hominid. Thus, with Andrews and Aiello (1984), we can see the ancestral species as forest-living climbers of increasing body size. As they became heavier they increasingly engaged in below-branch feeding and climbing postures, developing that short trunk and highly mobile forelimb of the living hominoids.

Humans do, of course, bear certain characteristics in the forelimbs that are adaptations to erect posture and that result from the downward-hanging position of the arms. Though we are a much more recent arrival on the earth than the kangaroo or the dinosaur, it is interesting that we share with them some reduction in our forearms. We can interpret this as a correlation of bipedalism and the development of manual skill.

Finally, we need not be surprised to find that the muscles that carry the weight of the body in climbers are, with their areas of attachment, reduced in humans (Ziegler, 1964). Although humans use the same muscles for carrying as the apes do for climbing, these muscles do much less work.

Though we bear the marks of the apes in our bodies, we are profoundly different from any of them. As noted earlier, we actually share more of our locomotor morphology with *Alouatta* than with any living ape. This demonstrates with some force how far both we and the African apes have evolved from the condition of our ancestors.

Suggestions for Further Reading

Few authors have paid much attention to the evolution of the forelimb. As a general introduction that runs parallel to this discussion, see the appropriate chapter in W. E. Le Gros Clark, *The antecedents of man*, 3rd ed. (Chicago: Quadrangle Books, 1971). An important study of the forearm is to be found in J. R. Napier and P. R. Davis, *The forelimb skeleton and associated remains of Proconsul africanus* [London: British Museum (Nat. Hist., 1959)]. For reference to the primate hand, see J. Biegert, "Volarhaut der hande und fusse," *Handb. Primatenk.* II/1 Lieferung 3, pp. 1326 (Basel and New York: Karger, 1961), and R. H. Tuttle (1969), "Quantitative and functional studies on the hands of the Anthropoidea," *J. Morphol.* **128**, 309–364. The forelimb of *Australopithecus* has been discussed by J. T. Robinson, *Early hominid posture and locomotion* (Chicago: Univ. of Chicago Press, 1972), and by J. T. Stern and R. L. Susman (1983), "The locomotor anatomy of *Australopithecus afarensis*," *Amer. J. Phys. Anthropol.* **60**, 279–319.

For a more general treatment, see R. H. Tuttle, "Parallelism, brachiation and hominid phylogeny," in W. P. Luckett and F. S. Szalay, *Phylogeny of the primates* (New York and London: Plenum, 1976).

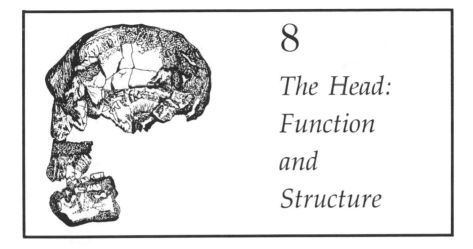

8

The Head: Function and Structure

I. Functions of the Head

In approaching the subject of this chapter, we first consider the evolution of the head as a structure that has come to assume an increasing number of functions in animal evolution. To understand the evolution of the human head we must examine these different functions carefully. In order to reveal their phylogenetic origin, we shall consider the structure of a primitive living creature, the lancelet (*Amphioxus*), which is believed to be very similar to the ancestor of all vertebrates. This small marine animal looks like a minute fish and measures about one inch long (Fig. 8.1).

In the lancelet (as well as in even more primitive invertebrate forms) the head can be recognized by three characteristics. First, it may be defined as the part of the body of the animal that precedes the rest in locomotion; second, it is near to, or incorporates, the mouth; third, it contains at least one sense organ. In preceding the body, the head is able to investigate the environment that the organism is approaching, to direct the organism toward food (in many instances by moving toward light or up an attractive chemical gradient), and to receive food by the mouth. Since movement toward food is basic and essential behavior, we find the nervous system developing between the head receptors and the effectors (the swimming muscles), which lie behind it along the animal's body. In the course of the evolution of both invertebrate and vertebrate animals, a small knot or ganglion of nerves appears near the receptors and the mouth. Within this small ganglion of nervous tissue, the primitive brain, information about the environment (or about the food) is analyzed and an appropriate pattern of behavior is effected by the locomotor system. In the evolution of vertebrates, the extent of sensory investigation of the environment has vastly increased; increased sen-

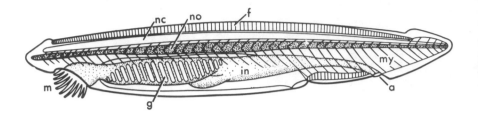

Figure 8.1. The lancelet, a primitive animal still found in shallow seawater, believed to be very like the ancestor of vertebrates. *m*, Mouth; *in*, intestine; *a*, anus; *g*, gills; *no*, notocord (the primitive spinal column); *nc*, nerve cord; *my*, myotomes (the lateral swimming muscles); and *f*, fin (from Colbert, 1955).

sory input relating to environmental conditions has characterized the appearance of more complex species. The evolution of multiple sense organs has been accompanied by the evolution of a relatively massive central nervous system, which analyzes the vast input from the receptors and generates a wide range of behavioral activity.

In vertebrates, therefore, with the brain evolving near the sense receptors in the head, we find that the head serves the animal in two separate ways: first, as a complex organ of interaction with the environment; second, as a protective case for the central nervous system—the brain. The functions of the head may be summarized as follows:

I. Relations with environment
   A. Metabolic
      1. Food intake through mouth and masticatory apparatus.
      2. Inspiration and expiration of water-borne oxygen or air through the mouth, with which the gills or lungs are connected. (They arose as outgrowths from the alimentary canal.)
   B. Sensory
      1. Deployment of sense receptors concerned with input from the environment, both distant and contact (especially food recognition, mate recognition, and predator recognition; see Table 8.1).
   C. Motor
      1. Communication: a secondary function of masticatory and respiratory structures; sound production through trachea and mouth; movement of facial muscles.
      2. Defense and offense: a secondary function of the dentition; evolution of horns on the skull in some reptiles and mammals.
      3. Grooming and gripping: a secondary function of the masticatory apparatus in many animals.
II. Protective case for the brain

TABLE 8.1. The Sensory Receptors of Mammals

Organ	Sense	Stimulus
Eyes	Vision (optic)	Radiant energy: light
Nose	Smell (olfactory)	Certain volatile chemicals
Ears: cochlea	Hearing (aural)	Vibrations in air (or water)
Ears: semi-circular canals, etc.	Balance and movement	Direction of gravity and rate of change of direction of movement, i.e., acceleration
Taste buds	Taste	Certain nonvolatile chemicals[a]
Skin, lips, and tongue	Touch, temperature, and pain	Local pressure change on skin, temperature change, damage to body tissues

[a]See Chapter 9, III. For discussion all other receptors, see this chapter.

The evolution of these specific functions of the head will be considered later. In Chapter 9 we will consider feeding and digestion and related behavior; Chapter 11 refers to the evolution of communication.

The form of the head is a product of these functions and their underlying structures. It is also a product of the extent and direction of gravity and the medium through which the animal moves. In many marine and freshwater animals, for example, the head is streamlined for swimming, and its shape is dictated primarily by hydrodynamic factors. In birds aerodynamic factors predominate. In terrestrial vertebrates, however, the overall shape of the head depends on structural factors: (1) the masticatory apparatus, (2) the eyes, (3) the nose and nasal chambers, which comprise the muzzle, (4) the brain, and (5) the position of the skull on the vertebral column.

The evolution of the masticatory apparatus will be discussed later (see Chapter 9, IV). In the next section of this chapter we shall consider the influence of this apparatus on the form of the head as a whole.

II. The Masticatory Apparatus and the Head
The size and power of the masticatory apparatus obviously affects head form. A large and powerful dentition requires large and powerfully built jaws and jawbones, the *maxilla* and *mandible;* together they will make an animal appear *prognathous* (its jaws protruding) if the other parts of the skull are not expanded equally. Not only will the bony parts of the apparatus affect head form, but so will the related muscles and the extent of their attachment areas.

All the muscles of mastication have their origin on the skull, and the extent of their attachment areas has changed in evolution according to the degree of their development. This fact of skull architecture is particularly relevant when we study the *temporal muscle*, whose origin lies on the braincase (*neurocranium*) in the area of the *temporal* and *parietal* bones (Figs. 8.2 and 8.3; also Figs. 9.1 and 9.2).

The extent of this area of attachment varies in accordance with the size and the power developed by the masticatory apparatus. In the large-jawed primates, such as the gorilla, the area of attachment of the temporal muscle is so large that the areas on each side meet on top of the cranium, and the bone grows to form a ridge or crest at this meeting point to increase the anchorage still further (Fig. 8.3A). The height of the crest varies according to the development of the jaws, so that in the lighter-jawed female gorilla it is smaller than in the male, or absent altogether. This crest, called the *sagittal crest*, is seen in primates whose jaws are proportionally larger than their neurocrania—those in which the temporal muscles are so large that they cover the neurocranium and actually meet in the sagittal plane along the top of the skull. It has been shown by allometric analysis that brain size is less geared to body dimensions than jaw size, so that larger animals will have jaws larger in relation to the neurocranium than smaller animals. This applies both within (where females and juveniles are smaller than adult males) and between species.

A second crest, the *nuchal crest*, develops in heavy-jawed primates where the extended temporal attachment area meets the proportionately large attachment area of the nuchal muscles (see Figs. 5.4 and 5.15). Since the size of the masticatory apparatus is controlled not only by the individual's genes but also by the use it receives, it follows that these two factors also affect the size of the attachment areas of the temporal muscles and, ultimately, the size of the sagittal and nuchal crests.

Many of the fossil skulls of the larger robust specimens of *Australopithecus* carry a marked sagittal crest (Fig. 8.2C), but in the smaller gracile lineage the attachment areas of the temporal muscles were relatively reduced in evolution along with the masticatory apparatus, and the neurocranium was expanded, so that in *Homo erectus* the areas have retreated down the sides of the skull. Here the areas of attachment are bordered by the *temporal line,* which shows distinctly on the surface of the skull bones, and in human evolution this line moves down the side of the skull as the relative size of the jaws is reduced and the size of the neurocranium increases (see Figs. 5.15 and 8.3).

Since the activity of all the powerful masticatory muscles occurs within the perimeter of the head, the stresses developed by their action must be entirely absorbed by the skull architecture. When the muscles contract and the jaw is closed, the force exerted for chewing pulls the mandible and neurocranium together. The force is considerable, and the bony structure between the roof of the neurocranium and the maxillary teeth (that is, the face and brows, all of which receive the force from the lower dentition) must be very strong. Thus, the thickness and form of the facial bones is related to the power developed by the masticatory apparatus (Endo, 1966).

Nearly all this force is transmitted by the molar and premolar teeth

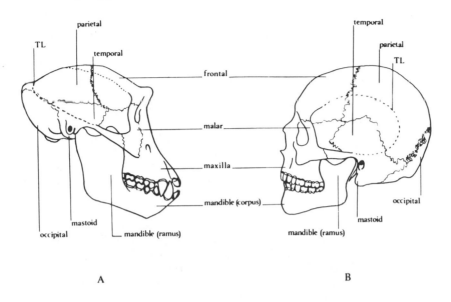

Figure 8.2. Skulls of a chimpanzee (*A*) and a human (*B*) showing the different bones mentioned in this chapter. *TL*, Temporal line.

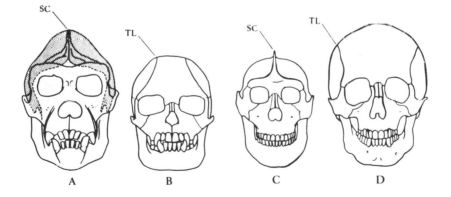

Figure 8.3. Facial views of gorilla (*A*), *Proconsul* (*B*), *Australopithecus* (*C*), and human (*D*); skulls drawn approximately the same size. Note the different development of the sagittal crest in each genus. The temporal muscles that originate on the sagittal crest and surface of the neurocranium and that are inserted upon the mandible have been superimposed on the gorilla skull. *TL*, Temporal line; *SC*, sagittal crest.

and passes up the side of the face to the cranial vault. Because of the position of the nasal cavity and orbits, the force must be carried around these openings by the maxilla and *zygomatic* bones (Fig. 8.4). The supraorbital torus (which is the anterior edge of the *frontal bone* of the skull) acts as an essential cross-member at the top of the face in species with a receding forehead in order to transmit the masticatory forces across the whole vault of the neurocranium rather than at each side only. Where present, the torus also resists other forces developed by the contraction of the temporal muscles. As their points of origin and insertion are closer to the midline and each other than the widest point between them, they develop a considerable horizontal compression at the sides of the frontal bone. When the forehead is vertical, as in humans and the orangutan, the vertical plane of the bone is able to take the stress of chewing without the development of a supraorbital torus. Thus, the size and form of the facial bones, the nasal aperture, and supraorbital torus are related to the size and power of the masticatory apparatus and the form of the neurocranium (see Fig. 8.3).

Finally, it should be mentioned that the cheekbones—each the *zygomatic arch* of the temporal and zygomatic bones—carry the origin of the masseter muscle, which is also an important masticatory muscle. These bones, which form a bridge across the temporal muscle, also vary in size and form according to the development and power of the jaws (see also Fig 9.2).

The masticatory apparatus and brain are certainly the two most striking determinants of head form (Fig. 8.5). Before we consider the brain, however, we must consider the part played by the different senses in determining the form of the head both directly and indirectly, through their influence on the evolution of the brain.

III. The Eyes   We have already described how vision evolved as an essential adaptation for an arboreal predator (Chapter 3, II) and, more generally, in response to the demands of the forest environment, where a sense of smell was possibly of less value. Receptive to radiant energy of a certain range of wavelengths (380–760 mμ), the eyes came to provide information at a distance about size, color, texture, movement, pattern, spatial relationships, and distance—information not accurately obtainable by other means. Here we will discuss three aspects of the evolution of vision: (1) the retina, (2) the neural correlates of vision, and (3) the orbital region of the skull.

Cats have relatively large eyes, but, apart from a few such carnivores, the eyes of some primates are outstandingly large. Cats are classic predators, and the large eyes of some of the smaller primates gives credence to Cartmill's visual predation theory (Chapter 3, II). The amount of light that will enter the eye depends upon the aperture of the lens, and discrimination and sensitivity depend upon the concentration and kinds

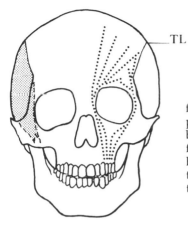

Figure 8.4. Diagram showing how the forces developed by the masticatory muscles, in particular the *temporalis* (shown on left passing behind the zygomatic arch), are transmitted from the lower jaw to the vault of the skull in humans. The lines of compression that pass through the facial bones are shown as dots on the right side of the skull. *TL*, Temporal line.

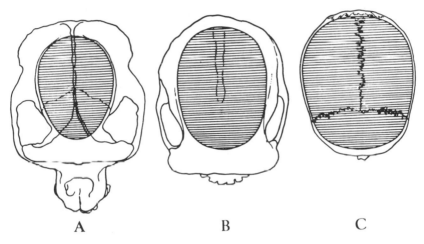

Figure 8.5. Skulls of gorilla (*A*), *Homo erectus* (*B*), and modern human (*C*) from above showing the relative sizes of the cranial cavity (shaded) and masticatory apparatus protruding below the neurocranium (from Weidenreich, 1939–1941).

of photoreceptors on the surface of the retina. A large eye will allow the entry of more light and permit greater sensitivity. The first stimulus to increased eye size may have been the nocturnal habits that are believed to have been characteristic of the ancestral primates (Martin, 1979). Today prosimian eyes are specially adapted for nocturnal vision with a light-reflecting membrane called the *tapetum lucidum*, which lies behind the layer of photoreceptors. There is a high density of photoreceptors in the retina of the very sensitive "*rod*"type. These are connected via the bipolar cells to the ganglion cells: one of the latter receives input from a number of rods and so is able to respond to low light intensities (Fig. 8.6). The amount of information received by the brain from the eyes is probably approximately related to the size of the optic nerves. The dog and cat have 150,000 fibers in each nerve; humans have 1,200,000 (von Bonin, 1963).

The eyes of the higher primates, however, are adapted for daylight (diurnal) vision, have no tapetum, are sensitive to color, and have a fairly consistent structure (Woollard, 1927). The night monkey of South America, *Aotus*, though nocturnal, has higher primate eyes without a tapetum. The disadvantages of such an animal having "diurnal" eyes are mitigated to some extent by their large size. Interestingly, the little tarsier, often considered to be a very unusual prosimian, also has eyes with no tapetum. This and other characteristics have caused taxonomists to place it among the higher primates. It is nocturnal and has the largest eyes relative to body size of any primate. We may conclude that these two species evolved from diurnal ancestors and are secondarily nocturnal.

The differences between the human eye and that of a monkey are slight. Both have a higher proportion of the so-called "cone" type of photoreceptor, which is specialized for discrimination, including color, rather than for sensitivity; each ganglion cell receives input from only a few cones (Fig. 8.6). Fine visual discrimination seems to have assumed great importance in the forest, for in higher primates and in *Tarsius* we find the evolution of the *fovea centralis* ("yellow spot") not found in any other mammals. It is a small area of the retina where the usual layer of nerves and blood vessels does not lie over the actual photoreceptor cells, which here contain no rods and form a dense layer of cones. At this point, therefore, the surface is pitted so that light will fall directly on the receptors. At the same time the ratio of ganglion cells to receptors is higher (1 to 1.3), while at the periphery it is 1.32 (Wolin and Massopust, 1970). In humans the *fovea* makes possible the kind of discrimination required in reading and other fine work. But the fovea had already evolved in the forest, and in fact is more marked in the monkey *Cercocebus* than in humans (Woollard, 1927). It is clear that the remarkable quality of color vision that we enjoy arose as a diurnal adaptation to forest life. There is no doubt, however, that changes were also necessary in the brain as a correlate of the greatly increased quantity of visual information that became available, and it is these changes that above all have carried the perception of higher primates beyond that which may be attributed to most other mammals.

We have already mentioned the evolution of stereoscopic vision among the primates (Chapter 3, VI). This important development has involved changes in the position of the eyes, in the structure of the optic nerve paths and brain, and, finally, in perception.

The evolution of the primate visual sense is correlated with a large expansion of the part of the neocortex that includes the portion of the brain concerned with vision, the visual cortex. In Fig. 8.7 the brains of

Figure 8.6. Diagram of a section through the human retina. Note that the light passes through the nervous tissue which overlies the layer of light-sensitive rods and cones. In prosimians the tapetum lucidum lies below this layer. At the fovea centralis, the light-sensitive cells are exposed directly to light, without the overlying nervous tissue (from Gregory, 1966).

the tree shrew and two primates are drawn the same size. In the tree shrew the visual sense has not yet replaced the olfactory sense, as can be seen from the relatively large size of the olfactory bulbs. In the higher primates such replacement has occurred, and, although the visual cortex gets smaller relative to the rest of the brain, it is still absolutely expanding. This expansion is even more true of the visual association area, the part of the cortex nearest to the visual cortex, which makes connections between the visual cortex and other parts of the brain. In primates the

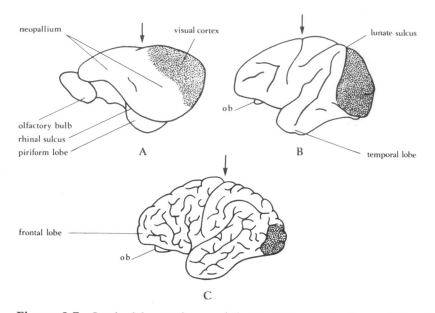

Figure 8.7. Cerebral hemispheres of the tree shrew (*A*), *Macaca* (*B*), and human (*C*), drawn the same size. Note the variation in extent of folding of the cortex and the relative size of the olfactory bulbs (*ob*) and visual cortex (shaded). The temporal and frontal lobes of the neocortex hide the piriform lobe of the paleocortex in the higher primates, which is visible here only in the brain of the shrew. The arrow points to the central sulcus, which separates the frontal and parietal lobes. The rhinal sulcus separates the paleocortex and neocortex, and the lunate or "simian" sulcus demarcates the visual cortex. (For a diagram showing the different lobes of the brain, see Fig. 8.12.)

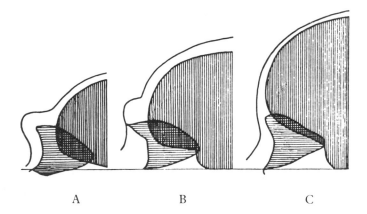

Figure 8.8. The relationship between the orbit and the frontal part of the cranial cavity in chimpanzee (*A*), *Homo erectus* (*B*), and human (*C*). Note the form of the supraorbital torus (from Weidenreich, 1939–1941).

expansion of the brain was led by the neocortex, which came to occupy an increasing proportion of the total brain volume. The visual cortex and the visual system as a whole became increasingly integrated with total brain function through the visual association area. All this involved an increase in total brain size, which determines the size of the neuro-cranium as a whole.

The size and position of the eyes have played an important part in molding the skull, in addition to the psychological and neurological importance of the eyes in human evolution. We have seen that primate eyes are often large and have come to face forward. The eye sockets— the *orbits*—are appropriately large and occupy a more central position in the face. The eyeball is a delicate structure, and its form is of necessity optically precise, so it must at no time be subjected to stress or pressure. It must also be at the surface of the body and under precise motor control. As a result, the eye is rather like the ball of a universal joint in a well-lubricated and well-padded bony socket, which gives it support and protection. The eye muscles move it within this socket, and with the eyelids, the facial muscles can be contracted across the socket to protect it from direct blows. Eyelashes protect the eye from sweat, rain, and dust.

The higher primates and *Tarsius*, alone among all vertebrates, have the eye socket totally enclosed and shielded in bone. The completion of the postorbital septum—the thin sheet of bone that separates the orbit from the temporal fossa—has been claimed to serve a number of func-tions, but probably the most important is that it protects the eyeball and its musculature from the intermittent pressures generated by the tem-poral muscle during chewing. The higher primate ancestor could chew and hunt simultaneously! The septum also provides an extra area of origin for the temporalis muscle.

As the eye sockets enlarged in evolution, the reduction of the muzzle made room for them in the face and allowed them a more central posi-tion. With the muzzle, they also moved back into the head, and in humans they lie closer under the anterior part of the brain than in other primates (Fig. 8.8). The bony rim of the orbit serves to protect the eye as well as to transmit the forces developed by the masticatory apparatus. The upper part of this rim assumes the greatest size (and gets the name of supraorbital torus) in big jawed forms in which the eyes have not fully retreated under the forepart of the brain. Such is the case in the gorilla and chimpanzee, while in humans and the gibbon the torus is less well-developed, since the jaw is smaller and less prognathous, and the eye is better protected by the forepart of the neurocranium. In the orang-utan, the relationship between the face and the neurocranium is dif-ferent; this also results in a much more lightly built supraorbital torus (Biegert, 1963). It is of interest that the torus is heavily built in all the Hominidae except modern humans, for it was only in the last stages of human evolution that the jaws finally retreated and the neurocranium finally expanded over the top of the eyes. This expansion of the brain has resulted in the flat vertical human forehead and the more or less vertical human face, which carries the deep eye sockets.

IV. The Nose The rhinarium, the sensitive skin around the nostrils, is an important sense organ in the prosimians, and its wet surface no doubt enhances its sensitivity to air currents, which can be interpreted directionally. With the ascendance of the visual sense with its strong directional character, the olfactory sense has come to be relatively less important in the larger, diurnal, higher primates (Fig. 8.9). In the daytime of the arboreal environment the nose supplies information about the quality and intensity of odor, but is perhaps less important as a source of directional information.

The air required for respiration is drawn into the nostrils and over the olfactory *mucoperiosteum*—the mucous membrane bearing olfactory receptors that covers the complex foldings of *turbinal bones* within the nasal cavity. In the higher primates, the complexity of the *turbinal bones* has been much reduced, together with the total area of mucoperiosteum. That both the muzzle and the jaws are reduced in modern humans is, however, coincidental. In some animals they have receded independently: in the gorilla the muzzle is slight, the jaws large; in the little elephant shrew (*Elephantulus*) the muzzle is far more developed than the small jaws.

It is the reduction of both the muzzle and the jaws in humans that has resulted in our flat faces. This change has altered the center of gravity of the head and contributed to the balance of the skull upon the vertebral column. The reduction of the muzzle came early in the evolution of the higher primates; the reduction of the jaws came late in human evolution (see Fig. 5.13).

A visit to the zoo, however, will reveal the human nose to be rather more prominent than that of most monkeys (except for the astounding proboscis monkey) and apes, perhaps partly due to the recession of the jaws and the expansion of the brain, which left only a small space for the nasal cavity. In fact, the olfactory epithelium covers only a small part (2.5 sq cm) of the lining of this cavity, so it is not unreasonable to suppose that some other function is performed by the human nose besides smelling. The fleshy surroundings of the nostrils contain fatty tissue and are lined with a thick, moist mucous membrane with many blood vessels, suggesting that the function of the nose is to warm and moisten the air on inspiration. The fatty tissue insulates the nasal cavity and the blood maintains the temperature within. The mucous membrane itself acts like flypaper to catch dust and small insects, and for this purpose there is a clump of hair near the opening of each nostril. Clearly, the warming of air could be important to the North Temperate peoples not only to protect the delicate membranes lining their lungs, but perhaps also to increase the effectiveness of their olfactory organ, which is more sensitive at body temperature.

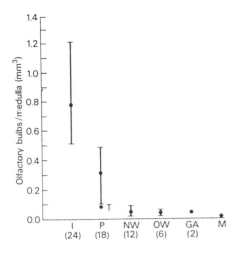

Figure 8.9. The volume of the olfactory bulbs expressed as a ratio of the volume of the medulla of the hindbrain in insectivores and primates. *I*, Insectivores; *P*, prosimians; *NW*, New World monkeys; *OW*, Old World monkeys; *GA*, great apes; *M*, humans. In the diagram, *T*, tarsier (after Passingham, 1982).

The moist mucous membrane also acts to humidify the inspired air. It has been shown that it secretes up to one liter of water per day and maintains the relative humidity of the inspired air at 95% (at body temperature). Its function is clearly to protect the delicate inner surfaces of the nasal and oral passages and the lungs from desiccation. The nasal index of modern humans (which indicates the shape of the nasal openings) is highly correlated with the absolute humidity of the air of the region inhabited (Weiner, 1954). The nasal breadth is correlated with temperature and humidity (Wolpoff, 1968). Thus, many of the tropical human races have flatter, more open noses than have those occupying dry cool areas, such as Mongolia. The prominent human nose is an organ for filtering, warming, and moistening the air and has perhaps evolved in size for that purpose since the movement of humans onto the dry plains and later into the cool temperate zones at the beginning of the Middle Pleistocene.

It has already been pointed out that the sense of smell is closely related to the basic drives of feeding and sex. In modern humans smell is an important stimulant to feeding, but individuals in which this sense has been destroyed can survive with little difficulty by depending upon the civilized preparation of foods, where testing of food quality by its scent is less essential than it is for traditional gathering societies. In its second function, sex, smell has again been replaced mainly by sight, but scent and perfume are still used by men and women as a sexual stimulant.

The relative size of the olfactory "bulbs" of the brain, to which the olfactory nerves run, can be seen in Fig. 8.9. Analysis of the olfactory impulses begins in the olfactory bulbs, and from them fibers pass to the lower lateral parts of the cerebral hemisphere (the *piriform lobe*) where further analysis presumably occurs. In mammals electrical stimulation at this point results in reflexes such as licking, salivating, and chewing.

From the piriform lobe connections run to the areas of the cerebral cortex, where scents are memorized and recognized. Though olfactory bulbs have been reduced in primate evolution relative to body size, the human olfactory sense is still capable of considerable discrimination, and the nose is capable of remarkable sensitivity. This has remained with us in spite of the fact that the importance of the olfactory stimulus in behavior is now much reduced; almost certainly a recent event.

V. The Ears Study of the evolution of the head must include consideration of the ear and the other soft parts of the face. The ear and the organs in the head associated with it form a complex mechanoreceptor that is sensitive to sound waves in the air, to gravity, and to movement. Like light, but unlike scent, sound waves travel in straight lines and therefore can be used in direction finding. For that reason the ears are separated on each side of the head (as are the eyes but not the olfactory organ) and function as an important stereophonic sound detector. They can supply data about the direction, frequency, and amplitude of sound waves.

The different parts of the ear serve the following functions among primates:

1. The *outer ear* is primarily a directional receiver of sound waves. As in many mammals, the external ear of most primates can be moved by both its own and associated muscles to receive the maximum amplitude of sound waves and to aid in discrimination between different sources of sound. The outer ear also plays some part in thermoregulation, as it does in mammals generally. There are a large number of blood vessels in the outer ear, and like the rest of the skin they are subject to dilation and contraction to regulate heat loss. This is important to animals with heavy fur and naked ears like lemurs. At the same time, ears held close to or away from the body can adjust the effective surface area for heat loss. The different sizes of the ears in different primates may be due as much to their function as thermoregulators as to their function as organs of hearing.

2. The *middle ear* receives the sound waves at a diaphragm (the ear drum) and transmits them mechanically with a reduction in amplitude via the three bones of the inner ear (which were made over from the reptile jaw) to the *cochlea* through a second diaphragm (the *fenestra ovalis*). This mechanism is peculiar to mammals and gives them more efficient hearing than other vertebrates, which are unable to hear the higher frequencies (Fig. 8.10).

3. Sound waves of reduced amplitude are transmitted to the liquid-filled cochlea in the *inner ear*, where the vibrations in this liquid are detected by the mechanically stimulated cells of the *organ of Corti*. The small hairlike processes attached to these cells respond selectively to movements in the surrounding liquid and send nerve impulses to the brain by the auditory nerve.

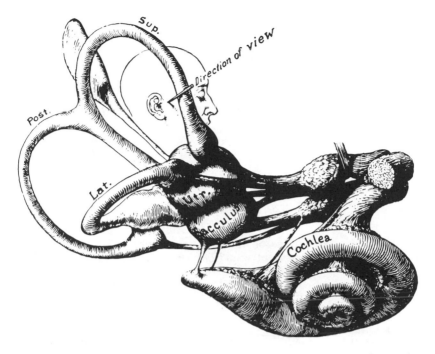

Figure 8.10. The mechanoreceptors of the inner ear are a system of interconnected liquid-filled tubes. They are shown here with their nerve supply. *Lat*, the horizontal semicircular canal; *Post* and *Sup*, the two vertical canals at right angles. The utricle (*Utr*), sacculus, and cochlea are labeled (after Hardy, 1934).

4. In connection with the cochlea are other liquid-filled vessels, the *utricle* and *sacculus*, organs that are lined with similar mechanoreceptors and transmit information to the brain about the posture of the head in relation to gravity.

5. Associated with this system of mechanoreceptors are the *semicircular canals*, which lie in the three planes of space. Sensory cells here detect the direction and acceleration of movement of the head.

It is certain that an arboreal animal like a primate needs a very efficient sense of movement and posture, but in fact the organs described are not very different from those found in fish, reptiles, and birds. Such organs are necessary to any motile vertebrate. In its function as an organ of hearing, the ear has changed very little during the evolution of the primates. We should note, however, that the range of sounds that can be heard has dropped in frequency as the primates have increased in size (Fig. 8.11).

Presumably, an acute sense of hearing is of less vital importance to noisy diurnal animals with few predators than to silent creatures that

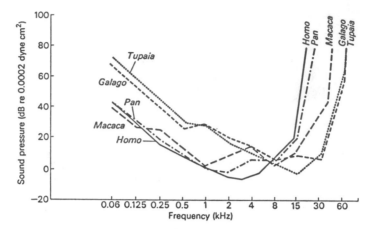

Figure 8.11. Audiograms of humans, three other primates, and the tree shrew (*Tupaia*). The intensity of sound (which can just be heard) is plotted in decibels (dB), while the frequency is plotted in kilohertz (kHz). Note that in the larger higher primates, the frequency range that can be heard decreases (after Stebbins, 1978).

hunt by night or walk in fear of death. The tendency in primate evolution has been to lose acuity of hearing of the higher frequencies but to gain discrimination in detecting lower frequencies and sound frequency patterns. For example, prosimians can hear frequencies up to 60 kHz, while humans can only detect at most, sounds up to 20 kHz. This change is correlated with an increase in body size; humans are little different in this respect from *Pan*. However, it also coincides with the development of vocal communication within the primate group, using sounds of widely different patterns.

In the most recent phases of human evolution the functions of the ear have not changed greatly. The outer ear still occasionally functions as an organ of thermoregulation when, as a result of the dilation of blood vessels, the ear turns bright pink (sometimes caused by blushing). The musculature of the ear has been greatly reduced, however, and few people can move their ears much. This reduction of motility begins among higher primates, and the reason for it can only be guessed, though it may be associated with increased motility of the head.

The high degree of auditory discrimination that we possess, which has been demonstrated by Stebbins (1976, 1978), enables us to understand the complex sounds of language and music. This discrimination may be due to some improvement in the cochlea itself where frequency is concerned, but is more likely due to the evolution of the auditory cortex, where the incoming signals are analyzed and sound patterns learned and recognized. Experimental work suggests that monkeys (and even cats and chinchillas) are able to make basic discrimination between vowels and phonemes almost as easily as we can

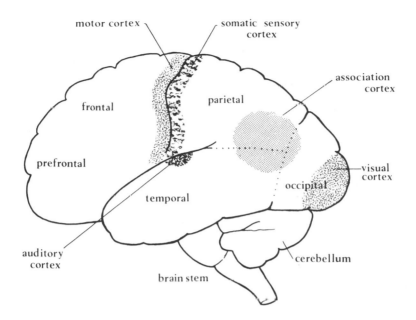

Figure 8.12. View of the left side of the human brain, showing the main subdivisions of the cerebral cortex, together with the somatic sensory and motor cortices. The dotted area in the inferior posterior parietal region shows very approximately the secondary association cortex discussed in the text. (Compare Fig. 11.7.)

(e.g., a/i, i/u, ba/da, da/ta). A review of these data by Passingham (1978) shows there is as yet not evidence of improved sound perception in humans.

The auditory cortex lies in the temporal lobe (Fig. 8.12), and it may be no coincidence that memory, which also appears to be localized in this area to some extent, is closely concerned with language as well as with visual images. Words and music may be memorized in an almost infinite variety of patterns. There is no question that the human brain has evolved a potential for sound discrimination and analysis that is greater than that found in any animal. Since, however, we cannot easily discover what animals can hear, we are at present unable to demonstrate what surely is human superiority.

In summary, it may be said that in primate evolution the ear has become concerned mainly with the perception of the part of the environment that consists of other members of the species, the social environment, rather than the general environment. That is, it is concerned with the kind of auditory discrimination required in the kinds of extensive vocalization found among higher primates, and finally in human linguistic communication. Appreciation of music is perhaps a by-product of the ability to analyze speech sounds.

VI. Changes in Brain Structure and Function

Our large, deeply folded brain is the most striking feature of *Homo sapiens*. It is not only absolutely large, but large in relation to our body size. Among mammals generally, the larger animals have the larger brains (though brains do not increase exactly in proportion to body size, but by a lower power). A larger brain is not necessarily an evolutionary novelty, or does it necessarily indicate special mental power (the brains of elephants and whales may reach 4000 and 6700 cc, respectively). In order to discover if a brain is unusually large in an animal, we must plot log brain weight against log body weight, as we did in our discussion of limb lengths. Such a display reveals that throughout the order primates there is a constant relationship between the two, the only exception being *Homo sapiens* (Fig. 8.13). We find from this data that the human brain is in fact 3.1 times the weight that would be predicted for a primate of our size (Passingham, 1982). The apes and monkeys follow, with the largest brains relative to their weight among all land mammals. The brains of elephants, large though they are, are merely average mammal brains when related to body size (Jerison, 1973). Only seals, dolphins, and toothed whales compete with the primates in relative brain size. However, the research sample of these animals is small, and the architecture and mechanics of the skull and brain have been profoundly modified during their extraordinary adaptation to a marine existence. The meaning of the relatively large, deeply folded dolphin brain is still one of the great mysteries of biology.

The extraordinary increase in the size of the human brain has been examined in relation to its different parts (for review, see Passingham, 1982). Although some of the lower brain centers have increased somewhat more than would be expected in a primate of our body size, the most spectacular increase by far has undoubtedly occurred in the *neocortex*—the part of the brain associated with the evolution of mammals that has assumed such important in primate evolution. The neocortex is clearly seen as the deeply folded surface layer of the cerebral hemispheres (Fig. 8.14); it is 3.2 times the size predicted for a primate of similar body weight (Passingham, 1982). It is so deeply folded that in humans 64% of the surface of the cortex is hidden in the sulci (or fissures); in apes 25–30%; in monkeys, only 7% (Fig. 8.14). This, above all, is the part of the brain that has given the human species its extraordinary capacity for learning, for culture, and for language.

By microscopically examining as well as electrically stimulating and recording within the living brain, it has proved possible to identify to some extent the functions of its different parts. Figure 8.12 shows some of the main functional areas in the human cerebral cortex. The evolution of these different areas in the primates may be related to the evolution of the sense organs we have discussed, and may be summarized as follows:

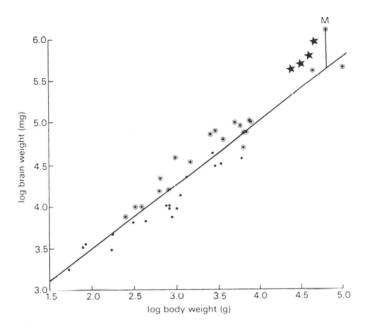

Figure 8.13. Bivariate plot of log brain and log body weight of 38 nonhuman primates and humans (*M*). The regression line is fitted through the points of the nonhuman primates. The human brain is 3.1 times as large as would be expected in a nonhuman primate of the same weight (vertical line). Dots, prosimians; larger circles, higher primates; stars, fossil hominids (after Passingham, 1982).

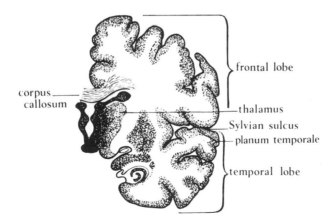

Figure 8.14. Section through one human cerebral hemisphere showing the extent of folding of the cerebral cortex.

1. The olfactory bulbs and piriform lobes are concerned, as we have seen, with the analysis of the olfactory stimuli from the nose. The piriform lobes form part of the *paleocortex*, which is the most archaic part of the cerebral cortex. The paleocortex is separated from the neocortex (which lies above it) by the *rhinal sulcus*. In the evolution of the primates the neocortex increases as the relative size of the paleocortex decreases (although the absolute size increases). As the neocortex expands, the rhinal sulcus moves down the side of the brain and the piriform lobes come to lie underneath the brain rather than to one side. These changes were shown earlier in Fig. 8.7.

2. The *occipital* (posterior) *lobe* of the cerebral hemispheres has been shown to be associated with the analysis of visual images, and the cortical layers here are, as we have seen, known as the visual cortex. As might be expected from what has been said of the importance of sight in primate evolution, the visual cortex has expanded as part of the neocortex, but its association area has expanded even more (see below). The visual cortex itself is clearly defined by a curved fissure called the *lunate sulcus*. The development of the visual cortex and its association cortex is one of the striking features of the primate brain.

3. As part of the neocortex, the *temporal lobe* has also expanded in primate evolution. A small area on its surface constitutes the *auditory cortex* (Fig. 8.12), but the rest appears to be concerned with the elaboration of visual analysis and with visual and auditory memory. These functions are obviously important in a group of animals that has increasingly come to rely on visual memory more than on other imprinted sensory patterns.

Clearly, the temporal lobes also play an important part in the human brain. Not only is memory to a major extent visual, but so is the stream of consciousness. Evidence suggests that in humans the temporal cortex is involved with the whole record of experience and with the reactivation of that experience (Penfield and Roberts, 1959).

4. As part of the neocortex, the *frontal lobes* have also expanded in primate evolution, an expansion that has probably overtaken that of most other parts of the cortex in the latter stages of human evolution. It is the enlarged frontal lobes that are always considered characteristic of *Homo sapiens*, and that have, in the last stages of evolution, given us a neurocranium that has expanded over the eye sockets (protecting them) and formed our vertical forehead.

The function of the frontal lobes is partly known, and partly obscure. Portions of the cortex lying in front of the central sulcus are clearly identified as *motor cortex* and premotor cortex (see Fig. 8.12). In these areas are located the transmitting neurons for the motor organs of the body's musculature. The "mapping" of these areas has been carried out by electrical brain stimulation in conscious human beings and other animals.

To the front of the motor cortex lies that part of the frontal lobes termed the *prefrontal area*. Here we find evidence of a speech center (see

Chapter 11, VI), but in other places no clear reaction can be obtained by electrical stimulation, and the only evidence of function comes from the striking but complex changes in personality that result from removal of various parts in human subjects. Evidence of this kind suggests that these areas are concerned with the maintenance of drive and with its inhibition and restraint. The prefrontal areas appear to make possible initiative, flexibility, and sustained attention in human behavior.

The human ability to sustain attention and concentrate on a particular goal may be an important development in our evolution. Kortlandt (1965) states that as a rule nonhuman primates cannot concentrate on one issue longer than a quarter to at most one-half hour. He contrasts this short attention span with that of carnivores like wolves, who may continue to lie in ambush, follow the same prey, or dig a den for many hours or even several days at a stretch. Clearly, a herbivore that persistently searched for one kind of food not readily available would be likely to die of starvation, whereas a carnivore diverted from its chosen prey would never succeed in running that prey to exhaustion. It would seem that insofar as we evolved as hunters and gatherers we necessarily developed the ability to work persistently toward a remote goal, and sustained attention was obviously a most important feature of early human behavior. We can see here, and we shall see again, how a change in human ecology and in the human way of life required the evolution of new kinds of behavior, of new mental processes, which in turn proved an essential basis for human culture.

5. The parietal lobes, which lie behind the central sulcus and extend back to the occipital lobe, have also expanded, possibly even more than the frontal lobes (von Bonin and Bailey, 1961; Passsingham, 1982). The cortex of these lobes is twice as extensive as might be predicted for a primate of our body size.

We know a little more about the function of the parietal lobes, and it appears that their expansion was also of the utmost importance in human evolution. In the anterior part of the lobes, just behind the central sulcus, we find the somatic sensory cortex, which receives inputs from the body's skin sensory receptors (e.g., pressure, heat, cold, pain). The mapping of these areas has also been carried out by electrical stimulation in conscious human beings as well as in other animals (see Fig. 3.9). Beside this narrow strip, parietal areas of the cortex are usually labeled the *association cortex*, a descriptive term arising from the fact that the area receives intracortical connections from the three sensory receiving areas: the auditory (in the temporal lobe), the somatic sensory (in the anterior part of the parietal lobe), and the visual (in the occipital lobe), concerned with hearing, touch, and sight, respectively. In monkeys and apes these areas connect directly with the surrounding cortex, so the nearby areas are labeled "auditory association," "somatic sensory association," and "visual association." But in humans we find an expansion of the parietal lobe between these three areas to form what has been described as "the association cortex of association cortexes" (Geschwind, 1964). This secondary association area is small in apes but immense in humans. All its connections are with the neighboring cortex rather than with the lower centers of the brain.

The secondary association cortex is the inferior posterior parietal region approximately delineated in Fig. 8.12, and it is of the greatest interest that part of it, at least, seems to coincide with a major speech area (see Fig. 11.7). In this area of the cortex, so particularly human, it seems possible that we may identify at least one place in which integration of the signals from the different sense organs takes place. In the speech area, too, there is evidence of abstraction. As we shall see (Chapter 11, VI), this so-called "speech area" is to be found only on one side of the brain; on the other side there is evidence of neural activity associated with the elaboration of notions of space and body movement. Such elaboration can again result only from the integration of sensory input.

The striking result of this expansion of the frontal and parietal lobes in the last stages of human evolution is that they have covered not only the paleocortex but even the large and newly evolved visual cortex. The simian sulcus, which demarcates the visual cortex in monkeys and apes, is almost obliterated in humans, and the visual cortex comes to lie for the most part underneath the brain rather than on top. It is thus hardly visible from above. The primary motor, somatic sensory, visual, and auditory areas, which bulk so large on the surface of the cerebral hemispheres of a monkey, are in humans to a great extent crowded down into the sulci.

While the primates have evolved advanced sense reception and motor control, it looks as though the specifically human achievement lies in the special and highly developed integrative functions of the brain. These functions appear to have developed with the anatomical correlates of cortical expansion in the association areas with extensive intracortical connections. These increased intracortical connections are particularly apparent in the greatly increased area of cross-section of the *corpus callosum*, the immense body of fibers connecting the two hemispheres of the cerebrum (Figs 8.14 and 8.15).

One of the most interesting and, until recently, least understood developments in the evolution of the human brain is the appearance of cerebral bilaterality. In our discussion of the primate visual system (Chapter 3, VI) we noted that each side of the brain is associated with the opposite side of the body, and this feature is characteristic of vertebrates generally. Where coordination of input and output is required between the two sides, it is achieved by means of fibers that connect the two hemispheres, the corpus callosum mentioned above. It has now been established that in human evolution, the redundancy in those brain functions of the two hemispheres that are not directly related to input and output has been reduced. For example, in the cat memory is duplicated—learned behavior is stored in both hemispheres. Such brain functions that do not require spatial duplication are present in humans

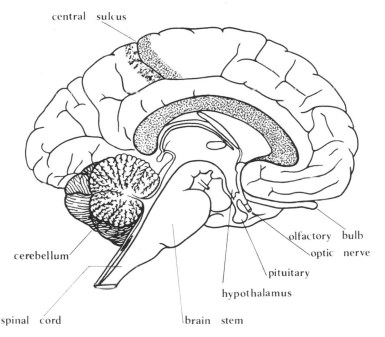

Figure 8.15. Median sagittal section of the human brain showing (*upper part*) the internal (medial) surfaces of the left cerebral hemisphere, below which lie the transverse fibers of the corpus callosum that connect the two hemispheres (shaded). The *central sulcus* separates the somatic sensory and motor cortices (shaded, as in Fig. 8.12), which continue on the median side of the hemisphere. The left olfactory bulb and optic nerve are shown at the base of the brain, together with the thalamus, hypothalamus, and pituitary. Like the cerebellum and brainstem, these organs lie medially and are cut in this view.

on one side only; a "division of labor" between the hemispheres has evolved. Speech and word memories involving symbolization and classification are a specialty of the left hemisphere in most people, while concepts of proportion and relationship in space and sound are stored and analyzed in the other side (Ornstein, 1972). This division of function is extensive in humans and possibly unique; by reducing redundancy it increases the capacity of the brain considerably without further increase in its size. It may have evolved when the expansion of the brain itself came to an end, toward the end of *Homo erectus* times. The differentiation of function is also coincident with differences in anatomy of the two hemispheres, which are no longer mirror images of each other (Geschwind, 1972). Thus, the human brain has not only increased in size and complexity, but has undergone a novel kind of bilateral specialization.

It would be wrong, however, to suppose that it is in the cortex alone that we must look for the neurological correlates of humanity, for the other parts of the brain, the so-called "lower centers," have evolved and expanded in conjunction with the cortex, with which they function in close reciprocal collaboration. Lying along the inner surface of the temporal lobes and toward the lower center of the cerebral hemispheres are a group of structures of extremely complex anatomy. They consist broadly of the *cingulum*, the *hippocampus*, the *thalamus* and hypothalamus, the more complex masses of the *basal ganglia*, the *mid-brain*, and the *amygdala* (Mark and Ervin, 1970), and constitute the limbic system (Figs. 2.8 and 8.16). As we have seen, the function of this system is to mediate certain types of sensory input and motor output in a manner recognized subjectively as emotional. This ancient part of the brain constitutes the "emotional brain," inherited from our early mammalian ancestors.

Emotional responses mediated in the limbic system serve motivational purposes in meeting crucial adaptive tasks: finding food and water, avoiding predators, achieving fertile copulation, caring for the young, and so forth (Hamburg, 1963). An emotion has several components: (1) sensory, (2) physiological, (3) subjective, and (4) motor. It is clear that the physiological components of emotion that take the form of nervous and endocrine activity have not altered significantly during the evolution of mammals. The activity of the limbic system and the production of the adrenal hormones with which, among others, its activity is associated, are common features of all mammal species. The expression of the motor component, however, varies a good deal, and in humans it is subject to inhibition and even repression by the cortex, the effects of which have been far-reaching. In this and many other ways evolution of the cortex has brought much of our behavior under conscious control and given it greater flexibility.

As in the case of the visual, somesthetic, auditory, and other regions of the cortex receiving input, the limbic system itself has an association cortex in the lateral and inferior parts of the temporal lobe. Through this association area the limbic system makes connections with the other association areas and brings to sight, touch, and sound the associations of fear, pleasure, rage, and pain, whichever is appropriate (Geschwind, 1964). Learning in animals that depends upon trial and error (and is equivalent to the experimental conditions of learning by reward and punishment) results in association being formed between sensory input and limbic response. Experience with a high emotional component will be recorded more deeply in the memory, and it is biologically important experience that is accompanied by emotion. The learning and memorizing of biologically useful experience is facilitated at the expense of biologically useless experience and behavior. Thus, all mammal learning, including human learning, carries a limbic component.

In humans, the secondary association area in the inferior parietal region of the cortex allows intracortical connection to be made without a necessary association with the limbic system—which may be only marginally involved. In this way associations between different external events may be recognized and memorized without much if any emotion,

cingulate   gyrus

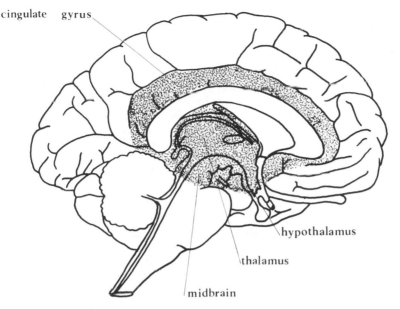

hypothalamus

thalamus

midbrain

Figure 8.16. Medial sagittal section of the human brain showing the limbic system (shaded). The cingulum is part of the cingulate gyrus. The amygdala and hippocampus are not visible as they lie on each side, on the medial surface of the temporal lobe.

and this possibility forms the basis of *detachment*—the specifically human potential for memorizing events that are not of immediate biological importance. Detachment from the limbic system also allows some human responses to be relatively free of emotional overtones; together with the potential for inhibition, it is the basis of cool reasonableness.

Another important structure is the cerebellum, which like the cerebral hemispheres, can be separated into a central medulla and cortex. It, too, has become deeply fissured in evolution (Fig. 8.15). Its function is the regulation of the action of all skeletal muscles. The pattern of muscular activity is determined elsewhere; the cerebellum, like a servomechanism, controls the strength and timing of contraction of the separate muscles involved in any movement. Damage to the cerebellum disrupts all basic activities and skills, even walking; movements become clumsy. It is also involved in the learning of automatic skills. Its action is unconscious but of immense importance to any vertebrate with rapid movements requiring accurate control. Clearly, primates of all mammals depend on the cerebellum to integrate and control their muscular activity, and in human evolution it has expanded in size second only to the neocortex. We have inherited a highly evolved cerebellum from our primate ancestry, which no doubt plays an important part not only in locomotion but also in the delicate motor control required in toolmaking and in more complex and precise activities such as piano playing.

From this brief summary we can see that different parts of the brain have expanded in varying amounts in human evolution. The most striking expansion has occurred in the neocortex and cerebellum. However, if we compare the volume of the neocortex with that of the brain as a whole in primates, and plot the figures in the usual way, we find that the increase in human neocortex size is predictable for a primate of our brain size. This means that the brain proportions set up in the more primitive primates have held, even though the brain size itself has trebled in relation to body size. So it seems that we have a very large "standard" primate brain (Passingham, 1982).

Each part of our large brain seems to have played an essential role in human evolution, but it is the neocortex that has contributed most to making us the creatures we are. The temporal and parietal lobes appear to have given us new possibilities of integrating our experience, the frontal lobes have given us new control over our behavior. Bilateral function has increased the effective capacity of the brain and reduced redundancy. The new association areas of the cortex have made possible some freedom from emotion in the learning experience. It is important to stress, however, that in no sense did the brain *cause* the evolution of human behavior. The human brain and human behavior evolved together as necessary correlates in response to environmental change and natural selection.

VII. The Brain as a Determinant of Skull Form

The importance of the brain in the maintenance of life at all homeostatic levels necessitates its protection in a bony case, the neurocranium. In order to give maximum strength for a given amount of material and weight, the neurocranium has tended in evolution to be spherical, mechanically the most efficient form for protective material. The brain itself is soft and does not appear to dictate greatly the shape of the neurocranium, only its volume. The neurocranium is in fact normally far from spherical, since the actual form is dictated by the development of the various bony appendages to it that we have discussed, in particular, the orbits, muzzle, and jaws, which make up the face. The relative size of the neurocranium and these components, together with the posture and the neck musculature, establish the shape of the head.

The neurocranium consists of two layers of bone, the so-called *inner* and *outer tables*. The inner table surrounds the brain itself; the outer table follows the structural demands of the other components of head form (Moss and Young, 1960). Between them is a lightweight, webbed bony structure, the *diploë*, and, where the inner and outer tables do not closely coincide, air spaces are formed called *sinuses*. Any section through the neurocranium shows the diploë, which can be seen later in Fig 9.2. According to mechanical principles, the strength of such a structure to resist bending can be increased, without increasing the weight of the sructure significantly, by moving the two outer components apart, and that is what has happened in the evolution of the neurocranium; skulls with a thick diploë, such as those of *Homo erectus*, weigh only a little more, yet give much more protection to the brain than do skulls with a thin diploë. It is also apparent that in *Homo erectus*, the tables of the bone themselves are thickened, and this may well increase protection against local forces of compression such as might be caused by a blow. The thickening may also be the product of a general trend in the species toward increased robusticity of the entire skeleton.

The size and shape of the brain is well known in the living primates, and in the fossil forms it can be estimated by making a cast (an *endocast*) of the inside of a complete fossilized skull (Fig. 8.17). It should not be supposed, however, that the cranial capacity is equivalent to the volume of the brain: because of the presence of associated structures, such as membranes, blood vessels, and ventricles containing cerebrospinal fluid, the human brain itself occupies only two-thirds of the total cranial volume, and in other primates the proportion may be even less (Mettler, 1956). The cranial capacity varies a great deal in modern humans and will have varied in more primitive forms as well, so that the figures we have, which are based on a few fossils, may be slightly misleading but are unlikely to be very far from correct (Table 8.2).

From our knowledge of the dating of fossil remains and their cranial capacity, it is clear that the great spurt in brain size began in early *Homo*. It is important, however, that there was at the same time an increase in body size, from a mean stature of about 4 ft and weight of perhaps 45 lb in *Australopithecus* to a mean stature of 5 ft 5 in and weight of 150 lb in some races of *Homo sapiens*. However, the sample size of the fossil hominids with sufficiently well-preserved crania is extremely small (see Table 8.2). It is significant that the increase in brain size coincided approximately with the elaboration of stone tool kits, the beginning of cooperative hunting, and the invasion of northern latitudes (see Chapter 9, X).

The expansion in endocranial volume has meant that in human evolution the neurocranium has enlarged more or less at the same time as the facial bones and jaws have been reduced. The effect of the jaw-bones on the form of the head is, therefore, less than that of the brain; as a result, the human head comes quite near to its "ideal" spherical form. We are "eggheads" because our cranial capacity is a prime determinant of skull form: the face and jaws are now a relatively minor appendage. Simultaneously, the reduced nuchal and masticatory muscles leave their areas of anchorage rather small, smooth, and no longer bordered by crests. With this change in emphasis of head function and structure, we see the rounded skull vault rising above the orbits in human evolution; this development has been quantified by Le Gros Clark (1971). (See Table 8.3 and Fig. 8.18.)

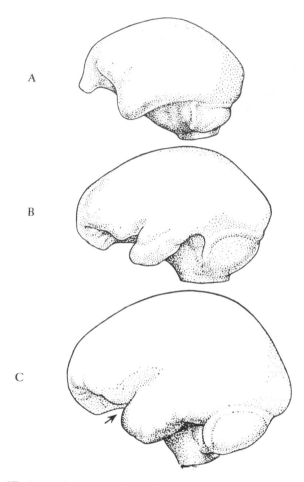

Figure 8.17. Lateral view of the endocranial casts of the gorilla (*A*), *Homo erectus* (*B*), and human (*C*). The arrow points to the notch that indicates the position of the sylvian sulcus between the temporal and frontal lobes (from Le Gros Clark, 1971).

When, therefore, we visualize the overall form of the neurocranium in the final stages of human evolution, we see that its shape becomes increasingly spherical: in the horizontal plane the cranium becomes less dolichocephalic (elongated) and more brachycephalic (rounded). This brachycephalization has been demonstrated in the most recent stages of human evolution as well as in earlier times, and it seems to reflect the continuing trend toward reduction of the jaws and face and improvement in cranial balance. As can be seen from Fig. 8.18, the supraorbital height index (FB/AB) measures the height of the skull vault above the level of the top of the orbit, and so indicates the height of the neuro-

TABLE 8.2. Cranial Capacities of Some Primates, Including Sample Ranges of Fossil Hominids[a]

Primate	Ranges (cc.)	Mean (cc.)
Macaca	—	100
Papio	—	200
Hylobates lar	82–125	103
Pan troglodytes	282–500	383
P. gorilla	340–752	505
Pongo	276–540	405
Australopithecus afarensis (*n* = 2)	385–450	418
A. africanus (*n* = 6)	428–500	452
Homo habilis (*n* = 5)	509–752	638
H. erectus (*n* = 14)	727–1225	930
H. sapiens (approx.)	1000–2000	1350

[a]Data from Tobias, 1971; Cronin *et al.*, 1981.

cranium relative to the orbital region of the face. From these figures it is clear that in this characteristic *Australopithecus boisei* forms a group with the great apes (index 47–52) and *A. africanus* forms a group with *Homo* (index 61–77). There is a considerable gap between them, reflecting the difference in the relative sizes of the neurocranium and face and probably in the total body size between these two species of *Australopithecus*.

The relative expansion of the neurocranium is also clear when the skull is viewed from the rear. As the brain becomes larger, the height of the point of maximum breadth on the side of the skull rises (see Fig. 5.15).

In summary, we see in human evolution a threefold increase in relative brain size, in cranial capacity, and in the size of the neurocranium, compared with what we would see in an ape of similar body size. This

TABLE 8.3. Supraorbital Height Indexes[a]

Primate	Mean supraorbital height index and source	
Pan gorilla	48	
P. troglodytes	47	Le Gros Clark (1950)
Pongo	49	
Proconsul africanus	55	Davis and Napier (1963)
Australopithecus boisei	52	Tobias (1967)
A. africanus	61	Robinson (1963)
Homo erectus	63–67	Ashton and Zuckerman (1951)
Modern Human	64–77	

[a]A useful discussion of this index is to be found in Rosen and McKern, 1971. For meaning see also Fig. 8.1.

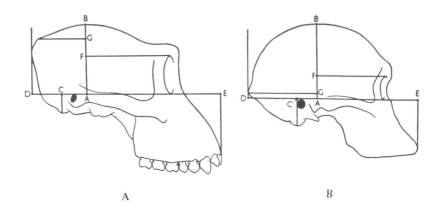

A   B

Figure 8.18. Outlines of skulls of female gorilla (*A*) and *Australopithecus* (*B*), with construction lines to calculate the skull indices introduced by Le Gros Clark: *supraoribital height index FB/AB, condylar position index CD/CE* (from Le Gros Clark, 1971).

rapid evolutionary change is accompanied by a reduction in the masticatory apparatus. In this way the human head is fundamentally remolded in the form of its skeletal structure.

VIII. The Face

We owe the characteristic features of the human face both to its underlying bone structure and to the form of the soft parts, which now require consideration.

The eyes have been discussed; their sensitive and delicate movements give the face a liveliness that no other feature can impart. The surrounding muscles serve to protect the eye and at the same time add to overall facial muscular activity in this region—particularly that associated with frowning and smiling. The form and function of the nose, a prominent feature of the face, have already been described. The muscles that move the nostrils are less developed in humans than in other primates, as are those of the ears.

The lips serve both a sensory and motor function. As a tactile sense organ they carry a high density of sensory nerve endings; the surrounding skin also carries hairs, around the follicles of which are wrapped further sensory nerve endings, a condition reminiscent of the facial vibrissae of mammals. This great sensitivity relates to the function of the lips in food investigation and is reflected in the area of the sensory cortex devoted to them. The area of motor cortex related to them reflects their importance not only in facial expression but in speech (see Fig. 3.9).

The facial musculature as a whole is not only concerned with the masticatory apparatus and the protection of the eyes but also plays an important role in social communication through facial expression. The differentiation of the facial musculature is peculiar to mammals and strikingly advanced in primate evolution. In many mammals, facial muscles are involved in behavior patterns not directly concerned with feeding. Teeth are often displayed to frighten away intruders, as well as in response to pleasure. The faces of the baboon and mandrill carry features that are not merely mechanical but *sematic* (that is, concerned with signaling, with communication within the group); they contribute effects that may be imposing and even terrifying in appearance. There is little doubt that natural selection has determined the mechanical function of the head as well as its appearance through shape, color, and distribution of hair, which is very important in a social context.

In the higher primates, facial expressions function primarily as signals between members of the same species. Most animals perform a courtship ceremony, which is a signaling device to bring the sexes together with proper timing and stimulation; it is hard to say exactly how large a part facial expression plays in these rituals among primates. Among humans, although dance is often a preliminary to sexual communication, speech and facial expression are of paramount significance.

Studies of primates have revealed that in monkeys and apes facial expressions are varied and significant in communication, and their increasing importance seems to be correlated with a reduction in facial hair. Examples are manifold: part of the range of expression visible in the face of the chimpanzee is reproduced in Fig. 8.19. The origins of facial expression in humans are discussed by Andrew (1965); they are an important means of communication in all social primates.

It seems clear, then, that in assessing the determinants of head form we must not omit the soft parts that affect facial appearance and expression. Natural selection operates on the form of the head to evolve a structure with varied functions. The function of the head as an organ of communication in a social primate is not by any means less important than the functions that determine the form of its bones.

IX. Hair and Skin

Our nakedness is one of our most striking features. The loss of body hair that has occurred in human evolution powerfully suggests that our origin is tropical, but nakedness cannot be the result of a tropical origin alone, since the other tropical primates have retained their body hair. The minority of tropical mammals that have lost their body hair include the elephant, rhinoceros, and hippopotamus, large animals with thick, heavy skin, and subcutaneous fat, which forms an effective layer of insulation. The remaining majority of mammals are protected by their hair or fur from the sun's tropical heat as well as from cold in temperate regions. The advantage of nakedness must lie elsewhere.

The evolution of nakedness has not involved any major evolutionary

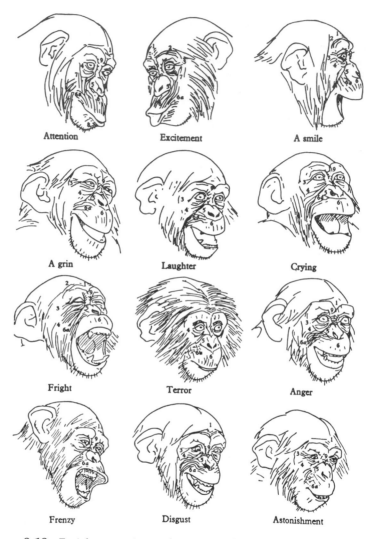

Figure 8.19. Facial expressions of a young chimpanzee in various moods. Some of the creases are marked with numbers to emphasize that each is by no means confined to one expression (from Kohts, 1935).

novelty. We have about the same number of hair follicles as the great apes have; the difference is that the majority of our hairs grow very little (Montagna, 1965). The fact that we retain certain areas of body hair is perhaps even more surprising than the loss of it.

The glands of the skin have probably assumed greater importance as a result of the human loss of body hair. The *sebaceous* (oil-producing) glands are more numerous and active in humans than in most other mammals, but their function is still obscure. While they may protect the hair and skin from drying out in the tropical heat, or from the effects of constant moisture, their remarkable distribution on the human body (such as inside the cheeks and on the inner edges of the eyelids) suggests a more complex function that is not presently understood (Montagna, 1965).

The sweat glands fall into two groups: the apocrine and eccrine glands. The *apocrine glands* secrete the odorous component of sweat and are primarily scent glands that respond to stress or sexual stimulation. Before the development of artificial scents and deodorants, they no doubt played an important role in society. In modern humans these glands occur only in certain areas of the body, in particular in the armpits, the navel, anal and genital areas, the nipples, and the ears. Surprisingly, glands in human armpits are more numerous per unit area than in any other animal. There is no doubt that the function of scent in sexual encounter is of the greatest importance even in the highest primate.

The *eccrine glands*, which are the source of sweat itself, have two functions in primates. Originally they probably moistened friction surfaces (such as the volar pads of hand and foot), improved grip, prevented flaking of the horny layer of the skin, and assisted tactile sensitivity. Glands serving that function are also found on the hairless surface of the prehensile tail of New World monkeys and on the knuckles of gorilla and chimpanzee hands, which the latter use in quadrupedal knuckle-walking. Glands in these positions are under the control of the brain and adrenal bodies, and as we know sudden stress may produce sweaty palms.

The second and more recently evolved function of the eccrine glands is the lowering of body temperature through the evaporation of sweat on the surface of the body. The hairy skin of monkeys and apes carries eccrine glands, but, except in one species of macaque, they are neither as active nor as numerous as in humans. In most other primates they may be considered to be no more than organs of excretion. We are equipped with two to five million active sweat glands, which play a vital part in cooling the body. The heat loss that results from the evaporation of water from a surface is enormously greater than that which could be expected to occur as a result of simple radiation. The fact that sweat contains salt necessitates a constant supply of this mineral, as well as water, if we are to survive in a tropical climate.

It has been observed that like almost all mammals, primates sweat very little. Even hunting carnivores such as dogs lose heat by other means (e.g., panting). Sweating has evolved as an important means of heat loss in humans, a fact that is surely correlated with the loss of body hair. The apparent significance in human evolution of achieving an effective means of heat loss would appear to indicate that hominids were subject to intense muscle activity, producing much metabolic heat. A light coat of hair is reasonably effective at shielding an animal from environmental heat — from the warm air, and the sun's rays. The human cooling adaptation is surely effective not only in protection from environmental heat but in assisting the loss of metabolic heat. Our ancestors could not afford even the smallest variation in body temperature. With such a highly evolved brain the maintenance of a constant internal environment was a requirement of prime importance in human evolution.

The above considerations seem to suggest that the human skin, with its poor covering of hair and millions of sweat glands, evolved as an adaption of a savanna-living primate that underwent considerable physical exertion during the heat of the day. Most carnivores hunt by night, but hominids, being diurnal primates, must have gathered their food resources by day. The main argument against this hypothesis is that sweat seems inappropriate as a means for a savanna animal to lose heat because it requires the ingestion of large amounts of water, which is not easily obtained on the savanna. However, all primates require reasonably pure water to drink at least once if not twice a day, and the archeological evidence makes it clear that those sites in which hominids settled were always near a freshwater stream. If some water was available, surely that would have been sufficient for the modest requirements of a sweaty hominid?

For the present the most likely conclusion is that the forest-living protohominids might have already lost much of their body hair, because radiation is the most effective means of heat loss in a humid forest where the evaporation rate is very low. On moving to more open country and taking up a more energetic lifestyle, their successors would have come to adopt evaporative cooling under the circumstances we have described.

The loss of body hair is, of course, by no means complete. The hair on the head gives protection to the brain from radiant solar heat in tropical climates. It is possibly relevant that baldness in humans in north temperate climates is stated to be on the increase (Montagna, 1965), which may be because hair serves a less vital protective function in cool climates. Yet it will serve to keep the head warm, and thermoregulation in the brain is very important. There seem to be other factors at work here, especially when we remember that among women baldness is rare, yet they have to survive the same climatic conditions as men. Man, incidentally, is not unique in his baldness; one species of macaque and the orangutan both go bald in middle age, while the uakari monkey has a completely bald head when it becomes an adult.

In fact, it is certain that the presence and absence of hair on the head as well as elsewhere is controlled by other than purely mechanical factors. Hair has been retained around the eyes (brows and lashes) as well as the ears and nose, where its function is primarily protective, but elsewhere it appears to be mainly sematic in function, acting as a signal to other members of the species (Goodhart, 1960). A glance at Fig. 8.20 suggests that this is indeed the case, for many species of monkeys have

Figure 8.20. *Cercopithecus* monkey (the guenon) has striking patches of distinctive hair corresponding to man's head hair, mustache, and beard. These patches are important as signals in sexual competition, just as they are in humans (see Fig. 8.22). (Courtesy Belle Vue Zoological Gardens.)

striking patches of hair that correspond to our own. In monkeys it is *epigamic* in function, that is, it is used in signaling in relation to sexual dominance. The guenon illustrated here has not only a very smart hairstyle but also a white beard and mustache. The beard is quite common among species of *Cercopithecus*, and a flowing mustache is found in the South American emperor tamarin. Such hair patches, which develop at puberty and are found in males only, are likely to have some social function and are probably related to intermale rivalry; it is not that the female selects the most impressive male, but more probable that the most impressive-looking male frightens away his rivals. This hierarchy of dominance in social prestige is based on appearance as much as physical force, hence the evolution of epigamic hair (see Chapter 10, VIII). There is considerable variation among different human races in the development of facial and head hair. Figure 8.21 suggests that facial hair has an important and recently evolved function in humans. Some human races have much more facial hair than the chimpanzee and gorilla. Shaving is a cultural convention that uses the presence and absence of facial hair as a means of modifying its epigamic function, presumably to lower the level of threat in a cooperative social group.

Pubic hair and *axillary* (armpit) hair must have a different function, however, since they are found in both sexes, and axillary hair is hardly visible unless the arms are raised. These two hair patches coincide with areas of skin containing apocrine scent glands, so it seems probable that

Figure 8.21. Although clothes have augmented the epigamic function of hair in both men and women, hair styles still retain importance in the establishment of sexual dominance and in effecting sexual attraction in men and women, respectively. Both head and facial epigamic hair patches vary in form and appearance among the human races. These paintings were made of individuals from Eastern Europe (*top left*), West Africa (*top right*), New Guinea (*bottom left*), and North America (*bottom right*). (Courtesy Hans Friedenthal and Gustav Fischer.)

the hair has some function relating to scent. Goodhart (1960) points out that the scented apocrine secretion is at first odorless and develops a scent only on exposure to air. He suggests that pubic and axillary hair serve to provide a surface for oxidative reactions and to facilitate the dispersion of scent in the air. However, there is reason to believe that pubic hair also has an epigamic function in men, especially when it grows up the belly, and it may have been selected as an attractive adornment in women too. There is a useful confirmation of the theory of epigamic hair patches to be derived from warrior dress. The British guardsman's bearskin, the field marshall's plumed hat, and indeed the Highlander's sporran—the ornament of animal fur worn over the pubic region—all serve to accentuate our natural hair distribution. Perhaps the most recent to evolve was the pubic hair, which tends to be sparse in primates and which would be visible to other individuals only in a bipedal primate.

Skin color is a striking human characteristic that attracts much attention. The basic racial distinctions in this characteristic are slight; they are due not to a difference in the number of pigment-forming cells (*melanocytes*) but to the amount of pigment manufactured by these cells. The melanocytes inject pigment into the surrounding cells of the *epidermis* (the outer layer of skin), where it forms a shield to protect the cell nucleus from the harmful ultraviolet rays of the sun. Hair is pigmented by similar injection into the cells of the hair follicle. Active melanocytes may be selected for camouflage (as in prosimians) or, in naked humans, for protection from the sun's rays. The differences of skin color in human races are probably due to the advantage that fair skin offers in the temperate regions in allowing the sun's rays to activate the synthesis of vitamin D in the skin. The evidence has been recently reviewed by Williams (1973). The activity of the melanocytes is an example of a characteristic subject to both developmental and genetic homeostasis.

The actual quality of the skin varies probably as much within a single population (if not in a single individual) as it does in the whole human species. Different kinds of skin occur in different parts of the body, and the form of the skin appears to be genetically determined. Modern humans appear to possess fine smooth skin, but this is probably due to its relative hairlessness and the protection that clothing and modern comforts offer. Again, a smooth skin may be a product of sexual selection, as Darwin believed, particularly among women (Crook, 1972).

The skin is a very important and complex organ. Besides enveloping the body and maintaining an effective boundary and container for the body's tissues, it is a strong waterproof membrane and a highly complex sense organ, responsive to pressure, temperature change, and damage. It is also an organ of thermoregulation. It is strong, hardwearing, and generates claws, nails, and hair, as well as protective pigments. Human skin has evolved with the rest of human body as an adaptation to our changing environment.

X. The Human Head

The human head can be seen to have been formed by the evolution of a number of separate functional complexes. How these complexes have influenced its internal and visible structure will now be summarized:

1. The evolution of the visual sense, accompanied by the expansion of the neocortex, was the first factor in the evolution of the human head from that of an early primate. In particular, the large eyes facing forward

for stereoscopic vision changed the appearance of the face and its bony architecture, which was modified to protect the delicate eyeballs.

2. Recession of the muzzle and reduction of the turbinal bones followed, with a resultant flattening and even hollowing of the face between orbits and jaws. The olfactory sense took second or even third place as the primate's means of perceiving its environment.

3. Erect posture and later bipedalism brought a forward movement of the occipital condyles and foramen magnum, which, together with the recession of the face as a whole, changed and improved the balance of the head on the spine, resulting in the reduction of the nuchal crest and nuchal area.

4. At a late stage in human evolution came the recession of the masticatory apparatus and a reduction of the power and extent of the grinding surfaces involved in the mastication of food.

5. The brain had been increasing in size throughout primate evolution; with the reduction of the masticatory apparatus came a final great spurt, and the brain trebled in a little over two million years. This final enlargement filled out the forehead and contributed to the vertical face of modern humankind.

6. Hairlessness revealed a face that contained not only the speech apparatus but also a musculature that was increasingly used in visual communication by facial expression. With the reduction of facial hair, especially around the eyes, the face became an important means of communication and the seat of beauty.

7. Finally, we find the whole body changing in appearance as a result of small modifications in hair growth, sweat gland development, and melanin production. The face and body assume an important role in communication through facial expression and body language. We note that in many mammals and especially primates hair patterns have a sematic role characterized especially by epigamic features.

In this chapter we have tried to subdivide the evolution of the human head into the evolution of a series of separate functional parts and then, as it were, to fit them all together again. It is important to emphasize that, since the head is so highly integrated a structure, the evolution of each part must have affected the morphology of every other part. A change in size of any portion will have effected the balance of the head on the vertebral column and so the development of the nuchal musculature. At the same time, the amount of development of any muscle (and in particular the masticatory muscles) will produce tensions and stresses in the bony architecture and in turn will affect its form.

In practice, the interactions of the different functional units of the head are very complex and can be fully grasped only by considerable familiarity with the skulls themselves. The basis of such interaction has, however, been presented here and should form a means of approaching a functional study of the cranial morphology of the hominids.

Suggestions for Further Reading

A useful account of the structure and function of the mammalian sense organs and the brain is to be found in J. Z. Young, *The life of mammals* (Oxford: Oxford Univ. Press, 1957). More recent texts that develop some of the ideas referred to in this chapter are by H. J. Jerison, *Evolution of the brain and intelligence* (New York and London: Academic Press, 1973) and R. E. Passingham, *The human primate* (Oxford and San Francisco, Calif.: Freeman, 1982).

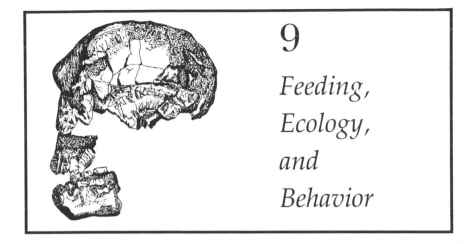

9

*Feeding,
Ecology,
and
Behavior*

<table>
<tr><td>

**I. The
Food
Search**

</td><td>

Before investigating the evolution of feeding in primates, it is helpful to look again at the sense organs and to separate them functionally into two kinds: the *contact receptors*, which supply data about the immediate environment (taste and touch), and

</td></tr>
</table>

the *distance receptors,* which measure intensity and direction (or extent of change) of external conditions (Fig. 9.1). We recognize three types of distance receptors: *chemoreceptors* measure the density of certain vaporized substances in the air (smell); *photoreceptors* measure the intensity and wavelength of certain electromagnetic waves (vision); and *mechanoreceptors* measure vibration, sound waves, and gravity (hearing and balance—see also Table 8.1). The evolution of these senses has been discussed.

Together with mating behavior, the search for food is a basic activity of all animals. An animal may spend most of its waking lifetime doing this, and only a surplus of food in the presence of other limiting factors on population growth will allow for the development of other behavior patterns less immediately essential. It is an important fact that in their adaptation to an arboreal environment primates were able to exploit a rich source of food that had not been tapped before to any great extent.

It is now clear that among mammals generally there is a close correlation between body size and basal metabolic rate. This means that small animals with a rapid basal metabolism need high-energy foods (such as meat and seeds), while larger animals can get along with low-energy foods, which require a longer period for digestion and assimilation (Martin, 1981). With their considerable range in size, the living primates have been able to exploit both the plant and animal products of the forest, and this feature characterizes their adaptation.

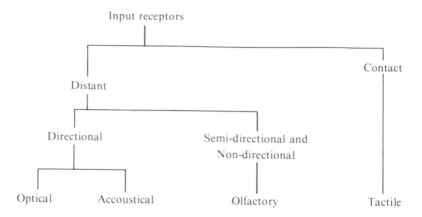

Figure 9.1. Schematic classification of mammalian sense receptors.

An enormous proportion of the energy received from the sun is converted into trees, as distinct from herbs, and the primates were able to consume not only the foliage but also the fruit and seeds of the forest, together with many of the smaller animals that feed on them. Fruit and seeds have high nutritive value, and arboreal animals obtained access to food resources of quite a different order of richness from those available to the grazing mammals of the plains. Edible fruits have evolved parallel to their consumption by animals, who in turn serve the plant by distributing the seeds in their feces. Thus, though the higher primates are broadly herbivorous, it is clear that the type of food they eat is different in texture and quality from that available to the terrestrial quadrupedal mammals that browse or graze.

Primates identify their food by sight, smell, touch, and, finally, taste. The preeminent importance of primate vision among the higher primates has already been mentioned, and it is certain that such an acute sense is of value in recognizing fruits and seeds, especially the former, which are often brightly colored. Birds and higher primates are the two groups of vertebrates known to have a well-developed color sense, and the colored fruits of the forest have evolved, no doubt, in conjunction with this developed color sense.

Although the visual sense is highly evolved in the carnivores and some ungulates, the sense of smell is by far the most important distance receptor in most mammals other than the higher primates, for whom its importance, however, should not be underestimated. Smell is still of primary significance among the nocturnal prosimians. The early mammals (which were probably nocturnal) and most recent mammals depend almost entirely on distance chemoreception in food- and mate-finding. For that reason, the olfactory sense is a primary source of information used in the generation of behavior. While sexual behavior is probably predominantly innate (see Chapter 10), a great deal of food-finding behavior is known to be learned, and the olfactory lobes are closely connected with the cerebral cortex (see Chapter 8, VI). It is significant that the cortex itself, with its vast memory store, evolved from the part of the brain originally concerned with smell (the *rhinencephalon*). It is not surprising, therefore, that, although in the evolution of the primates the visual sense overtakes the olfactory sense in overall importance, the latter plays some part in deeply rooted behavior patterns of feeding and mating, and this importance is not altogether lost in humans.

When food has been located and approached, the evolving primate hand plays an important part. Though the hand first evolved as an organ of arboreal locomotion in a small branch niche (Chapter 3, II), we can suppose that at an early stage in primate evolution it came to assume other roles. The evolution of an organ of manipulation has not greatly affected the diet, but it has enabled the animal to carry food to the mouth, rather than having to pluck food with the mouth alone. Therefore, the neck of primates has never lengthened as an adaptation to food-gathering; the prehensile and sensitive hand evolved instead.

When food has been identified by the eyes and tactile senses of the hand it is passed to the lips, where again we find a tactile sense organ of great importance. The sensory and motor areas of the cerebral cortex that in humans relate to the lips (see Fig 3.9) are very large; the mobile lips of the higher primates are a highly innervated organ of investigation. As the food is brought to the mouth it is held close to the nostrils, which can thus obtain reliable olfactory information about it. (This was especially important among nocturnal prosimians in which the olfactory sense and rhinarium are so important.) Innate physiological responses follow olfactory stimulation, such as the secretion of saliva and other digestive substances. Any food that has passed the lips and been subjected to so much investigation is unlikely to be falsely identified.

II. Diet

A review of primate diets shows that meat in one form or another is a basic primitive primate food. Ancestral primates evolved from insectivorelike animals and almost certainly depended on animal food, whether worms, insects, lizards, frogs, or eggs. We have seen that as mammals increase in size and their basal metabolic rate drops, they are able to exploit a higher proportion of vegetable foods. This correlation of body size, metabolic rate, and diet holds very well in the primates, from the tiny insectivorous lemurs to the giant gorilla, which is entirely vegetarian (Schaller, 1963). The slightly smaller chimpanzee, in contrast, enjoys termites and even small mammals (Goodall, 1968). The fully terrestrial baboons also enjoy meat when they catch a small animal, though they are predominately herbivorous. The large savanna baboon, for example, lives mainly on grass shoots.

The important conclusion for our studies is that primates have at all times been able to consume and digest animal food; indeed, they appear to enjoy it, but with their increase in size the higher primates have become well-adapted vegetarians. Humans are an exception, having reverted to a much more omnivorous diet with a considerable component of meat. The human diet varies a great deal; in the western world it is probably more variable than that of any animal species. Traditionally, meat represents a large portion of the diet only in North Temperate peoples who live in an environment short of vegetables during the winter. For that reason, Eskimos are almost completely carnivorous. Tropical humans also eat meat, and some will go to a lot of trouble to obtain it by hunting. However, most agricultural peoples, and even most pastoralists, have a primarily vegetarian diet, though East African Masai are an exception, for they traditionally subsisted mainly on blood and milk obtained from their domestic herds. Even hunter-gatherers consume only about 35% meat to 65% vegetables (Lee and Devore, 1976). It appears that, although meat is always highly valued, humans have a primarily vegetarian diet except in regions where arctic or desert conditions reduce its availability.

It appears, therefore, that there was no fundamental change of diet between our primate ancestors and ourselves. An analysis of animal bones on the living floors at Olduvai in Bed I times (about 1.7 mya) shows that the meat diet of *Homo habilis* at that stage included not only a variety of small animals, but some large mammals as well, even an elephant (M. Leakey, 1971). The change that occurred, therefore, in the diet of evolving hominids was to the consumption of more meat than might be predicted by our body size and to the utilization of larger animals—the plains mammals—which could be caught and butchered only with the technology and social organization associated with hominid hunting. The ability to obtain meat in this way was obviously of great survival value in an environment such as the savanna, where vegetable foods are indeed limited through much of the year and may be greatly reduced by intermittent droughts. It should be noted, however, that the archaeological evidence shows that these early hominids remained dependent to a great extent on smaller insects, reptiles, and birds.

It is important to note that every new evolutionary radiation has involved the exploitation of a new major source of food. Of course, the herbivorous mammals of the wide grass plains had already been exploited by carnivores, such as the lion and leopard, but the natural balance between those two groups appears to be about one carnivore to every hundred herbivores. The effectiveness of social hunting among humans enabled them to tap this huge reserve more effectively; there was enough food for both humans and lions.

The great cats and hyenas hunt either at night or during dusk and dawn; only the African wild dogs hunted by day with the hominids. Diurnal hunting was something new to Africa, as the dogs only entered Africa about the time that the hominids started hunting (ca. 2.0 mya). As Cachel has suggested (1975), it may have been the competition between these two species that triggered the final spurt in hominid evolution.

Meat is a concentrated form of food comparable to seeds, which have been exploited by the rodents and are no doubt a contributory factor to their great success. Meat contains a high percentage of protein and, when digested, will release the whole range of amino acids necessary for the synthesis of body tissues. Essential proteins are already synthesized by the herbivorous mammals and, though broken down in the process of human digestion to their component amino acids, are available in the correct proportions for resynthesis. Meat also contains vitamins (particularly in the liver) that are not readily available in a vegetable diet. There is little doubt that the final stage in human evolution (since the Lower Pleistocene) was correlated with the exploitation of the large terrestrial mammals. It is interesting that the baboon, another very successful plains-living primate, has been observed to kill and eat mammals occasionally in both East and South Africa. Our immensely successful evolutionary radiation must be associated, then, not with a fundamental change in diet, but with an important change in emphasis from a diet that was mainly vegetarian to one that was increasingly omnivorous. The change was attributable not to internal evolution of the alimentary canal and masticatory apparatus but to a change in ecology, in the species' whole environment, and in human behavior. In leaving the forest for the woodland and savanna, our ancestors changed not only their diet but their whole way of life.

III. Taste and the Tongue   The testing and identification of food does not end when it is placed in the mouth. The inside of the mouth is rich in tactile sensory nerve endings, as is the tongue. In addition, the tongue carries numerous papillae containing nerve endings sensitive to taste (the taste buds), which in humans (and presumably other primates) can distinguish the flavors sweet, bitter, acid, and salt. Such particular characteristics of food are not readily detectable by the nose (the sweetness or saltiness of a dish has no scent), for the relevant molecules are not easily vaporized. But the identification of these flavors is important: sugar is a valuable source of food, and salt is an essential mineral. Our marine origin has left us with saline body fluids, the salinity of which must be maintained exactly. Primates have been observed licking rocks containing salt and obtaining it from the salty sweat deposits in each other's fur. Bitterness and acidity are the characteristics of some natural poisons.

The human sense of "taste" is of course for the most part due to the activity of the olfactory organ, the nose. The smell of food in the mouth enters the nose by the *pharynx,* the internal passage from the mouth to the nose. There is no doubt that, although the sense of smell is now less important to us than it was to some of our ancestors, we still have the power of great chemical discrimination (in a "trained palate"); while the total volume of the brain involved with this sense may be relatively reduced, it is still of considerable magnitude among prosimians and the few nocturnal higher primates. In modern humans, the importance of smell has been more strikingly reduced as a determinant of sexual behavior than of feeding behavior.

Since the tongue lies between the teeth, it is able to sample food during mastication. Stimuli interpreted as undesirable can still result in rejection of the food. It is interesting that the nerve fibers from the taste buds lead to the brainstem and are strongly tied to innate reflexes such as salivation or rejection. The significance of taste is primarily innate, that of smell is to a great extent learned.

The tongue has other functions besides bearing the taste buds. In prosimians as well as in many other animals it has a roughened horny surface and aids in grasping the food in the mouth. The tongue initiates the process of swallowing and is important as a means of removing food particles from the teeth and in keeping them clean. In monkeys and apes this function is still important, ranking after tasting and swallowing, though the rough surface has been sacrificed to increased sensory discrimination.

The musculature of the tongue is an important consideration in a study of human evolution because the tongue makes possible the act of speech. This musculature is well developed in most mammals for the various functions described above, and little anatomical change was needed to turn the tongue into an organ of speech. It is anchored to the skeleton at four points, which include the inner surface of the *mandible* or jawbone at the *genial tubercle* (see Fig. 9.7) and the *hyoid bone* (see Fig. 11.6). The tongue also carries its own internal muscles, which make possible its complex changes of shape. It can expand backward into the pharynx and alter its shape, which will have the effect of modifying the vowel sounds produced in vocalization and language. However, the important changes that occurred in the evolution of language capability were not so much in the tongue's anatomy as in the motor control of the tongue by the brain (see Chapter 11, VI).

IV. The Masticatory Apparatus

Mastication is an essential process in the realization of the total value of foods. The enzymes and other substances that effect digestion operate on large masses of food very slowly. It is remarkable that some reptiles (such as snakes) are able to digest entire animals without any form of mastication, but the resulting rate of output of digested food substances is low and the animals undergo a period of postprandial sluggishness, or even coma. The warm-blooded mammal and the bird require a constant and high rate of digestion, and each has developed its own masticatory apparatus; among mammals it evolved from the jaws, and among birds the gizzard evolved from the alimentary canal. Besides breaking down large pieces of food, mastication also destroys plant cellular structure and frees the proteins, fats, and sugars from the insoluble cellulose cell walls that enclose the living cell contents. Thus, mastication releases an immense quantity of nourishment for immediate digestion, a development clearly essential to the herbivorous mammal.

The evolution of the primate dentition is of great interest insofar as the structure of the teeth reflects different dietary adaptations. In addition, of all the parts of the body the teeth have been most successfully preserved as fossils, and these fossil teeth have enabled us to understand something of the evolution of the primates. Jawbones, too, are very strong and have also been preserved in considerable numbers. The evolution of the human masticatory apparatus is probably more completely documented than that of any other part of the body, and deserves a detailed description.

The mammal jaw itself has evolved in accord with changes in its function. As we have seen, the reptile jaw was primarily a food trap, so it needed to be large, quick-operating, and escape-proof. It was not necessary for the jaws to develop powerful compressive forces between the teeth. In the python, for example, all the bones of the forepart of the skull and mandible are loosely connected to allow the swallowing of large food animals without mastication (see Fig. 2.3). The most primitive primate jaw was perhaps like that found today among the tree shrews, where little more than grasping action is required. As crushing evolved, more power was required, so the musculature developed and the jaw changed shape. Before consideration of these important shape changes, however, it is necessary to review briefly the musculature of the primate jaw.

The jaw is operated by paired muscles that move the mandible about its pivots, the *mandibular condyles*. The maxillary bones (or maxillae) that bear the upper teeth are firmly fixed to the skull, and the compressive force used in crushing and grinding is achieved by raising the mandible against the maxilla. The muscles that raise the mandible and close the jaw are the *temporalis* muscles, assisted by the *masseter* and *medial pterygoids* (also called *internal pterygoids*). The mandible is moved forward and sideways by the *lateral* (or *external*) *pterygoids*. Figures 9.2, 9.3, and 9.4 show the arrangement of these muscles. As we have seen, the skull forms a framework around the nasal passages and orbits to transmit these forces of mastication (imposed upon the maxillary bone) to the top and sides of the skull. The supraorbital torus and the vertical forehead are alternative ways of spreading this force; the former may be considered a product of the evolution of the masticatory apparatus.

The important changes that took place in the evolution of the primate jaw altered it from an insect trap with molars capable of some crunching to a powerful apparatus for the mastication of plant food. They occurred slowly during the earlier phases of primate evolution but were complete by the time the higher primates appeared. Two components of this change require consideration:

First, the *occlusal plane* is the line on which the teeth meet when the jaw is closed. In prosimians, with their mainly insectivorous diet, the point of pivot—the mandibular condyle—lies more or less on this plane; in higher primates it has moved up, well above this plane (Fig. 9.5). The result is

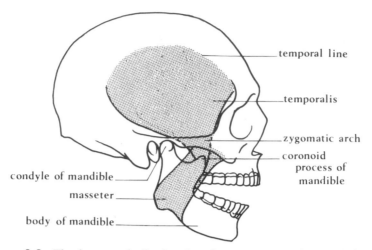

Figure 9.2. The human skull, showing the masseter and temporalis muscles, which raise the mandible to bite. The medial pterygoids (which also raise the mandible) are not shown, since they lie within the skull.

that, while in the lower primates the molar teeth meet first and the incisors later (like a pair of scissors), in the higher forms the whole dentition occludes simultaneously, which means that all the molars are equally effective as crushers and grinders. At the same time, the increased distance between pivot and teeth allowed the lateral movement necessary for grinding. The overall form and musculature of the jaw is modified accordingly. Thus, the jaw becomes an angled bone instead of a more or less straight one, the *ascending ramus* (the part bearing no teeth) being vertical rather than horizontal, as its name implies (Fig. 9.5). The displacement has been achieved in the skull itself by lowering the floor of the nasal chamber and the dental arcade in relation to the braincase and orbits, and in some species the palate is vaulted. The prosimian alignment is shared by the carnivorous mammals; the anthropoid alignment is shared by other herbivorous mammals.

The second primate development relates to the power generated between the molar teeth. This power is proportional to the relationship between the length of the power arm and the load arm of the mandible lever, for a given volume of muscle. The shorter the load arm (that is, the distance between the molar teeth and point of pivot at the condyle projected onto the occlusal plane), the greater the power of compression developed between the molars for grinding food (Fig. 9.6). In the evolution of the higher primates the relative shortening of the load arm has been achieved to a great extent by bringing the dental arcades backward and under the braincase, the resultant added advantage being that the maxillary bone can transmit the forces of compression more directly to the skull vault. Development of a short and powerful jaw was made possible

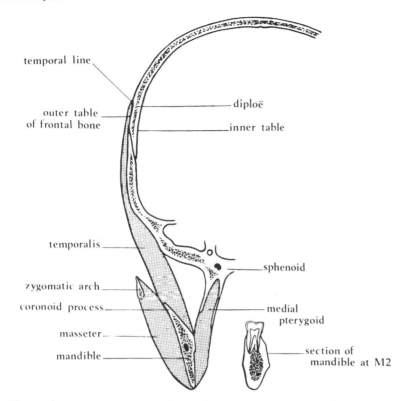

Figure 9.3. A section through the right side of the skull and mandible at the coronoid process showing the arrangements of the three biting muscles. *Bottom right*, a section of the mandible at the second molar showing the bone structure in which the tooth is embedded.

when the need for snapping action by the mouth became of secondary importance to the need for crushing action in the evolution of a herbivorous diet.

These two trends have operated simultaneously in the evolution of the higher primates and continued into the human dentition. Only in the last stages of the process was jaw size reduced relative to body size, and this reduction now deserves consideration.

**V. The
Human Jaw**

The concept of biological efficiency suggests that if an organ is larger than necessary it will be reduced by genetic adjustment in the course of evolution. For humans there was a distinct advantage in having a smaller jaw: a reduction in the size of the masticatory apparatus in an erect animal will help toward achieving a better balance of the head upon the vertebral column. A balance has been almost achieved in mod-

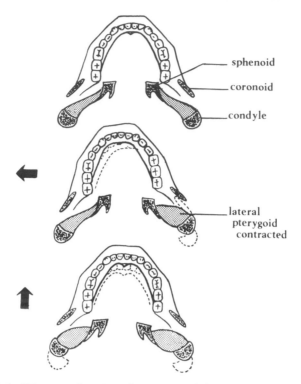

Figure 9.4. Diagram showing the action of the lateral pterygoid muscles, which are attached medially to the sphenoid bone of the skull and laterally to the mandibular condyle on the inner surface. In the middle drawing the mandible is pulled to the left by the action of one muscle only—the right lateral pterygoid. In the lower drawing the mandible is pulled forward by the contraction of both lateral pterygoid muscles (redrawn from Testut, 1928).

ern *Homo sapiens,* and the reduction in the size of the masticatory apparatus (a reduction of subnasal prognathism) is a factor of prime importance in this evolutionary development (see Chapter 5, IV). Experiments on animals have demonstrated conclusively that in ontogeny the final size of the mandible and masticatory muscles depends on the amount of use the masticatory apparatus is put to during growth—an example of developmental homeostasis. On the other hand, the size of the teeth is not affected in this way by environmental influence but is under more or less direct genetic control. The crowns of the teeth are, of course, subject to wear, but the length and breadth of the crown is not modified unless the tooth is damaged.

Two other more detailed features must concern us in a study of the later stages of human evolution: the form of the *mandibular body* and the evolution of the *mental eminence* or *chin.*

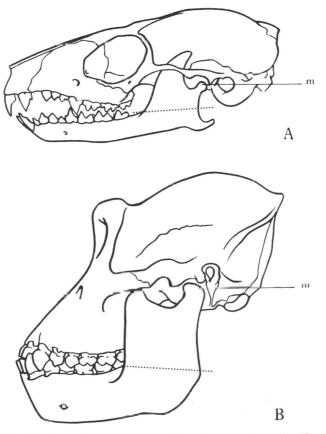

Figure 9.5. Skulls of the tree shrew (A) and a female gorilla (B). Note the relative sizes of the jaw and neurocranium and their relative positions. In particular, note the height of the mandibular condyle (*m*) above the occlusal plane of the teeth (dotted line) (from drawings in Le Gros Clark, 1971).

The mandibular body or *corpus* is the part of the mandible that bears the teeth. Its function, besides bearing teeth, is to transmit to the teeth themselves the forces put upon the mandible by the contraction of the masticatory muscles. Large teeth, such as the gorilla's canines, have large roots as anchors, and it follows that a deep mandibular body is necessary to support large teeth. In humans with much-reduced teeth, especially reduced canines, the body of the mandible may be quite shallow, but the bone structure has to transmit various forces and so must be of a certain cross section beyond that necessary to house the roots of the teeth. The depth of the mandibular body allows it to transmit the vertical forces involved in closing the jaw when the temporalis, masseter, and medial pterygoid muscles contract; the body of the mandible can be considered

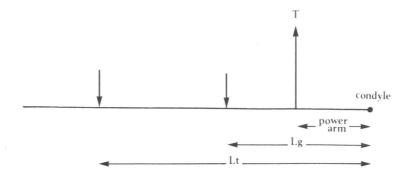

Figure 9.6. Diagram showing the length of the load arm on the mandible lever in relation to the power arm in the tree shrew and gorilla. The horizontal line represents the occlusal plane, and upon it are projected the condyle or pivot of the mandible, as well as the position of the force developed by the temporalis muscle and transmitted through the coronoid process (T). The relative length of the load arm is calculated on the basis of a constant power-arm length and is defined as the distance of the first molar tooth from the condyles. This length is indicated by Lt for the tree shrew and Lg for the gorilla.

as a girder, and its depth is directly related to the vertical bending stress developed at each point along its length. In effect, the teeth make it a form of U-girder with internal webbing (Fig. 9.3).

Forces in the vertical plane are not, however, the only forces acting on the body of the mandible. The lateral motion of the mandible is an important masticatory grinding movement and is brought about by the lateral pterygoid muscles (Fig. 9.4). Lateral movement is caused by the action of each muscle alternately: the right lateral pterygoid pulls the mandible to the left; the left lateral pterygoid pulls it to the right. The power for lateral grinding action, therefore, comes from one side only and is transmitted through the body of the mandible to the molar teeth on the other side. The bending stress on the mandible in the horizontal plane can thus be great, especially when the jaw muscles are strongly contracted and the friction between the molars is at its maximum.

In species that have evolved a powerful lateral grinding action of the molar teeth we find a thickening of the body of the mandible in the horizontal plane. Thickening is most apparent at the point at which the mandible is most curved, at the *symphyseal region* (the midline between the first incisor teeth). In the mandibles of higher primates, this point is strengthened by the development of internal buttressing, which may occur either at the lower margin, when it is called a *simian shelf*, or halfway up the body of the mandible, when it is called a *mandibular torus*. In some forms both kinds of buttressing occur together (Fig. 9.7).

In human evolution both kinds of buttressing have been lost, although their traces can be identified in some mandibles. The internal buttress has

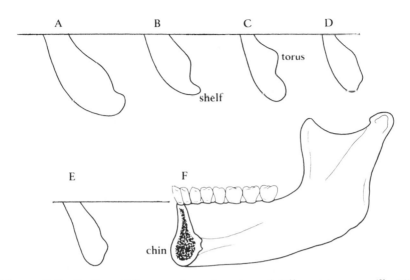

Figure 9.7. Section of the symphyseal region of different jaws: gorilla (*A*), chimpanzee (*B*), *Australopithecus africanus* (*C*), *Homo erectus* (Heidelberg) (*D*), *Homo sapiens* (early form from Krapina) (*E*), and modern human (*F*). Note the different means of horizontal stiffening: the simian shelf, mandibular torus, and chin. On the inner symphyseal surface of (*F*) are to be seen two small protuberances, the genial tubercles, to which are attached the muscles supporting the tongue.

been replaced by an external buttress, the chin. Clearly the stresses set up in the mandible of *Homo sapiens* are not as great as in earlier hominids, but, as can be seen from the wear of human teeth, the grinding action has by no means been lost and some strengthening is still necessary. The reason for this change of structure in the buttressing of the mandible of *Homo* has been revealed by a functional analysis by DuBrul and Sicher (1954).

Both the recession of the dental arcade and the movement of the head backward on the vertebral column (through the forward movement of the occipital condyles to achieve a better balance) have brought the mandible into very close proximity to the neck. If the form of the lower margin of the mandible had not changed, it would certainly have constricted the windpipe, larynx, and soft viscera of the neck, including the vital veins and arteries to the brain, which lie just behind the angle of the body of the mandible. DuBrul and Sicher have shown how the lower margin of the mandible has become everted to avoid this (Fig. 9.8). The human chin is the result of the eversion of the lightly buttressed *symphysis*; it is a necessary correlate of the reduced masticatory apparatus.

It is interesting to study the skull topography of the howler monkey in this context. In order to accommodate the laryngeal sacs in the space under the jaws, the occipital condyles have moved back under the skull; the jaws have come forward and a more "primitive" alignment with the

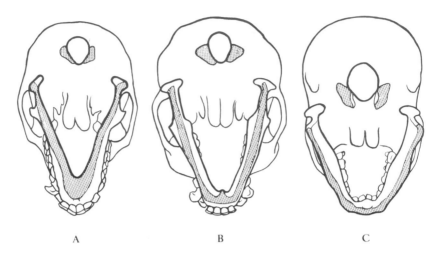

A   B   C

Figure 9.8. Basal views of the skulls of a monkey (*Cercopithecus*) (*A*), the gibbon (*B*), and human (*C*) drawn the same size. Note that the lower margin of the mandible (shaded) remains a more or less constant distance from the occipital condyles (also shaded), which support the head upon the neck. The buttressing of the jaw in humans takes the form of a chin in place of a simian shelf, visible in the monkey (after DuBrul, 1958).

skull has resulted. This development appears to be an alternative means of protecting the neck viscera.

The changes in the primate jaw that led to the human condition can be summarized as follows: (1) evolution of the ascending ramus, at right angles to the mandibular corpus; (2) retraction of the dental arcade under the skull; (3) reduction of the jaws and dentition; and (4) eversion of the lower border of the mandible.

VI. Dentition   A. *The Incisors.* Before we consider the different kinds of teeth it is necessary to review their arrangement in the jawbones (maxilla and mandible). This arrangement is termed the *dental arcade,* since in humans it is seen to take the form of a parabolic arc. In other primates, however, the teeth lie roughly either in two converging rows or as three sides of a rectangle (Fig. 9.9). Fossil evidence suggests that human evolution involved a change in the dental arcade from (presumably) one with straight rows via a more V-shaped arrangement to a curved parabolic arc of teeth (Fig. 9.10). At the same time, as we have seen, the jaws shortened and became smaller in relation to the rest of the head.

Since the different kinds of heterodont mammalian teeth have evolved from the similar teeth of a homodont reptile (see Chapter 2, III) there are no fundamental historical differences among them, and they

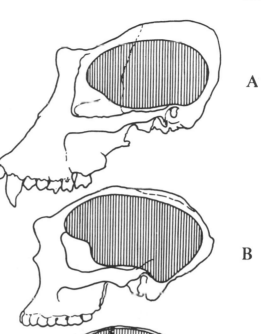

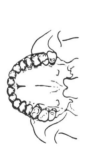

Figure 9.9. Upper dentition (*left*) and complete skulls (*right*) of gorilla (*A*), *Homo erectus* (*B*), and human (*C*). Note size and form of dental arcade and compare the size of the jaws and that of the cranial cavities (shaded) (from Weidenreich, 1939–41).

are in some species indistinguishable in form or function. In most primates, however, the four kinds of teeth remain distinct, though they are often reduced in number. They are the incisors (abbreviated as I), the canines (C), premolars (P), and molars (M). The teeth are conventionally numbered from the front backward, so the primitive hypothetical mammal that is shown in Fig. 9.11 had four kinds of teeth as follows: I1,I2,I3,C,P1,P2,P3,P4,M1,M2,M3, totaling twenty-two in each jaw.

The incisors, the most anterior of the whole dentition, are generally cutting teeth, as their name implies (though they grow into the tusks of

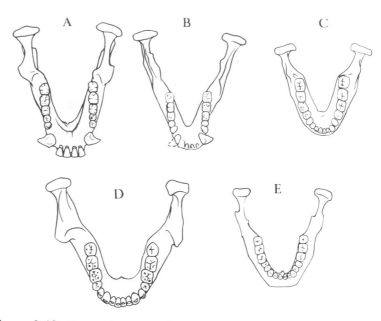

Figure 9.10. Lower dentition of various primates: chimpanzee (*A*), *Proconsul* (*B*), *Australopithecus africanus* (*C*), *Homo erectus* (*D*), human (*E*). Note the different forms of the dental arcade, from U-shaped to parabolic, and the size of the canine teeth. The early *Australopithecus afarensis* has an almost V-shaped jaw. (Not drawn to scale.)

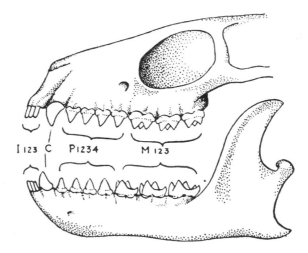

Figure 9.11. Diagram of the dentition of a hypothetical early placental mammal showing the arrangement and grouping of teeth from which it is believed that the various primate dentitions evolved (from Le Gros Clark, 1971).

elephants). Exceptions do occur among the primates, and in the lemurs we find a total change in function: the incisors of the upper jaw are very small pegs, and those of the mandible are procumbent and modified into pointed pegs to form a "comb." In some species these incisor combs are used for the collection of gums from trees, on which the animals feed. They are sometimes used in grooming. In the higher primates, however, the incisors are spatulate (chisel-shaped) and serve to cut food. They are used for biting into fruits, nuts, and shoots, for chopping leaves and stripping bark. Through their cutting function they have enabled many groups of mammals as well as the higher primates to utilize food that is too big to chew and swallow whole, and thus have made available a vastly increased food supply. As we know from our own experience with apples, broad upper incisors are particularly well adapted for fruit eating.

Among higher primates there are only four incisors in each jaw, two on each side (I1, I2). In most primates they tend to lie in a nearly straight line (Figs. 9.9A, and 9.10A) and form an efficient cutting edge. The incisors of the African apes are the largest and broadest and tend to make the dental arcade rectangular, though the lateral upper incisors do not share this enlargment.

In human evolution no further alteration has occurred in the function of the incisor teeth, so their form has remained more or less unchanged. Their size has decreased in accordance with the reduction in size of the whole masticatory apparatus. They lie in a curve instead of in a straight line (Fig. 9.9C, D) and lose their procumbency.

We have incisors of *Australopithecus* showing no very significant differences from the human incisors (Fig. 9.14). The incisors themselves tell us less about hominid evolution than do the other teeth, which we shall now consider.

B. *The Canines.* Throughout mammalian evolution as a whole the canine has tended to retain the pointed and rounded form of reptile teeth. Its function has been primarily that of grasping food, and its importance in this respect has been greatest among carnivores. It receives its name from its well-known shape seen in the dog (*Canis*). Among herbivores it has often assumed a different function: that of a weapon. The tusks of wild pigs are obvious examples, and this adaptation is also seen among the primates, e.g., the upper canines of lemurs and the canines of both jaws of the larger monkeys and the great apes (Fig. 9.12). In the absence of claws, the importance of large canines is probably enhanced. Since the animals are primarily herbivorous, the function of the canines in feeding is only one factor that may be responsible for this unusual enlargement.

Among the great apes, and especially in the gorilla, the canines are larger in the male animal than in the female, as is the supporting bone structure and skull. This is an instance of *sexual dimorphism*—of the form of a characteristic varying between the sexes. The size of the teeth is

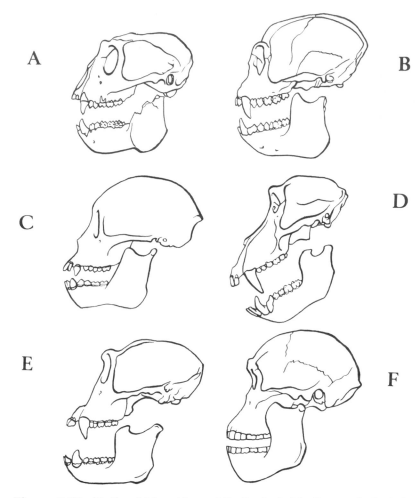

Figure 9.12. Skulls of *Mesopithecus* (*A*), *Presbytis* (*B*), *Proconsul* (*C*), baboon (*D*), chimpanzee (*E*), and *Australopithecus africanus* (*F*) (redrawn from Le Gros Clark, 1971). (Not drawn to scale.)

allometrically related to the body size, but is probably also adaptive in threat and combat. The evolution of sexual dimorphism is of social significance, since it affects the status of different individuals within the social group; it will be discussed in Chapter 10.

Primates with large canines generally have a *diastema*, a gap in the opposing tooth row into which the canine fits so that the jaws may be closed. The lower canines always fall into a diastema in front of the upper canines, and the upper canines fall behind the lower (Fig. 9.13). Le Gros Clark (1971) points out that the diastema is absent until the

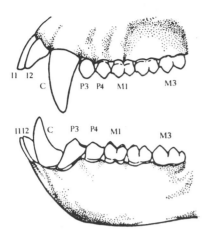

Figure 9.13. Dentition of the Old World monkey *Macaca*. Note the canine teeth (*C*), diastemata, sectorial lower premolar (*P3*), and bilophodont molars (*M1–3*) (from Le Gros Clark, 1971).

canines have fully erupted. It appears that the diastema is merely the result of active tooth occlusion (the way the teeth of the two jaws in fact interlock) rather than a genetically predetermined characteristic.

The *hypertrophy* (enlargement) of the canines has evolved as a striking feature of the living great apes. In human evolution, the opposite trend has occurred. The exact form of the canine varies among individuals (as the reader may easily verify) from a spatulate tooth, indistinguishable from an incisor, to a rather pointed tooth. It is significant that the root of the modern canine is longer than that of the neighboring teeth, suggesting that it may have been reduced from a larger tooth in the course of human evolution. The fossil evidence suggests that this was indeed the case.

Australopithecus afarensis has a diastema, and its canine teeth do project slightly beyond the surrounding teeth, especially the upper canines (Fig. 9.14). They show the ape kind of wear due to their contact with the lower canine and premolar as well as the human type of apical wear. Canines of *A. africanus* are a little smaller than these, are associated with no diastemata, and show the human wear pattern exclusively (Fig. 9.12). The modern human canine differs little from this condition. However, in the robust species of *Australopithecus* canine reduction has proceeded further than it has in the human line. In the very robust forms the canine teeth are relatively indistinguishable from the second incisors and have become very small. This trend is a complete reversal of that seen in the great apes, where canine size appears to be positively related to body size.

We must not assume that the human canine was reduced from a tooth as large as that seen in the living African apes. The Miocene *Proconsul* species from Africa varied a great deal in size, and their canines were not as large as in living apes. Toward the end of the Miocene we have the smaller apes of the *Sivapithecus* group, which also

Chimpanzee   Hadar   Human

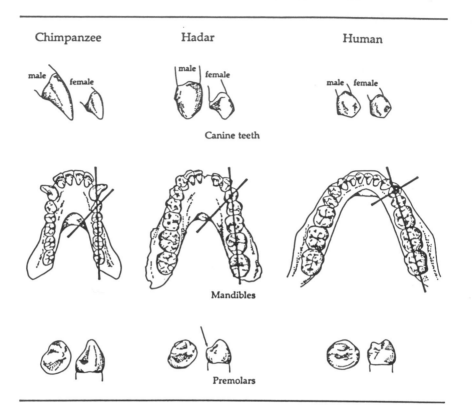

Canine teeth

Mandibles

Premolars

Figure 9.14. Comparison of the lower canine and premolars of *Australopithecus afarensis* with those of the chimpanzee and human. *Top,* male and female canines of each species. Note the reduction in both pointedness and sexual dimorphism in the human teeth. The middle line shows not only the shape of the jaws but the alignment of the lower P3. *Bottom,* the form of the premolar lingual cusp. In all these features the Hadar teeth are intermediate (from Johanson and Edey, 1981).

had smaller canines. Until we can find protohominid fossils we cannot know for sure the extent of canine reduction in the human lineage.

Finally, the human canine can still be seen to play some part in tearing food, especially meat, as in some individuals it is still a strong and slightly projecting tooth. It is likely, however, that in view of its small size it no longer plays any part, if it ever did, in fighting or display. Humans *in extremis* (e.g., with hydrophobia) will still fight by biting, but the value of the tooth as a weapon is today very limited.

C. *The Premolars.* The premolar teeth have served a variety of functions in mammalian evolution. In general they have slowly evolved from their peglike original condition as grasping teeth to flatter teeth for crushing and even for grinding; that is, they have tended to become increasingly molarlike in function, for the molars have also evolved mainly as crushing and grinding teeth. In the primates, however, the premolars have retained their peglike form to some extent, especially in some prosimians. Their evolution toward the molar condition has been termed "molarization," but the teeth become fully "molarized" only in the lemurs. In most species of higher primates the premolars and molars are still clearly distinct.

From an original four premolars in early mammals (see Fig. 9.11) we find only two surviving in catarrhine monkeys and hominoids (the third and fourth of the primitive series, known appropriately as P3 and P4— see Fig. 9.13). In Old World monkeys the premolars are typically *bicuspid* (with two *cusps* or points), which makes possible crushing as well as grasping, for the cusps interlock when the jaw is closed. One premolar has a special function, however, in this group. The anterior lower premolar is somewhat enlarged, has a single cusp, and shears against the upper canine when the jaw is closed so as to make an effective cutting blade. That is why this modified premolar (lower P3) is termed *sectorial* (Fig. 9.13). The same characteristic is also typical of the apes; the lower P3 is functionally correlated with the large upper canine.

This sectorial lower premolar is not characteristic of the Hominidae, which lack a large upper canine. In the early hominid *A. afarensis* we find a anterior lower premolar that is intermediate between the ape and the later human condition. It is basically sectorial and wears against the upper canine, as well as apically, as in humans. The smaller specimens are single cusped and in this characteristic the teeth are apelike; but the larger specimens carry a small second inner (lingual) cusp. The anterior lower premolars of the other species of *Australopithecus* and later hominids are all bicuspid, with the cusps more or less equal in size. Wear is always apical. In modern human lower P3s the *lingual* (inner) cusp is slightly reduced compared with the outer *labial* cusp (see Fig. 9.14).

In the past the shape (i.e., length and breadth) of the premolar crown has been considered an important characteristic in the classification of early Hominidae. However, in both the known species of *Australopithecus* and modern humans the shape is variable, and canine and molar teeth are of greater interest in this connection.

With the possible exception of the fossil ape *Gigantopithecus*, the Pleistocene hominid *Australopithecus boisei* has taken molarization of the premolars to its most advanced condition in the higher primates. As a result of the species' almost total dependence on the grinding action of the jaw, the premolars have become large and flattened like the molars. The absence of protruding canines makes possible a full rotary movement, which is most effective for the mastication of tough plant material (Fig. 9.15).

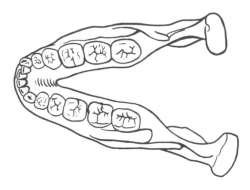

Figure 9.15. The heavily built *Australopithecus boisei* mandible. Note the heavily built jaw, the small incisors and canines, and the immense premolars and molars.

D. *The Molars.*   The evolution of the mamalian molar tooth has been the subject of a great deal of study and has been well summarized elsewhere (Le Gros Clark, 1971). For our purposes, we need only note that a more or less quadrilateral four-cusped tooth is found in both jaws of all higher primates, though a fifth cusp is present in the lower molars of the apes and some hominids. Apart from the exceptional New World marmosets there are always three molars in each jaw, and their general form is surprisingly constant.

In the Old World monkeys each of the molar teeth is marked by a "valley" or constriction between the anterior and posterior pairs of cusps (illustrated in Fig. 9.13). This type of molar shows what is called the *bilophodont* condition, and produces a deeply interlocking type of occlusion (Kay, 1977). While the teeth are perhaps most effective in crushing vegetation, the jaws do allow a lateral grinding motion, and the teeth move from side to side in relation to each other (Fig. 9.16). Full rotary grinding is inhibited by the canines, which allow some lateral movement but no movement in the backward and forward plane. It is significant that in this feature these herbivorous primates differ from the herbivorous ungulates and rodents; in the latter, *rotary* grinding is the function of the molar teeth. The plants of the open plains are much tougher than the tender fruit and shoots of the tropical forest; crushing and lateral grinding alone will serve to release the nutriment only from the succulent forest foods, and in the case of the savanna baboons, the succulent stems of grass.

In the apes and hominids the bilophodont condition is not found. Instead, the cusp pattern is more complex, and a fifth cusp has been retained in the lower molars called the *hypoconulid*. Ape molars are perhaps slightly more effective grinders, for in occlusion the teeth do not interlock so deeply, though rotary movement of the jaw is still to some extent restricted by the huge interlocking canines. The dentition seems to be adapted mainly for tearing and crushing.

Among the Hominidae there are two functional trends in the evolution of molar teeth:

Figure 9.16. Occlusion in an Old World monkey, *Macaca*. The first two maxillary molars are drawn with a heavy line, the mandibular molars with a thin line. The first molar is shown on the left. The central drawing shows the resting position; in the other two drawings the mandible is moved from side to side. The cusps are paired and alternate and only side-to-side movement is possible (from Mills, 1963).

1. The evolution in the robust species of *Australopithecus* is of relatively flat and very large grinding molars and premolars, accompanied by rotary action of the jaw and apical wear of the incisors and canines. As stated, this was made possible by a reduction in the size of canines to the level of the other teeth and was accompanied by molarization of the premolars. This trend appears most clearly in *A. boisei* and is believed to be an adaptation to a tough vegetable diet, such as would be found in a plains-living animal that had not widely exploited animal food (Fig. 9.15). In all species of *Australopithecus* M2 and M3 are larger than M1.

2. The reduction of the molar series as a whole and reduction of the third molar (the wisdom tooth, M3) in both jaws, in particular, is a trend seen in the later stages of human evolution. It is probably associated with the preparation of food by cooking, which lessens the need for powerful and prolonged mastication, and results in a smaller jaw. The large jaw and dentition were retained by the Neandertal people long after reduction among other groups had begun (Trinkhaus, 1983). In modern humans, the upper third molar is always considerably smaller than the other two and both third molars are sometimes absent, especially among Mongolians. The lower molars bear either four or five cusps (Fig. 9.10).

A characteristic of some interest in studies of human evolution is the order of tooth eruption. Schultz (1935) has demonstrated the order in the permanent dentition of the apes and *Homo* as shown in Table 9.1. The absolute and relative ages of eruption are shown in Fig. 9.17. In apes, all three permanent molars appear before the first deciduous incisor is shed: grinding teeth are needed for the mastication of food immediately after nursing ends. In humans molars are delayed in making their appearance and all deciduous teeth are replaced long before the molar series is completed. These data together demonstrate the much slower development of the human dentition, as a whole (which is correlated with the slower growth rate of the whole body) and the slower

TABLE 9.1. Eruption Pattern of Permanent Dentition

	Order of eruption							
	1	2	3	4	5	6	7	8
Great apes	M1	I1	I2	M2	P	P	C	M3
Homo (and rarely among apes)	M1	I1	I2	P	C	P	M2	M3

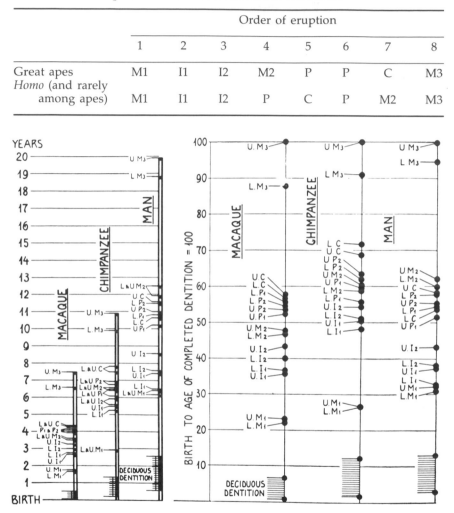

Figure 9.17. *Left*, absolute ages, and *Right*, relative ages of eruption of the teeth of the macaque, the chimpanzee, and modern man (from Schultz, 1935).

development of the molars, in particular (which is related to the reduction in importance of the molar series in humans). Unfortunately, information about the tooth eruption sequence of *Australopithecus* is limited, but what we have does appear to be within the patterns found in humans, which are distinct from those of the apes.

VII. Digestion

After mastication, the breakdown of foodstuffs by digestion and their assimilation into the body does not vary greatly among mammals. Studies in comparative anatomy (Straus, 1963; Martin *et al.*, 1984) suggest that no profound changes have occurred in the evolution of the human alimentary canal, but since there is no fossil record of these soft parts this cannot be verified by direct evidence.

There is, however, one variable in mammalian digestion that is of overriding importance: the extent to which specializations have evolved to make possible the digestion of cellulose. (Cellulose is a polysaccharide; though a multiple of a sugar molecule, it is normally indigestible because it is insoluble.) A specialization for cellulose digestion is typical of fully adapted herbivores, taking the form of an increase in the total volume of the alimentary canal to allow the development of a large population of microorganisms that are able to break down cellulose to synthesize protein and produce sugars. The microorganisms themselves and the sugars are in turn digested by the mammal. This development is typical of ruminants (such as cattle and sheep), which have multiple stomachs, and the herbivorous lagomorphs (such as the rabbit), which have a large *cecum*, a special side branch of the alimentary canal in which cellulose digestion takes place.

It might be expected that some extension of the alimentary canal would be apparent among the more herbivorous primates, and this is indeed the case. There is, for example, one unusual specialization in the mouth of certain primates in the evolution of cheek pouches. This characteristic, confined to the *Cercopithecus* group of Old World monkeys, allows food to be retained in the mouth for some time before swallowing, but while it presumably aids the digestion of starch by the salivary enzyme ptyalin, there is no reason to suppose that it is an adaptation for cellulose digestion. It is, however, an adaptive characteristic, in view of the high starch content of the diet of these arboreal forms and permits rapid harvesting of foliage.

The primate stomach is not greatly different from the human stomach, but in one subfamily of Old World monkeys the area of stomach lining is increased by subdivision into different compartments. This subfamily is the fully arboreal Old World Colobinae, whose diet is confined almost entirely to leaves and shoots with a high cellulose content. They may be considered folivores and are perhaps the only true vegetarians among higher primates. The multiple stomach and the presence of cellulose-digesting bacteria appears to be an instance of parallel evolution with the ruminants, and it is perhaps surprising that it is not more widespread in the order. It may also be noted that the monkeys appear to have been able to overcome to some extent the problem of toxic substances in foliage (sometimes described as the chemical defense of trees) more effectively than the apes (Janzen, 1978), who appear to be a more frugivorous group.

An alternative adaptation to that of the colobines is, however, found in the lemurs, where the cecum is often elongated and dilated—perhaps another case of parallel evolution, this time with the lagomorphs. In the Anthropoidea, however, the cecum is as small as in humans and appears to have no digestive function. There is one characteristic that separates the Hominoidea from the monkeys: the former have a vermiform appendix attached to the cecum. In the apes this tubular extension contains some lymphoid tissue, and in humans somewhat more (Straus, 1936). While its function is not fully understood, it does not appear to be vestigial, as has often been claimed, but rather is a hominoid specialization that has reached its greatest elaboration in humans. We can, however, have no certain knowledge of its evolutionary history.

The English 19th century wit Sidney Smith said that the secret of life lay in good digestion. While this vital process is under the automatic control of the lower centers of the brain, it also affects and is affected by the cerebral cortex. Bad digestion can poison our waking and sleeping hours, while our daily problems can equally upset our digestive processes. A digestion that can easily handle a wide range of foodstuffs has been of immense benefit to the species; it makes possible the wide human geographical range—its successful adaptation to so many different biomes and its great population growth. The variety of succulent food we can enjoy appears to be a direct consequence of our ancestry, yet it may carry with it some dangers, for there is no doubt that the early arboreal primates had to learn what fruit and leaves could not be eaten. Fruits poisonous to primates, which evolved with the fruit-eating birds (which thereby bring about the distribution of seeds), can be a disaster for any primate with too much initiative or too little ability to learn. Choice of forest foods is an important factor in primate adaptation: many fruits and leaves contain high levels of toxic substances. Humans, too, must have a knowledge of wild food qualities as part of their adaptation.

VIII. Ecology, Diet, and Behavior

The omnivorous nature of primates, and of their dentition and digestive processes in particular, made it possible for human ancestors to evolve from forest-living to savanna-living creatures. The flexibility in diet and behavior so typical of primates allowed an adaptation in food-finding behavior and food choice that made it possible for the early hominids to undergo a fundamental change in their environment.

On the basis of present conditions in East and South Africa, we can attempt retrospectively to discover the conditions in the late Pliocene. New geological evidence and faunal analyses from Olduvai suggest that rainfall may have been considerably higher in the Late Pliocene than it is at present. Furthermore, the living floors bordered freshwater streams and a saline lake, which are no longer present. The same conditions may apply to the important early sites at Koobi Fora.

Extensive excavations by the Leakeys at Olduvai show that *Homo habilis* consumed fish, reptiles, birds, and mammals, most of which were small (Fig. 9.18). Some sites carry remains of much larger game, and these are best considered butchery sites (Fig. 9.19). There is no doubt that *Homo habilis* had the capability for cutting up large mammals, even elephants, and the evidence of the living floor suggests that large quantities of meat were processed and shared. We can assume that as in all living hominids, roots, fruits, and other vegetable food formed an important part of the diet, though no evidence of such food is preserved. This sort of mixed diet is nourishing, yet hard to obtain. Like modern baboons, *Homo habilis* must have spent a large part of each day in the search for food. It is important in this context to note that when chimpanzees and baboons eat meat, they always kill it fresh; they never scavenge it. However, at the same time, it seems unlikely that meat which could be scavenged would be overlooked and indeed some of the bone fragments from butchery sites at Olduvai show cut marks made by flint flake tools superimposed on the tooth marks of other predators (Fig.9.20).

An important study of some of the social carnivores of Africa made by Schaller and Lowther (1969), sheds light on the possible adaptation of *Homo habilis*. One of their principle conclusions is that all the social carnivores are both hunters and scavengers according to the availability of meat. Meat is obtained in four different ways: (1) scavenging dead, old, or diseased animals; (2) driving predators off a kill; (3) catching newborn young and other small animals; (4) hunting healthy adults as a cooperative team. This, together with archaeological evidence, suggests that *Homo habilis* probably employed methods (1 through 3), but that method (4) only came into use with *Homo erectus*, a little over 1 mya. Under method (4) we find social carnivores hunting on a broad front (dogs and hyenas) and by ambush and stalk (wolves and lions). Early hominids added to this repertoire another method—the persistent chase, which is commonly seen among living hunter-gatherers. In this strategy an animal will be chased for many hours or even days and eventually be killed by spears. Persistence in the accomplishment of a task is a special human characteristic.

Gathering is an extremely important behavior pattern unique to hominids. It implies collecting small food objects (such as fruit, nuts, roots, eggs, or small animals) in a container—more than would be required for the individual concerned—carrying the food back to a larger group at a recognized meeting place and sharing it with them. Simple tools would also be required, such as, digging sticks and perhaps some kind of rake. We know virtually nothing of the development of food gathering, although the development of this altruistic behavior and the use of containers such as trays, bags, and baskets must have been very important steps in our behavioral evolution.

Homo habilis probably took gathering and food-collecting behavior to

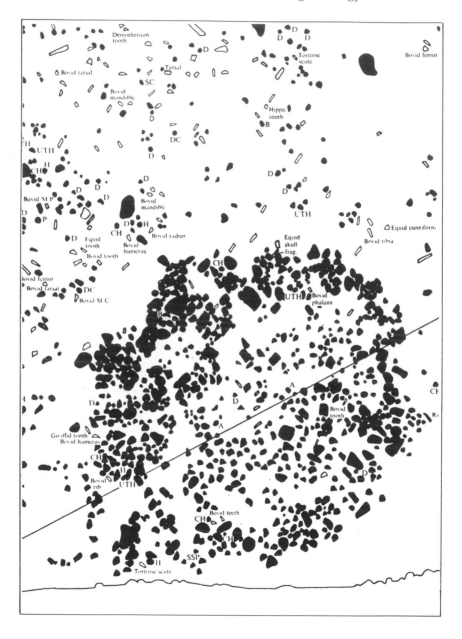

Figure 9.18. Plan of part of the excavated living floor—an ancient land surface—at Olduvai Gorge (Bed I), showing stone tools (solid black), food remains (bone fragments), and a circle of stones that suggests the foundations of some kind of shelter. The age is about 1.8 mya (from M. D. Leakey, 1971).

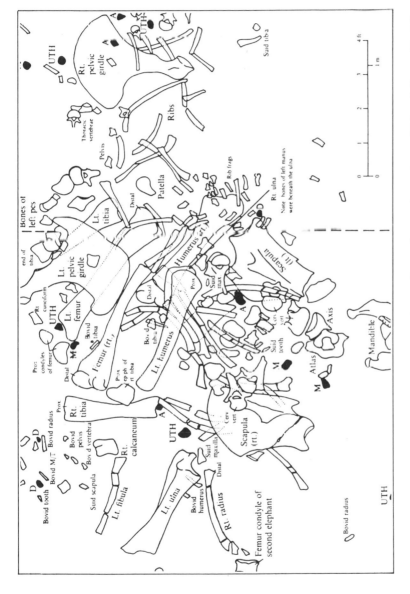

Figure 9.19. Plan of part of the excavated living floor of a butchery site at Olduvai Gorge (Bed I), showing stone tools (solid black) and almost the entire skeleton of an elephant, together with other food remains (from M. D. Leakey, 1971).

273

Figure 9.20. Reconstruction of *Australopithecus* living in the grasslands of the Transvaal in late Pliocene times, obtaining meat by scavenging and perhaps by hunting small or young animals. [By permission of the trustees of the British Museum (Natural History).]

an advanced level. Whether these hominids really developed the art of cooperative hunting we cannot say, but there is no evidence for the *regular* culling of large mammals at this time. A more probable hypothesis is that they scavenged widely and hunted and killed small animals, but probably only killed big game if they encountered a diseased animal or the wounded kill of some other predator.

Only in the Middle Pleistocene do we find clear evidence of a second major development in the human adaptation to living outside the forests, namely the adoption of *cooperative hunting*—a development in food-getting behavior of central importance to the story of human evolution. It seems that humans survived the climatic changes of the Middle Pleistocene by a new behavioral adaptation that affected fundamentally and irrevocably his psychosocial nature.

For mammals, food finding and feeding is normally not a social occupation, except among a few carnivores (such as lions, dogs, and killer whales). Most primates live in social groups that are not based on the need for social hunting and feeding. Social hunting, which requres active cooperation, is typical only of carnivorous mammals that feed on species larger than themselves. Wolves hunt in packs, particularly in winter when small animals are scarce and larger species are attacked.

It was in this way that early humans must have evolved cooperative hunting behavior. It seems clear that the prizes available as food make the evolution of cooperative hunting highly adaptive. If a group of individuals could cooperate to corner and kill an antelope, they would have enough meat for more than one individual and for more than one day. Cooperation was essential, and food sharing was a natural outcome of such cooperation, though it was probably first developed at a much earlier date by *Homo habilis* (Isaac, 1978). The sharing of a small kill serves little purpose; the sharing of a large kill is necessary and highly adaptive. It is clear that food gathering, cooperative hunting, and food sharing changed the social and physical attributes of early humans in a number of very important ways:

1. Food gathering required foresight and altruism as motives. It required containers and possibly digging sticks—the first tools.

2. Nursing and pregnant mothers would not have the endurance to carry their young long distances, so some females would have been left together, possibly guarded by a few males, thus involving the establishment of a home base and the division of labor.

3. Success in hunting required cooperation among hunters and concern for troop members left at home.

4. Traveling long distances and carrying heavy burdens of tools, weapons, and prey put strong selective pressure on the evolution of efficient bipedalism.

5. The exertions of the chase may have been responsible for the diminution of subcutaneous fat deposits, loss of body hair, and the considerable development of sweat glands, all of which aided the rapid diffusion of metabolic heat (Chapter 8, IX).

6. Catching and killing large animals required ingenuity and technological skill. The earliest weapons were probably stones and clubs (sometimes used by living apes). In the Middle Pleistocene, round stones are found that may have been used for throwing, possibly as a *bolas*, an ingenious and deadly device consisting of three stones strung together with leather thongs. On impact the thongs will wrap around the animal and probably bring it to the ground. From Torralba in Spain and Clacton in England we have the earliest preserved evidence of a wooden spear: an implement that was surely essential to the early hunters. Much later we have the introduction of the spear-thrower, or atl-atl, and finally the bow and arrow, perhaps 15,000 years ago.

7. The need to cut up large mammals must have stimulated the

development of better chopping and cutting tools; hominid teeth were clearly unsuitable for this function. Food preparation was an entirely new adaptation for a primate; evidence of it is clearly recorded in archaeological deposits where stone tools and bone fragments are abundant (e.g., Olduvai Bed I).

8. The meat would be in sufficient quantity to be shared by all hunters and carried back to base. Bipedalism makes hand carrying possible and frees the mouth for speech. In the absence of a suitable alternative diet during dry or cold seasons, bringing meat back to base would be strongly selected as a behavior pattern. Clear evidence for this can be found in the Middle Pleistocene, especially at Choukoutien.

9. Food sharing within a social group would certainly bring the group together in a very intimate way, a situation that probably encouraged language, especially for planning the hunt and relating accounts of it. As Roe (1963) has said, language was perhaps born of the need of intercourse as a result of the division of labor. For the first time, primates had something complex to communicate (see Chapter 11, VI).

10. Geographical knowledge of a wide area must have been important; the locality of useful water holes and herds of game must have become a vital subject for communication. Improved powers of perception, memory, and prediction were of immense value and probably evolved conjointly with speech.

11. The absence of individuals from the home base on hunting expeditions must have increased the overall division of labor between the sexes. The females with children would have taken over all other group activities at the home base.

It would be possible to list many other important changes that might have followed the evolution of social hunting, but it seems clear that the exploitation of the larger herbivorous mammals as a source of meat required cooperation, endurance, intelligence, foresight, and a precise means of communication. It was associated with two developments of cental importance, food sharing and the division of labor (Isaac, 1978). At the same time, it initiated the selection of an evolutionary trend toward further socialization together with the development of speech and technology, all of central importance in the story of human evolution.

We shall return to these socializing factors again later. Here it is necessary only to emphasize two things: first, that primate feeding habits preadapted bipedal hominids to a terrestrial existence (as they preadapted baboons); second, that the exploitation of plains mammals was made possible by, and also immensely stimulated, the evolution of hominid society. Thus, the final determinant of the direction of human evolution was environmental and ecological—not obscure, but striking and fundamental. Like all great evolutionary radiations of animals, the appearance of humankind on earth involved the exploitation of a new food source, and a very rich one.

**IX. Tools
and Resources**

Humans are characterized by being toolmakers. Tools must be clearly distinguished from artifacts, which are common in the animal kingdom— birdnests, beehives, and beaver lodges are among the finest examples. They are made by some creatures quite low in the evolutionary scale (such as the caddis fly larva that makes cases and nets). Tools are distinct from artifacts because they can be used to make other objects or to facilitate activities such as resource extraction. They are not by themselves of immediate and direct use.

Tool use has been identified among insects. Most striking is the use of a small pebble by the solitary burrowing wasp *Ammophila* to firm up the sealed entrance to its burrow. When the wasp has filled the burrow with eggs and a food supply of dead caterpillars, it seals the burrow and covers it with sand grains hammered firmly into place. In the end, all trace of the nest is obliterated.

The Egyptian vulture is also a tool user. It picks up stones in its beak, as large as it can hold, and drops them repeatedly on ostrich eggs until the shell breaks. The Austrialian black-breasted buzzard will drop stones on emu eggs and the English thrush will drop stones on snails until their shells are broken.

A somewhat different example is the satin bowerbird, which uses fibrous material as a brush to color sticks in its bower with paint made from charcoal or berries.

Among mammals there is the famous example of the sea otter (*Enhydra*), which collects from the seabed a stone and a shellfish, one in each forelimb, and then floats on its back with the stone on its chest opening the fish (such as abalone) by smashing the shell on the stone.

These examples of tool use imply that certain qualities belong to the species that uses them: the existence of manipulative organs (a beak and forelimbs in our examples), a good visual sense, and a behavioral flexibility that permits discovery and invention—although at a very simple level. The activity is in most cases a solution to a problem of obtaining resources, but in the case of the wasp and the bowerbird it presumably improves reproductive success. In the instance of the otter, important new food resources are made available as a result of tool use.

It is important to stress that motive must be present as well as potential. Tools allow these animals to tap food resources not otherwise available to them, and tools give them great advantage in that they may then have a monopoly or near monopoly of that particular resource. This freedom from competition is very desirable and unusual. Since the phenomenon of tool use is not confined to mammals, it apparently does not require a highly evolved central nervous system.

It is now well known that the *use and modification of tools* is also not confined to humans. Examples include the Galapagos finch *Cactospiza*, which uses a cactus (prickly pear) spine to poke out insects embedded in the branches or trunks of trees. Where there are no cacti, *Cactospiza* breaks off a short stiff twig of appropriate length from a tree. A second example is the British Greater Spotted Woodpecker (*Dendrocopos*), which pecks out a V-shaped cleft in a tree trunk and uses it to anchor pine cones, oak apples, etc., while it extracts seeds or insects. The African grey parrot will pry splinters of wood from pine posts (in a cage) and, holding them alternately in its two feet, will scratch the two sides and back of its head.

This second stage of technology, therefore, involves the natural object *specially modified* for tool use. The classic example is found among chimpanzees, which modify grass stems or twigs for termite fishing. The twigs are collected, any leaves are removed, and the length is adjusted to a standard. They are then poked into termite mounds. After a few seconds they are removed, and the termites clinging to them are eaten.

Less well known is the chimpanzee's preparation of leaves by chewing to make them act like a sponge, absorbing water for the animal to drink from awkward places such as a hole in a tree trunk. Modification of tools clearly implies not only discovery and imagination, but foresight. Thus, the preparation of a twig by *Cactospiza* is a remarkable feat. One might have expected that a highly evolved central nervous system was essential for this development. It is certainly very rare among animals. Among chimpanzees it is transmitted by learning, which is highly significant since it allows rapid modification of a technique and thus rapid development of it. Whether it has become a programmed behavior among finches as distinct from a learned one, which is how it must have begun, is not known.

Tool use and tool modification can be distinguished from toolmaking as *prototechnology*. In Table 9.2 these examples are shown as category 1a and 1b. Category 1a does not easily lead to 1b and does not logically imply such development. Category 1b, however, has immense potential, and while in the chimpanzee it has not led to further advances, there seems every reason to believe that it can give its practitioner the potential to develop true technology (category 2).

Humans, of course, are thought to be alone in the animal kingdom in the *making of tools*. The earliest evidence we have of humans making tools is, of course, lithic, which should not lead us to suppose that stone tools were the first human tools. Modern tribal peoples have only a small percentage of material culture that will be preserved in an archaeological site. The vast majority of the technology is biodegradable: besides stone, simple tools are made of bone, antler, horn, teeth, leather, wood, bark, and leaves and stems of plants. Biodegradable tools would have included digging sticks, trays, containers, shelters, thongs, threads, rakes, shawls, and many other simple objects.

The difference between categories 1b and 2a may seem slight at first, but in practice there is a world of difference between modifying a twig and removing its leaves (for the twig existed in the first place and could be clearly seen) and making a spherical pebble into a knifelike object for chopping or cutting. This surely requires far greater imagination. However, stone choppers and stone flakes do occur naturally and can be

TABLE 9.2. Categories of Technology

1. Prototechnology	a. Tool use
	b. Tool modification
2. Technology	a. Tool manufacture
	b. Stone technology (and secondary tools)
3. Pyrotechnology	a. Fire use
	b. Fire control
	c. Fire making
	d. Metal industries (smelting, casting, forging)
4. Facilities	a. Containers, cords, etc.
	b. Energy control
5. Machines 6. Instruments 7. Computers	See Chapter 12

produced by dropping one stone on another by chance. Sharp stone flakes were in our environment a billion years before hominids found a use for them. Once they did—where the motive was present and the brain sufficiently developed—the step from the use of existing objects to tool manufacture might have been rapid. Whatever other biodegradable technology such as digging sticks, rakes, and levers they may have had, and however much extra food in the form of roots, fruit, and insects they may have collected, the use of stone cutters and choppers would have opened up a vast new food resource (large mammals) not free of competition but otherwise unavailable to a hominid lacking a carnivorous dentition or even powerful canine teeth.

The manufacture of stone tools depended essentially on motive and manipulative skill, neither of which at first need have been highly developed. It led, however, to a most important development: because stones were so hard they could be used to cut softer (yet quite hard) materials such as wood and bone which, in turn, could be used as tools.

Thus, simple stone tools such as flakes and choppers could become, almost immediately, secondary tools (2b). In this way, human technology began its extraordinary development.

The essential factors involved in this development will be reviewed below:

1. A lead-in from the prototechnology of stone; that is, the use of naturally occurring flakes and choppers.

2. Observation of natural examples of stone toolmaking by percussion; that is, the effects of falling or dropped stones.

3. The existence of a motive related to cutting tool use; namely, the addition of large mammals as a food resource.

4. The availability of raw materials in certain areas.

5. The hardness of raw materials, allowing development of secondary toolmaking, as well as good cutting edges.

6. The existence in the central nervous system of a good visual sense, imagination and foresight, behavioral flexibility, learning potential, and motor skill.

From about 2.5 mya we can see the development of lithic technology in the archaeological record. As this development accelerated—and it accelerated incredibly slowly from our viewpoint—we find evidence of the unfolding of human skill, imagination, and ingenuity. The stone technology was no doubt accompanied by a wonderful proliferation of biodegradable tools. With few exceptions these have not been preserved from the earliest days, and for most of human prehistory the data of technological development are confined to stone tools. However, we should record the following: first known prepared bone point, Olduvai, about 1.75 mya; first prepared wood spear points, Clacton and Torralba, about 300,000 years ago; first bone tool kit, Choukoutien (China), ca. 475,000 years ago. Evidence is meager but suggestive of the extent of such tool kits.

Although stone tools are not known before 2.5 mya, tool manufacture must be very much older, and tool modification and tool use older still. We have no good evidence of this earlier phase of human prototechnology because we can only recognize stone tools under a limited range of circumstances: (1) when the tools are numerous and constitute an unnatural occurrence of shaped stones; where they have clearly been made in large numbers to a regular and recognizable pattern; (2) when the tools are made from a mineral not naturally present at the site and level where they were found; (3) when they are associated with other unmistakable signs of human activity.

Tools are so important in the story of human evolution that they form the basis of the science of archaeology. For the purposes of this section, however, it has been sufficient to show how the beginnings of technology—of tool use, tool modification, and eventually tool manufacture, came to release food resources in a revolutionary way that contributed so much to the peculiar evolving nature of humankind.

X. Food and Fire

Fire has many functions in human culture, and cooking is by no means the most important. However, since the use of fire has affected the digestion of food, it is appropriate for us to discuss the history and functions of fire in human evolution.

The ancient history of pyrotechnology (Table 9.2) is presently being pieced together by archaeologists. The oldest evidence for fire associated with human remains or archaeological evidence has been announced by

Chinese archaeologists to have been discovered at the ancient Chinese *Homo erectus* sites at Yuanmou and Xihoudu. At the first site ashes were associated with two incisors of *Homo erectus*; while at the second, with charred bones, antlers, horses' teeth, and stone tools. These sites may be as much as 0.7 million years old, but they may also be the sites of naturally occurring fires.

Even earlier evidence for fire comes from Chesowanja in Kenya, dated at 1.4 mya. Here the charcoal and burned clay may be due to the recurrent bushfires that occur so frequently on the savanna. There is no clear evidence of hearths, but the ash is associated with stone tools and charred animal bones.

The first certain traces of *hearths* come not from Africa but from Europe. In a cave in southern France (Bouches-du-Rhone) called L'Escale Cave we have traces of what appear to be human hearths that date from at least the earliest part of the Mindel period, and possibly earlier; they are, therefore, well over half a million years of age (Howell, 1967). However, their association with early humans is not certain. Somewhat later, at a site in Hungary called Vertésszöllös (Oakley, 1969), we have definite hearths together with hominid remains dating from the middle of the Mindel glaciation of Europe, that is, the very beginning of the Middle Pleistocene (see Fig. 12.10). Humans were hunting and gathering in a temperate environment with cold winters. The area is a butchery site. Much deeper and bigger hearths of about the same date have also been found in the great cave at Choukoutien in China (Black *et al.*, 1933), the site and level of which have also revealed the fossils of the Peking people, *Homo erectus pekinensis* now dated at 0.475 mya. Here the cave deposits are immensely deep and the hearths contain many meters of superimposed ash; evidently, the fires were permanently maintained throughout the year. There is much burned bone, which implies the use of roast meat.

At a much later date, hearths are recorded from South Africa at the famous Cave of Hearths (Lowe, 1954), but the oldest trace there is probably a natural combustion of bat guano. However, above the thick ashy layer formed by the burning bat guano, there is a succession of hearths dating from the final Acheulian period. At Kalambo Falls, further north, there are hearths a little older (about 60,000 years BP).

In summary, and on the basis of the present woefully inadequate evidence, it looks as if fire may have been captured and used intermittently for nearly a million years before humans learned to keep it alive. The evidence suggests that the regular and controlled use of fire (which is not indicated at Chesowanja) was first developed in temperate regions in the Middle Pleistocene. Without appropriate technology, humans must have relied at first on capturing fire from natural conflagrations caused by lightning, or possibly by seepages of mineral oil or gas, or even by deposits of coal revealed by landslide; all of these have been known to show spontaneous combustion. We can trace a somewhat similar sequence to that we found in discussing the prehistory of tools, from fire use through fire control to firemaking (Table 9.2). The myth of Prometheus describes how humankind stole fire from the gods; by means of such a myth, this achievement is still recorded in our racial memory and celebrated in some tribes. It was an achievement of vital importance to humankind.

Only after humans had learned to handle fire could they have learned to make it, possibly as a result of striking stones in the manufacture of tools. However it came about, our ancestors learned to create, conserve, feed, and handle fire, and it became a precious tool in the advance of culture. By the time of *Homo erectus* in the Middle Pleistocene, fire began to assume many important functions, for it was able to provide protection against predators, a center point at the home base, warmth, light, a means of cooking, and an aid to toolmaking.

It is interesting that according to the present evidence the regular use of fire arose first in Europe and Asia at about the time of the Mindel Ice Age but probably during the warmer phases of this icy period (Perles, 1977). It seems highly probable that fire made cave dwelling practicable because it gave humans warmth and protection at night from wild animals such as the cave bear, which in cooler climates were dangerous as predators and at the same time competitors for shelter. Fire is still used in parts of Canada for driving bears out of caves.

The achievement of handling and making fire would have resulted in a less nomadic existence, and the cave fire would have supplied vital warmth and light against the long winter evenings and the cool wet climate of the glacial period. The effective day would have been lengthened, and work—toolmaking and the preparation of skins and other animal products—could have been continued by firelight in the evening.

The established fire must have led to cooking, perhaps inadvertently at first. Heat will break down the tough structural components of vegetables and meat and release the nourishing juices. Complex organic compounds are in some cases broken down into simpler forms by heat, so that the process of digestion is already begun, as is the process of mastication. Roasting must have been the usual way of cooking for a long time; boiling is a much more recent development, since it involves a watertight hollowed rock or skin bag into which hot stones ("potboilers") were dropped; fireproof containers developed later. It is perhaps not surprising, then, that the overall recession and reduction of the human masticatory apparatus, which is continuing today, began more or less at the same time as the appearance of hearths in cave shelters. A powerful jaw is no longer necessary for the mastication of cooked food. In the process of evolution it has been modified accordingly.

Finally, fire was used in toolmaking. Large stones can be split by heating followed by sudden cooling; wood can be hardened in fire to form effective spear points. Only a few millenia ago, fire was first used to smelt ores, and the use of metal has resulted in the rise of modern civilization. The first evidence of smelting is of copper production about 4500 years ago in eastern Europe; it came some half a million years after the first certain indications of the control of fire!

XI. Summary

In this chapter on food and feeding we have touched on matters that are important in two ways.

First, teeth and jaws form the most abundant fossil evidence for human evolution, and, second, the environment determines the direction and rate of evolution of a species directly and obviously by the way the animal interacts with it in the act of feeding.

We have examined the structure and function of the primate jaw and dentition and have noted the lack of full herbivore specializations, except in a few groups. The omnivorous diet of primates has made them adaptable to a greater range of environments than most other groups of mammals; it allowed adaptation to a fully terrestrial life. We have seen that such adaptation among hominids has taken two different and distinct courses, both involving reduction in the canine tooth: in the robust species of *Australopithecus* there is a trend toward molarization for maximum grinding efficiency; in the *Homo* line there is a trend toward an omnivorous dentition, with some reduction in the molar series (especially M3) and finally in the whole masticatory apparatus.

Such a hominizing trend was accelerated by the evolution of food gathering with containers, by social hunting, by the making of tools for cutting and chopping meat and bones, and by the mastery of fire. The social correlates of these developments have also been briefly considered, and it seems clear that they constitute a major cultural advance in the final stages of human evolution. Human beings and their society are no exception to the general observation that animal groups evolve as homeostatic adjustments to changing environmental conditions, relating primarily to the interaction of organism and environment that is involved in getting food.

Suggestions for Further Reading

The best general account of the evolution of the primate dentition is W. E. Le Gros Clark, *The antecedents of man*, 3rd ed. (Chicago: Quadrangle Books, 1971). For a more detailed study, see the classic account by W. K. Gregory, *The origin and evolution of the human dentition* (Baltimore: Williams and Wilkins, 1922).

For information on primate foods and their digestion, see D. J. Chivers, B. A. Wood and A. Bilsborough (Eds.) *Food acquisition and processing in primates* (New York and London: Plenum, 1984).

The ecological and dietary implications of the hominid adaptation have been discussed by C. Jolly (1970), "The seedeaters: A new model of hominid differentiation based on a baboon analogy," *Man* 5 (1970): 5–26. The emphasis is placed on language in C. F. Hockett and R. Ascher (1964), "The human revolution," *Curr. Anthropol.* **5**, 135–68. A useful account of early hominid adaptations is found in G. Isaac (1978), "The food-sharing behavior of protohuman hominids," *Sci. Amer.* **238**, 90–108.

The adaptations of living hunter-gatherers are discussed in R. B. Lee and I. DeVore (Eds.), *Man the hunter* (Chicago: Aldine, 1968) and R. B. Lee and I. DeVore (Eds.), *Kalahari hunter-gatherers* (Cambridge: Harvard Univ. Press, 1976). For accounts of other hunter-gatherers, see also M. G. Bicchieri (Ed.), *Hunters and gatherers today* (New York: Holt, Rinehart and Winston, 1972).

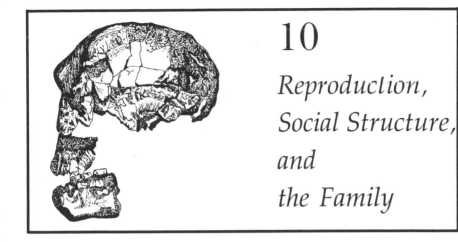

10

Reproduction, Social Structure, and the Family

I. Reproduction and the Placenta

To write a chapter on the evolution of human reproduction seems a project fraught with difficulty, for fossil remains tell us nothing of this all-important aspect of our subject. However, by comparing the reproductive processes of human and nonhuman primates, we can obtain valuable insight into the important changes that must have occurred during human evolution and the probable reasons for their occurrence. Because sexual reproduction is the mechanism by which a species both survives and varies in evolution, the form of the mechanism is of special interest to us. In this chapter we shall consider some significant changes in the behavioral aspects of the reproductive process, changes that have influenced the form of our society and the structure of our individual lives. Their study is a fascinating aspect of human evolutionary biology.

Although the development of the human embryo (the individual before birth) in the uterus, the process of gestation, is part of ontogenetic growth, it is convenient to treat it as part of the reproductive process. As we shall see, while the human lifespan has changed and lengthened during evolution (see Section VII), this first phase of human growth has not. Table 10.1, which relates the period of gestation to the total lifespan of different primates, shows that human gestation is relatively short compared with the postnatal period up to the completion of growth. While the total human growth period is twice as long as that of the great apes, human gestation is not significantly longer. The human infant is far more helpless than the infant ape or monkey, since it is born at an earlier stage in development, but its growth rate during gestation is much faster. When we consider (in Table 10.1) the figures for humans and, for example, the orangutan, it is relevant that the average increase

TABLE 10.1. Retardation in Humans[a]

Primate	Gestation (weeks)	Complete hair covering	Carpal ossification centers at birth	First dentition (months)	Second dentition (years)	Menarche (years)	Growing period (years)	Life span (years)
Macaque	24	During gestation	All centers	0.6– 5.9	1.6– 6.8	2	7	25
Gibbon	30	Onset during gestation completed after birth	2–3	1.2–?	?–8.5	8.5	9	33
Orangutan	39		2–3	4.0–13.0	3.5– 9.8	?	11	30
Chimpanzee	34		2	2.7–12.3	2.9–10.2	8.8	11	35
Gorilla	37		—	3.0–13.0	3.0–10.5	9	11	40
Homo sapiens	40	Never completed	0	6.0–24.0	6.0–20.0	13.7	20	70

[a]From Abbie, 1958; Schultz, 1956.

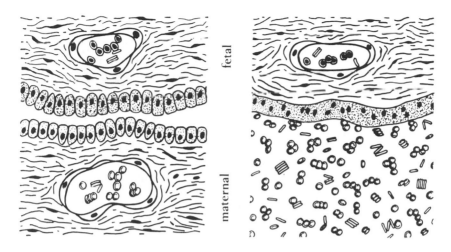

Figure 10.1. Diagrams of two types of placentae found among the primates. *Left,* the lemur; *right,* the human type, showing the relationship between fetal blood (*above,* corpuscles with dark centers) and maternal blood (*below,* corpuscles with light centers). In the human placenta the fetal tissues are bathed in maternal blood with no intervening maternal tissues. The cell walls of the fetal epithelium have dissolved.

in weight of the human fetus is 12.5 grams per day, while in the orangutan it is only 5.7 grams per day. The nature of the human placenta, among other factors, helps make possible this rapid growth rate.

Together with the monkeys, apes, and some few other mammals (insectivores and some rodents), humans share what would appear to be one of the most efficient types of placenta from the point of view of chemical interchange with the mother. In this *hemochorial* type of placenta, which may be compared with the lemur's *epitheliochorial* type (Fig. 10.1), there are no maternal tissues, only fetal tissues separating the fetal and maternal bloodstreams. The blood vessels of the endometrium—the lining of the uterus—are actually penetrated by the fetal vessels, and the former break down to form a spongy, blood-filled tissue, with the result that the fetal vessels are bathed in blood and chemical interchange between the bloodstreams is maximal (Luckett, 1975).

As a result of this intimate association between maternal and fetal tissues, the placenta at birth carries with it a considerable part of the endometrial lining of the uterus. The placenta is for that reason termed *deciduate.*

The differentiation and elaboration of the primate placenta promotes optimum conditions for the developing embryo. Not only is the supply of food and oxygen enhanced, but waste products are rapidly removed. Another advantage is the improved transmission of antibodies in the blood, which pass from the mother to the fetus and immunize the newborn against disease. There is, however, the concomitant disadvantage that, when the blood proteins of the fetus are different from those of the mother, antibodies may develop in the mother's blood, "protecting" her from the alien blood of the fetus, but in turn passing into the fetal bloodstream to cause clotting to the fetal blood. This dangerous effect, known as *isoimmunization*, probably tends to reduce genetic variability in blood proteins by removing variants from the breeding population.

A review of primate placentation from the primitive New World monkeys to the Hominoidea shows that the placenta develops at an earlier stage in the course of gestation in the Hominoidea than in the monkeys. The hominoid embryo does not begin its development in the *lumen* (open central space) of the uterus, but at a very early stage it enters the uterine wall and becomes implanted in the vascular tissue (densely filled with blood vessels) of the endometrium. This tissue is already being prepared for implantation from the time of ovulation (see Chapter 10, III), even though conception has not yet occurred. Thus, the hominoid fetus is able to draw the maximum amount of nutriment from the mother as soon as the fetal circulation develops, enabling it to grow at an exceedingly rapid rate.

II. Birth and Infancy

The overall reproductive rate in any species or population is controlled by a large number of factors, but only one of them can be considered here: the birthrate itself. It is characteristic of almost all primates that they give birth to one baby at a time. Among the higher primates only the South American marmosets regularly produce twins. The extremely low birthrate in humans (one child per year is about the maximum) is the end point of a process of reduced egg production (under *K*-selection) that leads to humankind from those forms of life (such as the salmon) that (under *r*-selection) produce many millions of fertile eggs per year. Evolution has selected in humans a reproductive process that enables us to maintain our numbers in a hostile environment, not by mass production but by prenatal protection and postnatal care.

The theoretical calculation of the reproductive potential of three primates is based on the assumption that the human and chimpanzee female produce one young every third year and the marmoset three sets of twins every two years, with a sex ratio of one male to one female (Table 10.2). It shows the striking discrepancy between potential and actual in the marmoset compared with that in humans. Of course, neither animal is able to utilize its full reproductive capacity, but it is clear that a human infant at conception starts out on a relatively safe journey. Its calculated chance of survival (in this example, with the population remaining constant) may be only 1 in 10, but in practice it is much greater than that, since the birthrate is never maximal. Observed infant mortality among chimpanzees is about 50% and the same figure may well apply to early humans (it still does in many parts of the world).

TABLE 10.2. The Reproductive Potential of Higher Primates

Primate	Period fertility of female (in years)	Fertility commences (years from birth)	Theoretical maximum number of offspring after 45 years
Marmoset	7	3	20 million
Chimpanzee	16	9	136
Human	28	17	21

We have stressed that competition is a correlate of the evolutionary process and that it is far more marked within species than between species, because members of one species require the same food and habitat while members of different species generally do not. This intraspecific competition is found not only between adults but also between infants and embryos. In animals that produce several young at birth, competition exists among the different fetuses for a limited supply of nourishment and space. Rapid development of the growing organisms may be favored for that reason, as well as for others that we shall discuss. The production of single offspring, which is typical of the higher primates, is important because it removes intrauterine competition and allows a slowing of maturation in the fetal stages.

Primates are *precocial* mammals: the young are able to move independently and crawl about within days of their birth. Their eyes and external ear channels are open and soon after birth the fur is well developed. There is an obvious advantage in such an adaptation, which is correlated with long gestation, long lactation, slow growth rate, and a long lifespan. Precocial mammals tend to be medium- or large-sized. In contrast, *altricial* mammals are usually small, have large litters of poorly developed young, and build nests to shelter them. The young are helpless and hairless at birth, with their eyes and ears sealed by membranes. Gestation and lactation are brief, growth is rapid, and the lifespan short. Precocial mammals are adapted for slow reproductive turnover, with a large parental investment of time and effort in the young (Portmann, 1962).

Like all other higher primates, humans are precocial mammals, but the maturation of the fetus is slow; it is born at a stage somewhat reminiscent of an altricial species. During human evolution there has been not only an acceleration of growth but a slowing of maturation, and as a result a change in the stage at which birth occurs in *ontogeny* (the development of the individual from egg to maturity). There are a number of factors that affect this stage.

In the first place, the fact that the birth canal passes through a bony ring in the mother's pelvis (see Figs. 6.3 and 6.10) means that its maximum size is fixed in any mature female, limiting the size of the fetus' head (which is the largest part of the fetus' body in its cross-sectional area). Since the brain is the highly evolved controller of all homeostatic mechanisms, including behavior, considerable development of this organ is necessary for the infant to develop to a level of independent activity. The size of the pelvic birth canal, therefore, limits the level of development of the newborn. The size of the birth canal itself is positively correlated with the overall size of the mother, so if its size is to increase in evolution (as it has), the overall size of the mature female will also be likely to increase, in spite of the competing advantages of small size (see Chapter 10, VI). Whatever other factors are involved in the determination of the stage of birth in monkey or human, the close relationship between the size of the fetus' head and the birth canal is a strictly limiting factor (see also Chapter 6, II).

The newborn human is helpless, and the bones of its body are incompletely formed. The latter fact, however, proves to have its advantages, for since the bones of the skull are not ossified or fused together they can survive compression and distortion without damage to themselves or to the soft brain within them. The head is, in fact, flexible and can be somewhat elongated during birth; it will subsequently and spontaneously regain its characteristic shape. Birth at an early stage in ontogeny, therefore, serves to provide a certain mechanical advantage, helping somewhat to overcome the limiting size of the birth canal.

Another advantage of early delivery may be the danger of iso-immunization if gestation is extended; it may be safer to feed the growing infant outside the mother's body rather than inside it.

But there are other, possibly more important by-products of this evolutionary change. Among humans, experience can influence the brain more directly and at an earlier stage in development than in other primate species. The human infant is soon adapting to its complex social and physical environment, and this may determine to a considerable extent the nature of the child's socialization and developing personality. On the other hand, it is surely disadvantageous to risk the infant's exposure to traumata sooner than absolutely necessary. Thus, we see the advantages and disadvantages of being born human. The helplessness of the newborn (including imperfectly developed homeostatic mechanisms like body temperature control) has been balanced to some extent by the evolution of a longer period of nursing and parental care.

It is noteworthy that the helpless human infant could not have evolved without a parent to carry it. The baby baboon or chimpanzee may be helped by its mother for the first day or two after birth, but otherwise clings independently to its mother's body hair during the daily search for food and water. Only a bipedal animal is in a position to carry its infant everywhere for at least a year. There is a close functional association between the evolution of bipedalism and the helplessness of the infant. The infant is being exposed to the vagaries of the environment—and in particular to the behavior of its parent—rather than remaining fully insulated in the constant environment of the water-filled fetal membranes within the uterus. The environment will start to mold the developing human at a relatively early stage in its growth, so that environmental influence will be greatly increased.

The influence of the environment upon the young animal begins with its relationship to its mother. The environmental input is in a sense filtered by the mother's behavior and especially by her responses to external stimuli. The close proximity of the mother's body and the comfort of the milk supply stabilize the first affectional relationship of the young creature. Suckling establishes communication between mother and child; the tactile sense is first in this respect. Close dependence makes possible the transmission of learned behavior patterns. The longer the period of dependence, the longer the period available for such transmission and the greater the long-term effect of the affectional relationship on the family group (see Chapter 10, IX). Eventually, at sexual maturity, the bond dissolves and the young begin to assume adult functions. The lengthened period to *menarche* (puberty) and the completion of growth is shown in Table 10.1. As we shall see (Section VII), this lengthened period of dependence on the mother is a factor of the utmost importance in human evolution.

Dependence on the mother has another and more alarming aspect. Among nonhuman primates, the infant can, after a few days at most, cling by its own hands and feet to its mother's fur on the ventral surface of her body. If momentarily separated from its mother, the infant can move toward her and cling to her without her aid. Later, in times of tension and anxiety, it can run instantly to its mother and climb onto her back to ride in safety. It can swing underneath her and feed from her at will. Thus, after about three days, the infant can secure its own survival and initiate and effect the satisfaction of its own needs from its mother.

In humans the situation is altered in a fundamental way. While the baby can still initiate the satisfaction of its needs by crying, it is entirely dependent on its mother to make an appropriate response. Unlike the ape or monkey mother, the human mother can choose how to respond to the cry of her infant—whether or not to satisfy it. It is now well established that human babies can be deeply and permanently disturbed by a continued failure by the mother to satisfy the child's needs, and especially by her failure to respond to the infant's cry in an appropriate way. In a young infant such response should take the form of intimate skin-to-skin contact between mother and young, which supplies warmth, nourishment, and security with the appropriate tactile experience. As Harlow demonstrated in a series of famous experiments (1959), the proper maternal response in monkeys does not appear to be fully realized if the mother was deprived of this experience in her own childhood. The same applies to humans (Bowlby, 1969).

Thus, the evolution in humans of infant dependency and its by-product, maternal responsibility, have had a profound effect on human nature. We find that social traditions dictate childrearing practices and thereby determine to a considerable extent the most fundamental atti-

tudes of individuals. Thus, different societies induce distinct social behavior in their members. The human species as a whole is distinguished
by learned cultural differences that can be quite profound (see Chapter
11, I).

III. Female Sexuality

The menstrual cycle is, of course, the human version of the estrous cycle of other mammals. The
most notable event of the menstrual cycle is the
monthly discharge of blood from the uterus; in the
estrous cycle, the most notable event is the period called estrus ("heat"),
in which the female both desires to copulate (is *receptive*) and stimulates
males for that purpose. In the higher primates (including women), vola-
tile aliphatic acids—*pheromones*—are produced in the vagina at estrus that
function as potent sex attractants. Ovulation occurs during estrus. This
period of heightened sexual attraction is seen in all mammals except
humans and a few other higher primates. Menstruation (loss of blood)
occurs at the end of the cycle, long after ovulation. Slight bleeding occurs
in many primates, but only in humans does it amount to a heavy monthly
flow. The term "menstrual" means "monthly," and the cycle does
occupy 28 to 30 days in most higher primates.

The full cycle, which is properly called estrous (rather than menstrual),
was shown earlier in Fig. 2.6 and occurs among all primates. Those that
have a breeding season, rather than breeding the year-round, may experience a period of *anestrus*, when the cycle subsides and sexual activity
ceases. Only a few of the prosimians and the Japanese macaques are
definitely reported to have such a nonbreeding season. The estrous cycle
is interrupted in the mature female higher primate only by pregnancy
and/or periods of *lactation* (milk production), which follow *parturition*
(giving birth).

The discharge of blood that is so typical of *Homo sapiens* follows the
stage in the cycle that has been described as "pseudopregnancy." Immediately after ovulation the lining or endometrium of the uterus begins
to develop its spongy blood-filled texture in preparation for the implantation of the embryo. If implantation and pregnancy do not occur, this wall
breaks down just as it does at birth, when the deciduate placenta is shed.
The increased blood loss in women would appear to be correlated with
the degree to which the endometrium is prematurely modified for implantation and the nourishment of the embryo. It indicates the speed with
which the human endometrium responds to the hormone progesterone
(see Chapter 2, IV), and the speed with which the embryo will come to
require nourishment from the mother and begin its growth.

In many primates estrus is accompanied by variation in the appearance
of what has been termed the "sexual skin"—a specialized area of skin on
the female contiguous with the labial and circumanal region. In these
genera, the skin undergoes cyclic variation in its external form and color.
The sexual skin is swollen and carries an unusually rich blood supply; at

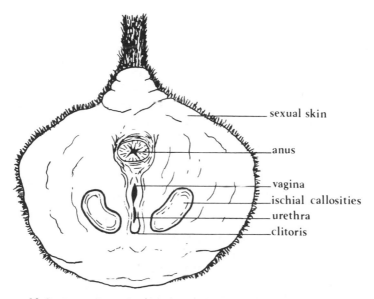

Figure 10.2. Rear view of a female macaque monkey at estrus, showing the great development of the hairless sexual skin. Note the position of the ischial callosities in relation to the other anatomical landmarks.

estrus it swells with intracellular fluid and may assume bright-pink coloration. The presence of sexual skin has not been reported among New World monkeys (except the marmoset) and varies a great deal among Old World forms, reaching its greatest development in baboons (Fig. 10.2). Among anthropoid apes it is only developed in the chimpanzee. The *tumescence* (swelling and thickening) and coloration of the sexual skin at ovulation is due to the increased level of the hormone estrogen in the blood. Its function apparently is to act as a visual sexual stimulant to males, thus ensuring that copulation occurs at ovulation.

As it occurs among nonhuman primates, therefore, estrus has three components: (1) its physiological basis in the hormone cycle and in ovulation, (2) the physiological signal of the female's pheromone and (in some species) the sexual skin's tumescence, and (3) the behavioral change in the female that makes her not only sexually receptive but even anxious to solicit the male's attention.

Among almost all higher primates we find that receptivity on the part of the female extends beyond the 1–3 days of estrus, which is the usual pattern in prosimians. In mammals generally, the phenomenon of estrus is short-lived (lasting perhaps only a few hours) and interrupts normal behavior, with the result that the female, and in turn the male or males, may be almost totally preoccupied with sexual activity. Among higher primates, with the exception of the gorilla and the New World squirrel monkey *Saimiri*, receptivity has now been observed to extend throughout the cycle and even into pregnancy (when estrogen levels remain high). There is a peak of mating activity at estrus in most species, but not all. The vervet monkey (*Cercopithecus*), for example, shows no outward signs of estrus, and mating continues throughout the cycle. This is the human pattern.

In women there is no sign of sexual skin, or is there any clear evidence of estrus with the possible exception of a very small rise in body temperature occurring at the time of ovulation. Though some evidence of a peak of sexual response among women at estrus has been recorded by Benedek (1952) and Udrey and Morris (1968), most of the available evidence suggests a maximum response just before menstruation, and not ovulation. This is probably due to the increase in fluid (edema) that develops at this stage of the cycle and increases the sensations of pelvic congestion (Sherfey, 1972).

In all mammals a period of anestrus necessarily follows the birth of young and accompanies lactation. Lactation is therefore of central importance in the control of fertility (Short, 1976a).

When a female baboon comes into estrus, it is believed that the dominant male of the troop will copulate with her at the peak of her tumescence. When a female chimpanzee comes into estrus, all or most of the mature males will copulate with her. Her sexual skin advertises her receptivity to all, and she welcomes all approaches. There is almost no element of choice on the part of the female in either species at estrus, or do males show much hesitancy in their sexual responses. Among humans, however, any man can theoretically choose to copulate with any woman and any woman with any man, so that what determines who copulates with whom is not a physiological mechanism but personal choice and, of course, social sanction. Among humans there is a possibility of choice of sexual partners by both sexes that did not exist before among nonhuman primates, except perhaps outside estrus. This opens up the further possibility of male–female friendship—the basis of a more permanent sexual relationship—which may account, at least in part, for the selection in human evolution of females who not only show a long period of sexual receptivity but little or no expression of estrus itself, which is hidden.

The female sex organs do not vary a great deal among the higher primates. The protective *labia* (lips of the genitalia; see Fig. 2.4) are rather more fully developed in women than in most other primates. On the other hand, the sensitive *clitoris* is much larger in nonhuman primates than in women, is exposed and often pendulous (except in the *Colobus* family), and even has a small *baculum* (stiffening bone) in *Cebus* (see Section IV). The small size of the clitoris in *Homo sapiens* is possibly due to one or both of two factors: (1) the absence of any swelling of sexual skin, which would obscure the clitoris and protect it from stimulation, and (2) a change in position in copulation (see Section V). It is so effectively placed for stimulation during ventral copulation that excessive size is unnecessary and could even be disadvantageous.

Finally, it is worth noting that the visual epigamic features that the female offers the male have moved in human evolution from the dorsal to the ventral side of her body. In place of the sexual skin, visible from behind, man responds most strongly to features on the ventral surface of the female, and these epigamic features have spread from the genital region to the face. We find the development of pubic hair, of rounded breasts with their *areolae* (the sensitive skin surrounding the nipples), and of smooth facial skin and lips (Goodhart, 1960). It is notable that only in women are the breasts permanently rounded, so that they appear (from an ape's viewpoint) to be lactating. Unlike other primates, they first become rounded at puberty, well in advance of the first pregancy. From Upper Palaeolithic times, perhaps 20,000 years ago, we have definite evidence from rock art of human interest in the epigamic sematic features of the human body (Fig. 10.3). Because today many of these features are hidden by clothing, their exposure and their use, like other aspects of human sexual behavior, comes under conscious control, which indeed, may be one important reason for the development of clothes. Thus, we find voluntary signaling by the female replacing the involuntary physiological signals of estrus. In both sexes sematic features are positioned for face-to-face communication, and the tactile receptors of lips, breasts, and other parts of the body have cortical connections to bring about erotic response.

IV. Male Sexuality   Male primate genitalia have been described by Pocock (1925), and those of the great apes and man by Short (1977). Although the male sex organs are basically similar, there are some very remarkable differences in size among the Hominoidea. The largest of the group, the gorilla, has the smallest penis (only 3 cm) and the smallest testicles (11.6 gm each). The smallest of the group, the chimpanzee, has a long penis (8 cm) and immense testicles (each weighs about 60 gm). Man, slightly heavier than the male chimpanzee, has by far the longest penis of any primate, which averages 13 cm and varies between 11 and 15 cm. His testicles are small compared with those of the chimpanzee, weighing 12 to 16 gm each.

A small bone or cartilage called the *os penis* or *baculum* is found in the penis of all monkeys and apes. It extends from the body of the penis into the left side of the *glans* (the enlarged tip of the penis), which it stiffens. This bone has apparently been lost in human evolution; the blood pressure in the glans during erection gives sufficient firmness for penetration without its support. The loss of the bone may be related to more gentle and careful copulatory behavior among humans.

In general, the length of the penis of Old World monkeys and apes seems to be related to its use in display and to the extent of the development of the sexual skin at estrus, which would otherwise serve to keep a short penis out of the vagina. Among those monkeys and apes in which the females do not have a developed sexual skin, the males have shorter

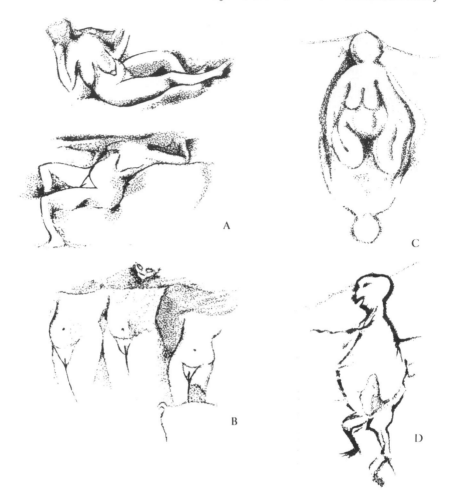

Figure 10.3. Some of the most ancient known representations of the human figure, of paleolithic age, (A) from La Madeleine, (B) from Angles sur l'Anglin, (C) from Laussel, and (D) from le Portel—all in France (drawn by Rosemary Powers).

penes. Man's relatively long penis may also be related to the long vagina in women, which with the labia protects the uterus from external infection. It may also be a product of erect posture and bipedalism. In many species of monkey, the penis is used, usually erect, for display. Frequently it is brightly colored and of considerable size. In some species, the scrotum is also brightly colored and enters into the displays. There is little evidence that it is used in display by any living peoples, though its presence and size has sometimes been accentuated in dress; for example, in the use of the penis sheath and the codpiece. There is no doubt that its size has always been considered important to men as an epigamic feature, if not to women, as can be seen in the erotic sculptures and paintings of many civilizations.

A free-hanging scrotum and testes are present in most of the higher primates, since the development and storage of sperm cells is more efficient at a temperature lower than that of the rest of the body. This has been experimentally demonstrated in different mammals, including man. The testes are obviously vulnerable in the scrotum, and in both the gorilla and gibbon they lie close to the body wall and cannot be easily observed.

The very different sizes of the testes is puzzling at first sight. However, Short has related testis size to body size, and then to the observed frequency of copulation in the African apes and humans. In 1200 hours of observation, Tutin (1975) observed 1137 copulations in a wild community of chimpanzees. Gorillas were observed to copulate 98 times in 2000 hours of observation (Short, 1977). Human (Americans), in contrast, are reported to copulate between 5 and 11 times per month (Westoff, 1974). It seems clear that testis size is correlated, not with body size, but with copulatory frequency: chimpanzees have a high frequency, gorillas and humans a low frequency.

However, when testis size is related to body size, we find that the size of the testes can be correlated with the mating system. This will be discussed in Section VI.

V. Sexual Behavior and Copulation

There is little doubt that the expansion of the cerebral cortex in human evolution and in particular of the frontal and parietal lobes has had as important an effect on human sexual behavior as on other activities. In mammals generally, sexual response depends on reflex behavior stimulated in the first place by hormones in the female, and certain olfactory, visual, and tactile sensations in the male. In human evolution these subcortical mechanisms have become subject to excitation or inhibition by the cerebral cortex.

On the one hand, cortical excitation increases the range of stimuli that can effect response, as by cortical association they may include many factors of no direct biological sexual significance. The rise of cortical control may also be seen in the behavior of immature individuals. As a rule, immature female animals indulge in no sexual play before the secretion of the appropriate hormones at maturity; immature males occasionally do. The young female chimpanzee does show some interest in sex before maturity, but only among humans will young boys and girls (under the less rigorous repression of Polynesian society, for example) indulge in coital play involving nearly all the elements of adult behavior except ejaculation (Beach, 1947).

On the other hand, the cortex may also inhibit sexual activity when a full range of valid biological stimuli is present; response to a real sexual situation may be totally inhibited. This inhibitory action of the cortex is so effective that men may easily become impotent; in women sexual desire related to estrus has almost disappeared. The estrogen still flows as before, but its effect is no longer clearly apparent; sexual desire is affected by stimuli that may operate at almost any time.

This development has been brought about by a change in the source of the sexual stimulus, which in the human female arises not from internal physiological factors (the hereditary hormone cycle) involving estrus, but from external factors (such as the advances of men), which may occur at any time but which are associated in the cortex with the arousal and satisfaction of sexual desire. Similarly, males respond not to pheromones and sexual skin but to features permanently present in mature women and under their conscious control.

A second factor relevant to the evolution of human sexual behavior is the adoption by humans of their characteristic (but not universal) face-to-face position in coitus—the so-called "ventral-ventral" position, with the female supine, the male prone. An approximation to this position is found in all hominoids, whose mating positions are variable, but, while the female gorilla or chimpanzee may lie on her back, the male will squat between her legs, not lie upon her body. This position was also reported to be common among the Australian aboriginees until it was pronounced sinful by western missionaries (Abbie, 1969). When the ventral position became common among humans (if it is the most common position) is uncertain. There are representations of human copulations dating from the late Pleistocene to Etruscan times (Fig. 10.3C is possibly an example) that seem to show humans behaving like animals in this respect, with dorsoventral entry from behind. However, pictures of animal copulations are equally common, and, in accordance with the magical interpretation of much early art, it seems possible that this "animal" behavior was specially adopted and represented by humans to increase the fertility of the game animals. It would not be unreasonable to assume the adoption of the ventral position as humans slowly became adapted to bipedalism in a terrestrial environment. A number of factors must have operated at that time:

1. The development of the buttock (see Chapter 6, II), especially the steatopygous buttock [which contains large fat reserves and occurs in San women], may be thought to make entry from the rear increasingly difficult. The longer human penis, however, seems to counteract this difficulty.

2. The adoption of a horizonal resting position in a terrestrial animal would lead naturally to the ventral position in copulation.

3. The ventral position might have been associated with the invention of speech and the development of friendship between men and women, and is surely associated with the ventral evolution of human epigamic features.

Consideration must also be given to the female orgasm, which has been claimed to be a uniquely human experience and which Ford and Beach (1951) suggest may be correlated with the extra stimulation that the clitoris receives in ventral copulation. There is, however, suggestive evidence of strong sexual response in female animals during *intromission* (full entrance of the penis) as well as before (Hrdy, 1981, Symons, 1979). What is perhaps uniquely human is the variability in a woman's sexual response from a regular absence of orgasm to its multiple expression. Again we can account for this by the increased extent of cortical control in sexual behavior generally.

There is no doubt that in some animals the function of the clitoris is to raise the level of sexual excitement to the point necessary for effective intromission. The gorilla is reported (Schaller, 1963) to stimulate the female's genital areas both orally and digitally (and in one case even to stimulate the breast). The male chimpanzee has also been reported to stimulate the clitoris orally, resulting in its erection (Ford and Beach, 1951), but during copulation gives it no further direct stimulation. However, the labia minora pull on the clitoris during dorsoventral copulation and stimulate it effectively in both animals and humans (Sherfey, 1972).

By this point in copulation a high blood pressure has been developed in the female, and the vulva and vagina are tumescent. This high pressure, associated with edema, has been measured in women and in the dog and been found to subsist throughout coitus (but much longer in the bitch, in which intromission is maintained long after ejaculation). Peaks of high pressure are identified by women as orgasm, which may be repeated: the experience is accompanied by contractions of the vagina and uterus as well as other subsidiary reactions. The female orgasm appears not to be uniquely human, but may involve an intensification of the sexual response seen in other mammals brought about by increased clitoral stimulation and associated sensations.

It appears that in human evolution the loss of the innate and predetermined sexual drive associated with estrus is balanced by the more effective stimulation of the clitoris that occurs not only before but during intromission. In mammals generally, the estrus of the female initiates copulation; the male responds automatically. Among humans and most higher primates, copulation is brought about at any time by mutual interest and stimulation. But only humans are released from the bonds of estrus, which brings about copulation and fertilization without the opportunity for a choice of partner. This gives the human sexes a unique equality not found among other animals. It makes the sexual embrace a mutual experience, possibly an important factor in the structure of human society.

Perhaps the most interesting result of this complex development was the individualization of sexual relations. Mate selection is, of course, common among some mammals and birds, but in humans the recognition of individual qualities goes much deeper. As a result of this and many other factors, more permanent bonds began to develop between man and woman. Relationships became based on friendship—compatibility—as well as sexual attraction. When female receptivity began to overlap the nursing of children, the man was brought into the mother–child relationship. Thus, more permanent sexual ties appear to underlie the structure of the family.

VI. Sexual Dimorphism and Secondary Sex Characteristics

Characteristics peculiar to the sexes but not directly part of the sex organs are present in the majority of animals. These differences usually develop at menarche and are called *secondary sex characteristics*. Among primates, differences in size are common: an example of *sexual dimorphism*. Among orangutans, for example, males may have twice the weight of females. In contrast, female gibbons are the same weight or may be somewhat heavier than the males.

In his book of 1871 Darwin wrote: "Many characters proper to the males, such as size, strength, special weapons, courage and pugnacity have been acquired through the law of battle." The development of selection for sexual dominance among males is based on appearance rather than physical force, and on threat rather than fight. This means that in *multi-male* social groups males with an impressive appearance (in the form of large teeth, epigamic hair, etc.) and a strong personality pass more genes to the next generation than do those less well-endowed. It is appearances that count. Among most primates and many other groups of vertebrates, male secondary sex characteristics are selected in general not by females, but as a product of intermale rivalry, which is reflected in the mating system. This is one classic form of sexual selection (Darwin, 1871).

Among Old World monkeys sexual dimorphism tends to be correlated with large body size and multi-male troops (Clutton-Brock and Harvey, 1977). Larger monkeys and apes such as the baboon and gorilla, are usually terrestrial, and among terrestrial species the gorilla shows the greatest dimorphism (Table 10.3). An exception to this rule is the orangutan, which is both arboreal and highly dimorphic; here ecological factors are also at work. The least dimorphic species are the monogamous ones (e.g., gibbons). It is likely that the bigger males in large terrestrial species are a product of sexual selection due to competition for access to females. Because females invest a great deal of time and energy in reproductive activities and their young, estrous females are always in short supply (Trivers, 1972). However, much of the variation in sexual dimorphism is still difficult to explain.

There is an interesting correlation between relative testis size and the mating system. Males of species with multi-male troops in which there is overt competition between males for females have larger-than-average testes; males of species with one-male groups or males that are monogamous have smaller-than-average testes (Harcourt et al., 1981).

TABLE 10.3. Differences in Body Weight Associated with Sex in Higher Primates[a]

Primate	Mean weight in pounds		
	Male	Female	Female as % of male
Baboon	75	30	40
Gibbon	13	13	100
Orangutan	165	80	48
Chimpanzee	110	88	80
Gorilla	375	198	53
Human (U.K.)	155	150	97
	(Height, 67.5 in.)	(Height, 62.5 in.)	

[a]The figures are approximate, since they are derived from small samples.

When we relate testis size to body size, we find that the solitary orangutan has very small testes while the promiscuous chimpanzee has very large testes. The small size of the human testis suggests that in our ancestry we enjoyed a one-male or monogamous mating system.

Sexual dimorphism, however, implies more than sexual selection through intermale rivalry. It may be advantageous to evolve a small female if food supplies are limited, and the distribution of food supplies in the terrestrial environment could be one factor that limits population density. With female baboons one-half the size of males, twice as many females can be maintained with a limited food supply than would otherwise be possible. Selection not only favors large males but probably also favors females as small as is compatible with their social and maternal roles.

It might be expected, therefore, that *Australopithecus* and early *Homo*, both terrestrial, would show considerable sexual dimorphism. It has, however, proved difficult to determine the amount of sexual dimorphism present in extinct species such as *Australopithecus africanus*. Since we cannot yet determine for certain the sex of fossil bones, we cannot know how much of the known variability is due to variation between the sexes. In theory, a large sample with considerable sexual dimorphism should give us a bimodal curve in the expression of certain characteristics, but in practice the sample size has not yet enabled us to achieve this degree of precision. At Swartkrans the fossils show considerable variability and appear to fall into two groups, which have been attributed to two different hominid species by a number of authors (e.g., Robinson, 1963; Tobias, 1971; F.C. Howell, 1967).

A somewhat similar situation exists at the Hadar site. The differences we see here, which are expressed in degree of robusticity as well as shape, are attributed to sexual dimorphism by Johanson and White (1979), while others maintain that two species are present. One thing is clear, however:

there is no evidence of a large canine in any skull such as we find in the larger male higher primates (see Fig. 9.14). (Although tooth size often tends to increase with body size, this relationship is neither invariable, nor is it allometrically expressed. It is certainly not characteristic of *Homo sapiens.*) Nevertheless, there is considerable variation in the robusticity of the skull and the development of the masticatory apparatus, and in the size of the skeleton; some of this variation almost certainly reflects a degree of sexual dimorphism. Heavier musculature and heavier bone structure suggest a heavier male, which is hardly surprising. What is surprising is that the heavy masticatory apparatus was already a thing of the past.

We can also make some predictions about the degree of sexual dimorphism that might be expected in *Australopithecus.* Since dimorphism is more pronounced in larger species, the extent of dimorphism in the small species *A. afarensis* and *A. africanus* would not be expected to be great. On the other hand, we have seen that there is some correlation of dimorphism with terrestriality, and on that basis we might expect more dimorphism in *Australopithecus* than in a chimpanzee of similar size. Finally, we have seen that there is more reason to expect substantial dimorphism within species with multi-male troops in which we can predict the existence of intermale rivalry for females; we may also predict little or no dimorphism in one-male or monogamous species.

We can conclude for now that we have no certain evidence of the degree of dimorphism in *Australopithecus* or in its postulated ancestor, but that it was not great. The fact that humans show a low level of body size dimorphism suggests that a reduction in this characteristic (together with a possible modification in the mating system) is likely to have occurred in human evolution.

Figures for average heights and weights of English men and women are given in Table 10.3 (Fig. 10.4). This size and weight difference is the most obvious secondary sex characteristic that differentiates modern men and women. In addition, we find in modern humans a number of minor variables such as shape of head, shape of pelvis, fat distribution, voice, body hair and skin, muscular strength, conception rate, infant mortality, growth rate, age at puberty, longevity, and energy utilization. Some factors, such as the shape of the pelvis, are directly connected with the reproductive function; others, such as fat distribution and age at puberty, are more remotely related to sexual function. Some may be associated with sexual selection (reduction of body hair and character of head hair), while others may be merely dependent on one of these factors (voice differences). Visible secondary sex characteristics are certainly selected in modern societies, where a small proportion of individuals who do not satisfy the socially accepted requirements for sexual attraction do not succeed in finding sexual partners. The reproductive rate of such individuals is reduced by sexual selection.

This second kind of sexual selection, also introduced by Darwin (1871), accounts for features selected by the opposite sex. If the choice of partner

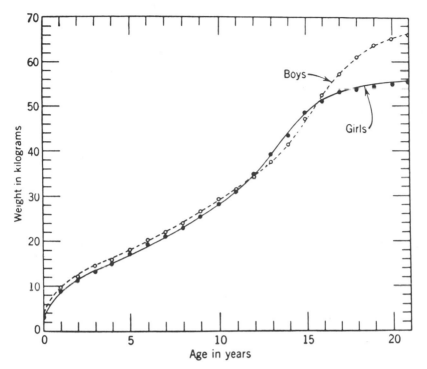

Figure 10.4. Sexual dimorphism in man: the growth curve and adult size of males and females is clearly distinct. Growth is arrested by the hormones produced at puberty, so the smaller size of women is a function of their earlier physical maturation (from Shock, 1951).

among humans was traditionally made by the male, it follows that only the physical characteristics of the female were subject to sexual selection of this kind. In particular, we may include under this heading the relative reduction of facial and body hair and the overall form of the face and body. Head hair, smooth skin, and well-rounded buttocks and breasts are today attractive to men and have almost certainly been subject to sexual selection; if they did not increase the chance of a particular female finding a mate, they did increase the chance of her mating with a male of high social status, so their joint descendents were perhaps more likely to transmit their genes to posterity (Dickemann, 1978). In practice, however, at least some selection is exercised in many human societies by both sexes, so the situation is more complex than this.

Having compared briefly the physiological and anatomical aspects of human sexuality with primate sexuality, we must now consider how the reproductive processes fit into the total lifespan of the individual (10.VII) and the structure of the society in which that individual lives (see Sections VIII–X). While mating activity in general is never, among higher primates, the main determinant of social structure, it is always closely linked to it. The mating system, which is almost always under social constraint, affects the size, nature, and success of the evolving primate population.

VII. The Human Life Span

One of the most important and far-reaching differences between humans and the nonhuman primates lies in their growth rate. A slowing of the rate of growth and maturity is already apparent in the evolution of primates as a whole and is characteristic of precocial mammals, but as we have seen, the trend is greatly accentuated with the coming of humankind (Fig. 10.5). The period of infant dependency, which averages just a few months in most precocial mammals, has already been extended to a year in monkeys and to two years in apes. In humans it is four to eight years. To these figures we must add a period of cultural dependency, which may continue even beyond sexual maturity.

As we have discussed, the helplessness of the human infant at birth distinguishes us from all other higher primates. In place of the rapid development of effective motor control common to most precocial mammals, this unusual helplessness continues during a period of fast growth. The brain, which is only about one-quarter of its final size at birth, doubles within one year; in contrast, the brain of apes is already half grown at birth (Fig. 10.5). Compared with nonhuman primates, the human brain forms a very high percentage of total body weight in the developing infant.

The human mother must protect and nourish her infant during its period of helplessness and carry it about with her for two or three years; in contrast, the gorilla infant clings to its mother for only 3 or 4 months. The long period of intimate contact between the human mother and child constitutes the appropriate environment for the child's early development. This maternal environment is now known to be essential to proper functioning as an adult for both human and nunhuman primates. There is no substitute for it that does not have some deleterious effect on the infant. The protracted immaturity of humans provides a unique basis for socialization, intellectual development, and individualization. Retardation of growth so extreme is only possible in a social context because of the protection offered by the social group. In turn, it is a central factor in the reinforcement of social structure and in the evolution of culture. This very slow growth rate is a unique and fundamental characteristic of humankind (Table 10.1).

Among girls puberty arrives in two stages. The first is marked by the completed development of the sex organs and the appearance of the secondary sex characteristics; it is termed *menarche*. The second is marked by the production of mature gametes (ova) and is termed *nubility*. The ages of menarche for some primates are shown in Table 10.1. In girls nubility follows two or three years later (about 15 years of age), but in monkeys and apes it develops more rapidly.

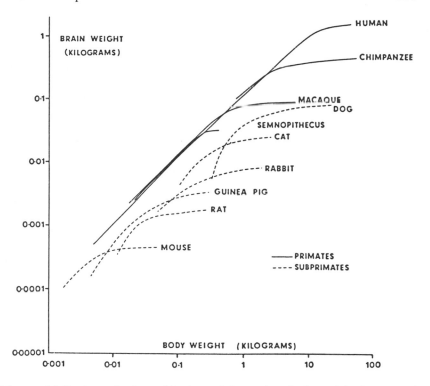

Figure 10.5. Growth plots of brain weight against body weight for several mammals through prenatal and postnatal life. Four species of primates (solid line) follow the same growth line, but humans extend the period of rapid prenatal growth well into postnatal ontogeny (from Holt *et al.*, 1975).

Most mammals now enter upon their full reproductive life. Among chimpanzees, for example, a young estrous female will accept the advances of any number of males in her group, from immature juveniles to dominant males (Goodall, 1968). Among many monkeys, however, the female may begin her reproductive life soon after menarche, while the male does not effectively do so until some time after puberty. The mating pattern among baboons, in contrast, appears to be one in which the younger males are permitted to copulate with the females when they first come into estrus, but after a few days one of the dominant males moves in and acts as consort to the female for the rest of estrus, when tumescence is maximal. The relationship between estrus and ovulation is not known for certain in all the primates, but in general ovulation occurs toward the height of estrus, so it appears that the most dominant males will father the young. The young males appear able to satisfy their sexual responses without serious competition from the older males, but they do not contribute much to the gene pool of the next generation until they too arrive at a position of seniority. It may be of only marginal importance, but this mating pattern also appears to be one factor that would reduce the incidence of son–mother matings that might result in conception.

The delay in the assumption of full reproductive life among baboons is correlated with their sexual dimorphism. The females complete their growth and are sexually and socially mature at about four years, but the much larger males complete their growth and reach social maturity about four years after reaching sexual maturity. For purposes of reproduction, therefore, there are in fact about twice as many females as males (although the actual sex ratio is 50:50). During the last four years of their growth the young subdominant males tend to live peripherally around the troop. They are, suitably, its most expendable members. A similar trend is detectable in human societies, for boys continue growth and social development for a longer period than girls (Fig. 10.4).

This social structure recalls in broad outline that of many African tribes in which the young men must spend their early maturity as hunters and warriors, while the elders remain in the village. Among present-day polygamous peoples the young men may have to wait many years to take wives, while the older men take their third and fourth. Though by no means universal, this kind of mating pattern allows natural selection a maximum opportunity to act upon the males. Only a male who has survived a long apprenticeship both in society and in hunting will be in a position to father children. There is maximum opportunity for a weeding-out of less healthy and less intelligent males. It is possible that natural selection will in this way tend to maintain the evolution of a mating pattern most readily subject to its action, since groups with such a mating pattern will be best adapted to their environment and so gain an increased probability of survival.

We can see, then, how natural selection, acting through society, can delay the full reproductive functions of adult male nonhuman primates as well as those of adult male humans. Delay in the full reproductive function of adult females is, however, an unusual and uniquely human characteristic. It is primarily related to the evolution of a complex culture, for, since one function of human parenthood is the transmission of learned behavior and culture, it is desirable that individuals themselves should have assimilated cultural traditions fully before they are called upon to pass them on to their heirs. This means that in western society education (at home and in institutions) may delay marriage beyond the advent of nubility. In most tribal societies, however, girls marry as soon as they are nubile.

Traditions of learning and culture are protected by specific (and particularly religious) institutions, and the possibility of cultural transmission is also assured widely by *rites de passage*; that is, social rites marking stages in the physical and social development of individuals. Initiation ceremonies are a case in point: they mark the attainment of sexual maturity and include a period of instruction in traditional tribal lore. In western society education is enforced by law, while marriage is prevented by a rule of minimum age. In England the minimum age has risen from 12 years in medieval times to 16 years today. Those in our own society whose cultural heritage imposes the need for a long period of education experience a long delay between the coming of nubility and the onset of a full reproductive life.

As we have seen, this delay seems to be nothing new among males, for a period of activity directed to the benefit of the social group has usually preceded a full reproductive role in nonhuman primates. On the other hand, the longer period of education now offered girls makes the assumption of parenthood at nubility inappropriate. The delay between nubility and marriage is greatly extended by the recent startling increase in growth rate, which has brought the age of menarche from 15 or 16 years down to 13 (Fig. 10.6). This trend, which is a reversal of the overall direction of human evolution toward a slower growth rate, is presumably due to changed ecological conditions, especially an improved diet, associated with western society and is perhaps an instance of adaptability rather than genetic change.

The trend is accompanied by a reduction in anestrus (or amenorrhea) following parturition, even during breast feeding, which in western countries is countered by the introduction of family planning and contraception. These important changes are summarized in Fig. 10.7.

One obvious trend in human evolution is the expansion in the actual amount of knowledge and skill that has to be transmitted from one generation to the next, an accumulation of information entailing an ever longer period of training and instruction. It follows that while the specialist aspect of this burden is taken from the parents' shoulders by the development of institutions of education, an immense responsibility still remains with the parents. The present trend toward earlier marriage in western countries appears to be a solution neither to this problem nor to that arising from overpopulation. The age of marriage is a cultural trait of great biological importance.

Humankind's "threescore years and ten" takes us through a slow period of development and growth and a long spell of childrearing into an old age beyond our reproductive years. The end of woman's reproductive term of life comes with the menopause, between the fortieth and fiftieth year; her survival beyond that age is an interesting and possibly unique phenomenon of the animal world. Postreproductive primates have been recorded in zoos and occasionally in nature but are certainly uncommon. It is not immediately easy to see how postreproductive life could be selected in evolution, since the fate of postreproductive individuals will not affect the composition of the next or future generations in terms of differences in the gene pool.

There are a number of reasons, however, why a postreproductive period of life could be of value to evolving humans. In the first place, the slow growth of children means that they will require the presence of their parents until about 15 years after their birth. Therefore, the mother who bears her last child at 45 will still be occupied in her function as a

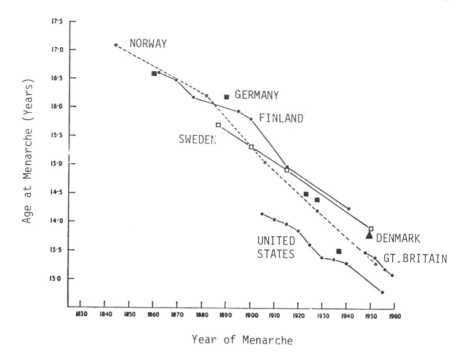

Figure 10.6. Age of girls at menarche, 1840–1960. This diagram shows clearly the remarkable change in the age at menarche that has occurred in Europe and North America during the last 120 years (from Short, 1976).

mother until she is at least 60; a postreproductive period will be essential for the survival of the last-born. There is surely an advantage here in widening the gap between the average age of death and the last ovulation. This is an example of selection through inclusive fitness (see Chapter 1, VII). It is significant that women outlive men (in the West) by an average of eight years.

A second possible reason for the development of a postreproductive period is that childbearing puts considerable strain on a woman and is likely to be more successful if it occurs when the mother is still in her prime. There is evidence that the reduction of egg-bearing follicles has been accelerated in human evolution (see Chapter 2, IV), so that by the advent of menopause there are none left in the ovary. It has been postulated that this is a response to savanna life, where the dry season (especially in lowland areas) imposes great nutritional stress (Donaldson, 1983). Unlike the sudden cutoff in egg production, the woman enjoys a much gentler slowing of her sexual responses, which continue well into her 70's and beyond.

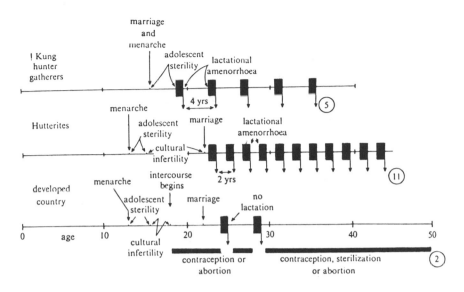

Figure 10.7. Changing patterns of human fertility. With the individual female's lifespan as the horizontal dimension, the top line shows the reproductive activity of a typical "hunter-gatherer" !Kung San woman from the Kalahari. These people do not use any form of contraception. Menarche is late (15.5 years—coincident with marriage) and adolescent sterility defers the birth of the first child until the age of 19.5 years. Lactational anestrus keeps births four years apart. Average maternal mortality probably resulted in a family size of five, three of whom might survive into their reproductive years.

The second line shows the situation among the North American Hutterites who also practice no form of contraception. Their high standard of health and nutrition has advanced the age of menarche to 13, but cultural taboos prevent intercourse until the average age of 22. Although all children are breast-fed, good nutrition means that anestrus is of shorter duration and births occur every two years, to give a completed family size of 11.

The bottom line shows an average lifespan in a typical developed country where menarche occurs at 13 years, intercourse begins in the late teens, and contraception or abortion are used to plan families. Lactation is of such short duration that it no longer induces anestrus, and birth control is necessary for perhaps 20 years after the second child is born. There is an enormous increase in the number of menstrual cycles the woman experiences (from Short, 1976b).

An important argument for the advantages of longevity follows the evolution of culture, and it applies to both men and women. The longer the period of life that must be spent learning, the more valuable to the population the adult individual will be. Among lower animals, especially altricial species, in which learning is not an important determinant of behavior, a quick turnover in population will make possible a faster rate of evolution and so greater flexibility in the face of environmental change. When behavior is innate it is known at birth, but when it is learned it has to be laboriously acquired by each generation and cannot be lightly thrown away.

Thus, the human dependence on learned behavior may be a reason for the selection of old age, as well as for an extension in lifespan. Old men will be greatly valued in a social group for their knowledge of hunting, toolmaking, and other masculine activities, and in particular for their experience of rare and occasional events such as flood, drought, and locust infestation. Similarly, older women will be valued for their knowledge and experience in childbirth, childrearing, food gathering, and preparation, and other household arts. In a literate society like our own the importance of the elders may not be so great in this respect, but in nonliterate societies the old make a unique contribution to the survival of the cultural animal, for they are the storehouse of knowledge and wisdom. It is hard to see how a female nonhuman primate could fulfill a socially useful role in old age after her young have matured, since her adult social function lies almost entirely in her reproductive capacity.

**VIII.
Primate
Social Life**
The higher primates are social animals, and the adaptive advantages of a social way of life are manifold. The prime advantage possibly lies in the fact that their social life facilitates learning. Clearly, infants can learn from their mother without the existence of social groups, or troops, as primate groups are usually called. But in social groups the amount of information received through observation can increase and broaden, and young individuals can benefit not only from the experience of their own mothers but also from that of their peers as well as the other older members of the troop. The troop retains a pool of experience and knowledge, which exceeds that of any individual female and is available to the growing infant. For example, the identification of edible foods, the recognition of competitors and predators (and what to fear in general) are vital pieces of knowledge. Such knowledge can be gained secondhand from other members of the troop and does not require individual experience. Since disaster to one individual is felt as shock throughout the troop, all members may learn from it. Young troop members may even imitate behavior patterns that have been taken up by the troop in the past and that result from such vicarious experience (Washburn and Hamburg, 1965). This socially learned behavior is the beginning of culture, and since it is of great importance in human evolution it will be considered further in Chapter 11; for the present, we need only note its existence as a fundamental adaptive determinant of the social life of primates.

Beyond the facilitation of learning, social grouping must have been selected in evolution as an adaptation for mutual defense against predators. The baboon troop tends to be centered on the females and juveniles; the males live peripherally and, as we have seen, are equipped

Figure 10.8. A baboon troop moving across country. Dominant adult males accompany the females with small infants and slightly older infants in the group's center. Some young juveniles are shown below the center, and older juveniles above. Other adult males and females precede and follow the group's center. Two estrous females (dark hindquarters) are in consort with adult males (from DeVore, 1965).

by their secondary sex characteristics with large canines that can be used to defend the troop. This pattern applies especially to terrestrial species. At the approach of predators such as lions, or leopards, the males may find themselves in position between the females and the danger and may act in concert to drive away enemies (Fig. 10.8).

Another important adaptive characteristic of the social group is its association with a recognizable area of land or forest that contains sufficient space, food, water, and safe sleeping sites for all its members. Behavioral spacing mechanisms operate to bring this about, reducing the possibility of overexploitation of food resources and conflict over those resources. The area in question—which may be partly shared with other groups and is certainly shared with other species—is called the *home range*. Within it, one can usually define a smaller area containing resources absolutely essential for survival, which is called the *core area*. Core areas do not overlap. Where this area is actively defended against intruders from neighboring groups is called a *defended territory*.

Among primates we find an overlapping home range pattern to be most common (Fig. 10.9); it is characteristic of most monkeys and apes. In areas where the food supply is rich and the area required for resources is small, the range may be reduced to a defended territory (as in the gibbons and a few monkeys such as the gray langur in Sri Lanka). In this case, which is the exception rather than the rule for primates, threats are used to maintain control of the territory and its essential resources. In all cases, the relationship between the troop and its en-

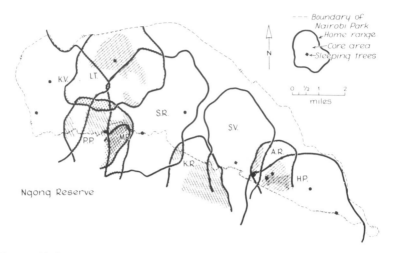

Figure 10.9. Plan of the home ranges and core areas of baboon troops in the Nairobi Game Park. Note the considerable overlap of home ranges but the separation of core areas that contain the sleeping trees (from DeVore, 1965).

vironment is ultimately determined by the distribution and density of the natural resources essential to the troop's survival.

Human hunter-gatherers commonly have a home range and core area pattern of territorial distribution. The ranges involved today are very large (Table 10.4), because these people are now generally confined to very poor environments where the population is extremely sparse. There is no evidence of defended territories among them, and it seems unlikely that humans were territorial in this sense before the development of agriculture, when resources became very dense and were a product of substantial effort, giving rise to a sense of property.

The primate troop is also characterized by an internal structure based on dominance, mother–child relationships, friendship, and sexual bonds. Cooperation and conflict are expressed daily in social interactions. Dominance is expressed through competition between troop members for limited resources. To understand how dominance functions in effecting a stable social structure, it is necessary to consider briefly the part played by conflict in the creation of the dominance hierarchy. Conflict can take three forms: play conflict, threat, and open conflict. The controlled and harmless conflict of young animals at play gives them the opportunity to develop precise motor control and coordination and at the same time to establish their social status. (In a similar way, a social structure will develop in a school, both according to seniority and within a group of peers.) A dominance hierarchy, which in some species is quite stable, is established for each sex on the basis of strength, agility, and personality, and similar factors appear to operate among humans. After completion of the period of play conflict, the

TABLE 10.4. Population Data on Higher Primate Social Groups[a]

Primate	Mean troop size	Mean population density (animals/km^2)	Mean home range (km^2)
Macaca mulatta	21.5	44	2.3
Papio anubis	34	4	24.3
Presbytis entellus	19	57	3.4
Hylobates lar	4	4.7	0.54
Pongo	2	2	5.0
Pan gorilla	10	1.8	6.2
Pan troglodytes	20	2	7.0
Homo sapiens			
San	20	0.01	1100–3000
Australian	35	0.03	260–1950

[a]From Clutton-Brock and Harvey, 1979.

social structure will usually be maintained by threat, which alone is usually sufficient and is obviously adaptive, since it avoids the risks of wounds and infections arising from open conflict within the troop. Every species has its own forms of threat signaling—an important form of communication in any social group (Fig.10.10).

Recent studies of chimpanzees in captivity have revealed the flexibility and complexity of their behavior and the extent of political in-fighting which can develop in the establishment and maintenance of the male hierarchy. At Arnhem zoo in Holland, a breeding group of chimpanzees has been kept for 12 years in a 2-acre enclosure and has been very carefully observed (de Waal, 1982). It appears that the males will use a variety of techniques to gain power and status, as well as covert influence, such as we find in modern human societies. Political activities (defined as social manipulations designed to secure and maintain power and status) and the subterfuges observed included dominance networks, subversive coalitions, power struggles, alliances, divide-and-rule strategies, arbitration, confiscation (of weapons), collective leadership, privileges, bargaining, and frequent reconciliations which usually take the form of kissing. Many of these complex social behaviors had been briefly observed or suspected at Gombe, but it was only at Arnhem, with its excellent facilities for observation, that they could be regularly and accurately recorded. It seems from this work that chimpanzees have the potential for *very* complex social behavior which can be evoked under appropriate circumstances. There is apparently a drive for power and prestige among them, and at Arnhem, power bestows not the best food rations (for these are fairly distributed to each individual) but access to estrous females. Thus we find a complete contrast to the less structured hierarchy of Gombe: at Arnhem the hierarchy is centered on

Figure 10.10. Threat is an important means of establishing and maintaining social structure in primate groups. The actual form taken by the gesture varies among different species, but the "yawn" and lowered eyelids of the baboon are typical (courtesy Irven DeVore).

power and its rewards—sex. It is clear that sensitivity, ambition, and social complexity in the chimpanzee differs little from that of humans and that these differences lie not so much in the level of social evolution, but in the existence and power of human language (Chapter 11).

Once established, the typical primate male hierarchy may remain stable for quite long periods. However, this hierarchy will be disturbed by the introduction of a strange animal into the troop, and open conflict may occur between adults to reestablish a modified social structure.

Conflict, usually in the harmless form of threat, is also very important in maintaining proper relations with other troops and even with other species. Where present, defended territory is maintained this way, and sufficiently threatening gestures on the part of the males will usually scare away intruders.

Founded on threat of conflict, the dominance hierarchy maintains the structure of the troop and orders its behavior and movement in a number of species. The dominant animals take precedence in gaining access to a limited food supply, so that in a famine not all animals will starve equally; the dominant will be more likely to survive. Normally among monkeys, where many species have large multi-male troops of mixed sex, the dominant males will also act as consort to the females at the height of estrus, for such females are a limited resource. In some species, separate hierarchies can be recognized according to the resource involved. For example, some baboon troops have one male hierarchy based on access to rare or desirable foods and another very different one relating to access to estrous females. While the young males determine their status through both their mothers and juvenile play, the females in many species inherit their status directly from their mothers, and in both sexes status usually increases with age. Female hierarchies tend to be very stable and status is inherited; male hierarchies are more labile and depend to a greater extent on individual size and personality. Females tend to remain in their troops throughout their lives, while males may move from troop to troop, first at adolescence and then at intervals of a few years.

Though rhesus macaques (M. mulatta) usually live in multi-male troops, in some situations they will form one-male groups, in which a single male associates with and controls 2 to 5 females that live in a strict hierarchy. Young females will remain in the group after adolescence and inherit their mother's social status, while young males regularly leave the group to join small all-male troops or start up new family groups themselves. Simpson and Simpson (1982) have observed that high-ranking females consistently produce twice as many daughters as sons, whereas mothers of lower rank produce more sons than daughters. This remarkable observation can be accounted for by the supposition that, since status can have a considerable effect on an individual's chances of survival under stressful conditions (such as when food is limited), it is clearly advantageous for a mother's Darwinian fitness to produce daughters of high status. High-status females will therefore produce a majority of daughters, and lower-ranking females (who cannot aspire to producing daughters of high status) will produce more sons, in the expectation that they will be able to succeed outside the existing group and perhaps eventually create their own one-male group. In other species, low-ranking females are simply less fertile (Dunbar, 1980). Although we have used expressions that seem to imply conscious choice on the part of the females, the strategy is, of course, not consciously undertaken. It is a product of natural selection, which has acted on the physiology and behavior of the monkey.

These are important observations because they show us that female reproductive physiology, status and dominance, and overt behavior are all closely interlinked. It is not presently understood how the process of determining the sex of the young is brought about, but the fact that it is possible demonstrates the extraordinary closeness between sexual physiology and behavior.

Any observer of primates will be struck by the amount of mutual grooming that takes place. There is no doubt that it serves an important function in the removal of parasites and the conservation of salt—both parasites and salt crystals are removed from the fur and eaten. But in some species the amount of time spent grooming is unusually great, and here it carries a second important function. It seems to be a gesture of friendship between individuals, for it is accompanied by a definite reduction in tension. We find grooming of this sort most common between mothers and their young and between siblings, but from time to time females may groom dominant males, so that grooming and sexual contacts intermingle. The behavior is pacifying, acting as a signal indicating friendly submission. Grooming has a bonding function within the troop and serves to hold members in affectionate interdependence. In some species, friendship is a very important component of primate behavior.

As a result of the early observation of zoo monkeys whose natural behavior was disturbed, it was for a time believed that sexual attraction between individuals was the primary bonding mechanism of the troop. We now know that many primate species have breeding seasons, and that in the Japanese macaque, for example—the most northerly monkey—there is no copulation from four to nine months each year, while stable group organization is maintained without interruption or change and there is no weakening of the social bond (Lancaster and Lee, 1965). Thus, sexual behavior brings about temporary bonds between individuals but does not in itself play a significant role in troop bonding as a whole.

Primate troops vary a great deal in size, from 2 to about 500, but the most common range is from 10 to 50 (Table 10.4). Considerable variation occurs even within a species in the same type of environment. There is clearly a relationship between troop size, home range size, population density, and available resources. The highest densities occur among the arboreal leaf-eating monkeys, the lowest are found in the terrestrial species (Clutton-Brock and Harvey, 1979). In the Old World higher primates, only the gibbons (seven species) and a species of langur, adapted as they are to a fully arboreal life, exist in small "monogamous" groups. The gibbon group usually consists of a mature male and female, with one, two, or three growing young of both sexes. This group has all the appearance of a human monogamous family, but, as we shall see, its adaptive determinants are different.

IX. The Nature and Evolution of the Human Family

It is clear that any extrapolations we may make from studies of nonhuman primates about the origins of human society are bound to be extremely tenuous and perhaps even misleading. Nevertheless, a comparative approach is the only one available to shed light on this fascinating problem. Because we believe that different human social structures are adaptations to particular environments and can readapt rapidly, we cannot conclude that one is necessarily more primitive than another. We shall, therefore, use the comparative method with caution and lay special emphasis upon the equivalence of environmental conditions between the societies being compared.

Among living Old World higher primates, one of the best known mating systems is that of "promiscuity" within a heterosexual group containing numerous males and females—the multi-male troop. This pattern is found among our closest relatives, the chimpanzees. In contrast, groups with only one adult male and a number of adult females occur in many species, especially the terrestrial monkeys; this is the one-male group, also known as the *harem*. In this category, the gelada, the hamadryas baboon, the patas monkey, and the blue monkey are probably the most well known (Crook, 1972). Beyond this we find even smaller groups among the gibbons in which one adult male and one adult female form a permanent "monogamous" bond.

Monogamy occurs among all the gibbon species, one species of langur monkey (*Presbytis potenziani*), the lemur (*Indri indri*), and a number of New World primate species. The advantages of monogamy are not obvious; on the contrary, there is likely to be strong selection pressure for males to breed polygynously whenever possible to increase their reproductive success. Monogamy is likely to happen only where the male leaves more offspring when he assists the female to rear their young (Trivers, 1972) or where females mate selectively with monogamous males who will invest more than polygynous males in their young (Vernor and Willson, 1969). Monogamy will, therefore, be likely to evolve where the male's assistance in the form of infant carrying and territorial defense is most necessary for the survival of infants. Monogamous primate species are territorial and are associated with constant intraspecific competition; evidently, the male can only manage to defend and support one female and her young.

Finally, the orangutan appears to be solitary, since the only recognizable permanent association is that between mother and young. Thus, we find a wide variety of mating systems among the higher primates, some of which are roughly paralleled in human societies.

Today, there are considerable variations in human mating systems, even omitting contemporary experiments in community living. In almost all instances, however, a nuclear family can be distinguished as a one-male unit with one or more adult females (*monogamy* or *polygyny*). In a world survey of human societies the majority are classified as polygynous, but the majority of marriages are probably monogamous, since polygyny is usually limited to the rich members of a society even where it is the accepted and desirable married state. The essential characteristic of human societies is the one-male group—the nuclear family. The one exception to this general rule is *polyandry*, a rare arrangement in which one woman is united with more than one man, often brothers. It appears to be an adaptation to limited food resources and is associated with a reduction in the proportion of nubile girls by female infanticide, a most effective if brutal means of population control (Hiatt, 1981; Beall and Goldstein, 1981).

In societies permitting polygyny, the number of females a man marries is based directly on his resources and status. In monogamous societies such as our own, marriage bonds may be strongly developed but serial monogamy is common and many men in the society may be supporting more than one woman (as lover or ex-wife), even if such a man is only legally married to one. The evidence is unequivocal that there is no overriding genetic component operating to bring about either monogamy or the degree of permanency in human relationships that we find in some other primates, especially gibbons.

How did the present human one-male group evolve from an ancestral primate mating system? If the ancestral system was that of the one-male group (as has been suggested by Eisenberg *et al.*, 1972), it is clear that human evolution introduced no startling novelty. Alternatively, the ancestral system may have been that of the chimpanzee, since they are our closest living relatives and occupy (in part of their range) a forest fringe/savanna environment that is probably very similar to the one occupied by the earliest hominids. Chimpanzees have an open and fully promiscuous mating system in which most adult males may have the opportunity to copulate with most adult females (Goodall, 1968). When a female comes into estrus (which probably occurs rarely since females are either pregnant or lactating most of the time), the males will gather and wait in turn to copulate. Surprisingly, in the wild they show almost no dominance in their sexual behavior. An estrous female traveling with a group of males may copulate 30 to 50 times in a day! Chimpanzees, however, also show a tendency to form short-term pair bonds. Sometimes a male and an estrous female will disappear for a few days into the forest together, out of sight of other males. This is colloquially referred to as "going on safari." Here they may copulate 5 to 10 times per day (Tutin, 1975).

Gorillas, who are about as closely related to us as chimpanzees, also have multi-male social groups, but the groups are small and *age-graded* (Eisenberg *et al.*, 1972). This means that the few males are under the dominance of a definite leader, recognized by his age and his silver-haired back, who controls all aspects of group life. The younger males, often his own offspring, will either usurp his position in due course or leave the troop to form new groups. Here we find a multi-male group that has much in common with the one-male groups we have already mentioned. The females may, however, solicit and mate with the subordinate "black-back" males when they get the chance (Harcourt, 1979).

We know that early hominids moved away from the forest fringe and came to exploit the more open country of the savanna, and we find that most living savanna terrestrial primates have a different mating system than the chimpanzee pattern. Among the savanna baboons the dominance hierarchy has come to determine an order of sexual privilege among males; consort pairs are formed by alpha males for a few days when a female is in estrus. Among other terrestrial monkeys in more arid regions and some forest species, one-male groups are a permanent feature of the social structure. Both these adaptations can be interpreted as appropriate to the resource distribution and predator pressure of the environment in which the species lives (Crook, 1972). However, a full understanding of the determinants of primate mating systems and group structures is still lacking (Clutton-Brock and Harvey, 1977).

There is no doubt that the need to carry a helpless infant would seriously impede the hominid mother's mobility in gathering food, suggesting that female dependence on males may be an early event in human evolution (Lovejoy, 1981). Bipedalism, which frees the arms and hands, may well have been associated with this development. We know bipedalism was the first major hominid development, dating from over 3.75 mya. It may be that the human one-male group is at least as old as this. The environment and the helplessness of infants may indeed have combined to bring about the beginnings of the human family.

Two pieces of evidence tend to confirm that the ancestral human group was the one-male harem type. First, the testis size of men is not relatively large and indicates a lack of intermale competition of the chimpanzee variety. Second, humans show limited sexual dimorphism comparable to that of chimpanzees, which seems to indicate small one-male groups (see Section VI). This is certainly the most economical hypothesis to account for the evolution of the human family.

The main difference between the *unconstrained* sexual behavior of humans and chimpanzees is in the regular (but not invariable) appearance of consort pairs, which are only rarely seen among chimpanzees. The absence of estrus and the permanent sexual attractiveness of women make long-term pair bonding a possibility, strengthened as it usually is by friendship. Society, in the institution of marriage, comes to reinforce such bonds in order to establish a relatively permanent nuclear family. The prime function of marriage appears to be to create an intimate and secure social setting that is appropriate for the development of children. It also tends to stabilize and bring order to social behavior in the area of sexual activity and consort bonding.

In fact, we find there is not a great deal of difference between the sexual behavior of humans and chimpanzees. These differences can be summarized as follows:

1. Loss of estrus and the permanent sexual attractiveness of women make the consort pair more common in humans than in *Pan*.

2. Loss of estrus means the greater possibility of choice in forming consort pairs, so that friendship (social compatibility) is a component of pair bonding.

3. Dominance and status play some part in the selection of sexual partners (but see Section X below).

4. Humans show a tendency toward relatively permanent one-male units not found among chimpanzees. This is ultimately reinforced by the institution of marriage.

The institution of marriage has been described by Miller (1931) as a device "to check and regulate promiscuous behavior in the interest of human economic schemes." The human family, bonded by marriage, has evolved as an economic and political institution, and we may suppose that it has arisen as an adaptation to economic and political demands rather than through the evolution of a biological tendency toward permanent pair bonding. With the development of cooperative hunting in the Middle Pleistocene, it is likely that females became finally and fully dependent on males for part of their food supply; this dependence would have become critical during winter months in the North Temperate regions. In turn, males would have become more dependent on the food-gathering activities of women. The children would have been dependent on *both* for their survival. This interdependence probably also coincided more or less with the evolution of a longer lifespan and an increasingly helpless human infant. In the course of time any mature female would have had to support two or three dependent young while pregnant with another. With the development of social hunting, therefore, we can see the one-male group bonded more closely, characterized by division of labor and economic interdependence.

The human family is the simplest social unit with a complete division of labor between adult individuals. It is to the fact that the roles of man and woman have been fully complementary that the family owes its continuance and stability. Any interchange of roles, such as we see today in western society, will threaten that stability. Economic independence for women is bound to have a fundamental effect on the economic basis of the family; its stability will then perhaps come to rest on sexuality and friendship rather than economic ties.

With the necessity for increasing cooperation between adult males, the unwritten contract of marriage between each man and woman became part of a political bond between descent groups. Such a bond may have proved necessary for social stability when humans could survive only through the resources gathered in the social hunt and other concomitant social institutions (see Chapter 11, VII). The social group was thus enlarged again, this time not as a promiscuous troop but as a far-reaching and complex political structure uniting descent groups broadly through intermarriage—all brought about by the phenomenon of exogamy (marriage outside the kin group).

Most primate troops exchange males, who often move to new troops during adolescence and establish themselves, with difficulty, in a new social hierarchy. The females tend to remain. Such troops can be described as matrilocal and patriarchal. Father–daughter incest has been observed, but it is rare; mother–son incest is very rare. The movement of males between troops is one of the factors that reduces the incidence of incest.

In human societies gene flow is maintained by controlled exogamy: the sanction against marriage with close kin, made possible by humankind's recognition of kinship and our engrossing interest in our relations. This, in turn, has probably arisen from the deepening and expansion of affectional relationships within the family. These relationships (father–son, mother–son, brother–sister, etc.), though derived from the sexual function of the parents, are reinforced by economic needs and services. The children are economically dependent on their parents for food, shelter, etc., and brothers and sisters have a subsidiary and complementary role to play in family affairs (food gathering, baby minding, and so on). Incestuous relations between members of the family would clearly upset the structure of the unit, since sexual and economic roles would be confused. It has also been supposed that they could be genetically dangerous in such a closely related group as a family, though the genetic basis of this claim is now being questioned.

One of the simplest types of exogamy is found today among the food-gathering tribes of Africa and Australia. In these peoples, families are usually united by ties of kinship into small local groups or bands of 20 to 50 individuals. A local band may occupy a well-defined area of land (comparable to the home range of primate groups) and live upon the plants and game found on it (Table 10.4). Members of the band hunt together and have obligations toward one another in sickness and old age. Being kinsmen, members marry outside the group yet within the tribe, so the incest taboo usually applies throughout the group.

This group of related families has superficial resemblances to a primate troop such as we find among the hamadryas or gelada monkeys. In these species the one-male groups come together for sleeping and when food is abundant, and split up when food is sparse and widespread. It seems possible that the one-male group may be the basic unit in all higher primates as Eisenberg *et al.* (1972) have proposed, that these groups may have banded together to form the common promiscuous troops in nonhuman species, and that in human society a group of related one-male troops came to form the band. Thus, the appearance of exogamy may be a direct consequence of the expansion of affectional relationships within a single descent group.

The existence of exogamy requires a social structure that stretches beyond the exogamous group. This factor results in another important difference between nonhuman and human society. Although home ranges overlap, primate intergroup behavior is generally unfriendly and antagonistic; there is, so far as we know at present, no supergroup social structure. Human intergroup behavior can be considerably more friendly, and a social structure generally exists that relates social groups. This broad structure is a result of exogamy and the political relationships developed between descent groups that exchange marriage partners. It seems clear that human social life is based on kinship and intermarriage (though this is no longer very striking in modern western society), for marriage is historically a contract not only between individuals but between different descent groups.

Thus, a local descent group may stand in a certain relationship to other local groups of the same tribe for purposes of marriage as well as *rites de passage*, where these ceremonies exist. Such a relationship involves periodic social gatherings, especially for potential marriage partners to meet. The gatherings commonly involve feasting and dancing, wherein nubile girls and marriageable men are able to meet in an intimate manner under socially controlled conditions. In many tribes supergroup structures take the form of clans. In such circumstances, however, descent-group exogamy may expand and enclose the whole clan, in which case the social structure will relate different clans within the tribe for purposes of marriages.

On the other hand, a reduction in the extent of the incest prohibition is characteristic of Eskimo society as well as modern western society. Because of the special conditions of Eskimo life (a very sparse, static population) and modern city life (a very dense mobile population), a more extensive exogamy is either impractical or unnecessary.

The final stage in the evolution of human society and the family was certainly dependent on the evolution of language and social institutions, which are discussed in the next chapter.

X. The Rise of Human Society

In studying the evolution of human reproduction we have encountered a number of factors that are important in understanding the development of human society. These are now summarized:

1. Improved conditions for individual growth, both prenatal and postnatal, result in an acceleration in growth rate from conception to two years, accompanied by a delay in maturation and a slowing of growth to maturity. A long lifespan evolves with a substantial postreproductive period, which may be associated with the development of culture and the need for the protection of a mother's investment in her children.

2. Increase of cortical control of sexual behavior is accompanied by an increase in the range of effective sexual stimuli as a result of response by association. Receptivity remains continuous, the phenomenon of estrus

is lost, and sexual behavior comes under more conscious control. Biological options in sexual relationships remain open, but they are probably based on the one-male pattern that may be a basic and ancient higher primate mating system.

3. Highly versatile creatures such as the early hominids can be expected to exhibit striking individual differences in personality and behavior. Since loss of estrus allows choice in sexual partners, social compatibility and friendship become important in the formation of consort pairs. This, together with the frontal position in coitus, emphasizes individualization in sexual relationships and strengthens pair bonding, which comes to have affectional overtones.

4. The human family evolves as a one-male group in response to economic needs, and this arrangement is reinforced later by social sanctions. Division of labor arises by age and sex.

5. Recognition of kinship is accompanied by prohibition of incest, a universal characteristic of human societies. Local descent groups become exogamous.

6. Economic competition is replaced by economic cooperation; sexual bonds and kinship are factors in such cooperation. Competition for food is replaced by food sharing, and individual status is maintained not so much by greed as by generosity. These changes arise from the move into a new terrestrial environment with an increasing dependence on mammal meat. Hunting necessitates cooperative group activity, which evokes a new, broader social structure.

7. The dominance hierarchy (if it were ever present) is replaced to a considerable extent by kinship as a mechanism of social organization, though a status hierarchy persists in human societies among males if not among females.

8. Broader social groups develop with an economic and territorial basis, and, with the further expansion of exogamy, social structures spread beyond the immediate descent group to clan and tribe.

The biological needs of living organisms are, broadly, food, sex, and safety. Among primates these needs are satisfied within the organization of the troop. In human evolution we see a new social grouping arise according to these needs in response to the difficult environment of the open savanna plains. Here the need for food was perhaps satisfied first by widespread gathering of sparse resources and later by close teamwork in hunting. Therefore, we find that economics determines a wider social structure, which is reinforced by political and sexual relations. Modern human society has evolved out of this economic-political-sexual structure, but it has been strengthened further by other developments. The basic framework that has been outlined in this chapter was enriched and controlled by the rapid development of culture, which has brought humankind to its present state.

Suggestions for Further Reading

An invaluable textbook on primate adaptation is H. Kummer, *Primate societies: Group techniques of ecological adaptation* (Chicago: Aldine, 1971), and a useful review of social behavior can be found in T. Rowell, *Social behavior of monkeys* (Harmondsworth: Penguin Books, 1972). The question of mating system and sexual selection is discussed in B. G. Campbell (Ed.), *Sexual selection and the descent of man 1871-1971* (Chicago: Aldine, 1972). The behavior of the great apes is reviewed in D. A. Hamburg and E.R. McCown, eds. *The great apes* (Menlo Park, Calif: Benjamin/Cummings, 1979,) and primate politics are discussed in F. de Waal, *Chimpanzee politics* (New York: Harper & Row, 1982).

A valuable review of sexual behavior among animals especially primates has been translated into English: W. Wickler, *The sexual code: The social behavior of animals and men* (Garden City: Doubleday, 1972). For a useful discussion of primate sexuality, see D. Symons, *The evolution of human sexuality* (New York and Oxford: Oxford Univ. Press, 1979), and on female sexuality see S. B. Hrdy, *The woman that never evolved* (Cambridge: Harvard Univ. Press, 1981).

For a well-illustrated article comparing human and nonhuman societies see M. D. Sahlins (1960), "The origin of society," *Sci. Amer.* **203**, 76–87. On the social structure of tribal society see (for example) R. Linton, *The study of man* (New York: Appleton-Century; London: Peter Owen, 1964). Examples of tribal kinship structure can be found in A. R. Radcliffe-Brown and D. Forde (Eds.), *African systems of kinship and marriage* (London: Oxford Univ. Press, 1950).

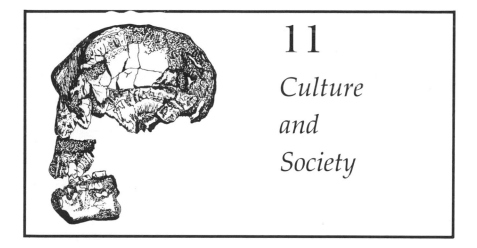

11

Culture
and
Society

**I. Culture
and
Protoculture**

Culture has been described, discussed, and defined at great length by social and cultural anthropologists, but has received scant consideration by biologists, for it has been generally assumed to be a uniquely human phenomenon and thus outside the field of biology. However, human biologists cannot avoid consideration of the origin and function of this most pervasive human attribute.

In the literature culture is almost always defined as a purely human possession; Dobzhansky calls it a species characteristic of *Homo sapiens* (1963). Most cultural anthropologists define culture as specifically human and at the same time dependent on symbolic thought or language. Kroeber and Kluckhohn (1952) define culture as consisting of "patterns explicit and implicit of and for behavior acquired and transmitted by symbols, constituting the distinctive achievement of human groups, including their embodiments as artifacts." L.A. White, too, states culture to be "dependent upon symboling . . . trafficking in meaning . . . which cannot be comprehended by the senses alone" (1959).

Evolutionary biologists must approach culture differently. Our definition must not be designed *a priori* to distinguish modern human society from nonhuman primate society; it must try to define in its simplest form that phenomenon which today we call culture, so that we can then see whether it can be found elsewhere in the animal kingdom and what function it performs. E. B. Tylor, the father of cultural anthropologists, defined human culture as "capabilities and habits acquired by man as a member of society" (1871). More recently, Keesing (1958) has defined culture as "the totality of learned socially transmitted behavior," and Oakley (1963) has defined it as "the sum total of what a particular society practises, produces, and thinks." These definitions may seem very broad, yet at the same time they are more precise. We are not concerned, like White and Kroeber and Kluckhohn, with trying to grasp an abstraction. We are concerned with what humans and animals do and how they come to do it. We must examine how human behavior and society differ from animal behavior and society so as to discover what behavioral and social novelties, if any, appeared during human evolution.

For us culture must be defined as a kind of behavior acquired in a particular way. We shall describe it broadly as the totality of behavior patterns of a social group of humans or animals that are passed from generation to generation by learning. Thus, if culture is socially determined, learned behavior, it obviously is not a specifically human phenomenon. Cultural behavior patterns are the outcome of versatility; they arise from behavioral flexibility and a capacity for observational learning. It seems clear that behavior patterns that become part of the culture of a group must be adaptive, though in what way is often not apparent. But there is no doubt that cultural behavior is an evolutionary development of the greatest importance, since a new adaptive behavior pattern developed by one member of a group may be quickly taken up by others through observation, and a behavioral response to the environment on the part of the population is much faster than a response by genetic mutation.

From the reports of Japanese monkey behavior (Imanischi, 1960; Kawai, 1965) there is no doubt that cultural behavior patterns can be easily developed by social primates. During the period of careful observation of these animals since 1952, they have been observed to adopt a number of novel kinds of behavior, which have been invented by one individual, imitated by other members of the group, and then passed from generation to generation (with, interestingly enough, varying degrees of success). Food washing and swimming in the sea are both patterns of behavior that were invented and became part of the troop's culture during the period under study. Similarly, it seems probable that the use of sticks by chimpanzees to extract termites from their nests, reported by Goodall (1968), is a similar type of cultural behavior pattern.

Haldane (1956) has reviewed the considerable evidence for cultural behavior among other mammals and among birds. A particularly interesting example of a novel cultural development among birds is the tradition of pecking through milk-bottle tops to consume the cream, which began in 1945 among Titmice (or Tits, *Parus* sp.) in those parts of western Europe where milk bottles with aluminum-foil tops were placed daily on doorsteps (Hinde and Fisher, 1951). The tradition spread throughout the area and continues to this day.

Culture has also been described as a "nongenetic adaptation" and as "an organ of human behavior" (Huxley, 1958). The first description will apply to all learned behavior, while the second will apply to material culture only. For us culture is more accurately seen as a set of behavior patterns that are learned *by* members of society *from* members of society and that are societal adaptations to the environment. Culture is a product of society and a property or characteristic of it. Through culture, therefore, behavior has reached a "supraorganismal level" (Huxley, 1958).

If cultural behavior is to be passed around a social group, its spread must depend on observational learning if not instruction (Chapter 1, VI). It follows, then, that culture will be limited to social animals that can learn in this way; but, given this possibility, any novel behavior pattern with a selective advantage will, under appropriate circumstances, be incorporated into the behavior of the social group in which it arises. Culture will spread by diffusion.

It is characteristic of culture that cultural techniques or traditions are passed down within an evolving group, and during their passage through time may change in accordance with the needs of the group and their changing environment. Cultures, in fact, "evolve," not according to the mechanism of organic evolution, but independently of it. While adaptive organic mutations (that is, new genes and gene combinations) may take generations to become established and spread through a population, new cultural adaptations may spread in one generation; they may be acquired very quickly and passed on by learning. Cultural adaptation is also flexible, and will allow numerous specializations for different ecological niches.

When, however, we come to examine the whole range of cultural behavior from birds to humans, we will need to recognize a clear-cut distinction between symbolic and nonsymbolic culture. This will enable us to distinguish human *symbolic culture* from the animal cultural traditions such as we have mentioned, which we shall call *protoculture*.

The manufacture of artifacts and tools is a particular sort of cultural behavior, involving objects taken from the environment and an appropriate organ of manipulation. As we have seen, birds use thorns, stones, and twigs as tools, and so do primates. Chimpanzees prepare their sticks for termite extraction. Humans fashion stones and many other objects with great skill to achieve a particular end. Such objects are a remarkable product of human cultural behavior. *Homo sapiens* is often defined as a toolmaker and, therefore, as "the cultural animal," but toolmaking is only one category of cultural behavior. It is not really different in kind from other cultural behavior except in its product, called *material culture*.

Evolution is opportunistic, and any novel behavior pattern with a selective advantage will be incorporated into the behavior of the evolving population. The use of external objects as tools or weapons increases the effective exploitation of the environment and decreases the danger of predators. The beginnings of material culture arose when objects were fashioned from raw materials—when sticks were intentionally sharpened and stones carefully shaped. Where it opened up extensive new food resources the new behavior would have been widely and rapidly adopted.

Techniques of toolmaking are not typically learned by trial and error at many different times but are invented by intelligent behavior on rare occasions and then transmitted as learned behavior from one generation to another. Children learn from their parents not merely by observation but also by instruction involving speech. Simple protocultural traits were certainly developed without speech, but the significant cultural advances that we associate with the Hominidae must almost certainly have awaited an adequate language. Speech alone made possible levels of abstraction that were necessary for the development of a more advanced material culture and what we think of as human society. Thus it is that social anthropologists such as L. A. White insist on the symbolic nature of culture, because most of human culture is indeed dependent upon *symbols*. (A symbol is an object or act the meaning of which is not self-evident or innate, but socially agreed upon. Meaning is bestowed upon it by those who use it.)

Humankind's immense capacity for the elaboration of symbolic culture, which is dependent on our increasing educability, is clearly a characteristic of great selective value; it is in some ways surprising that its extensive development is unique in the animal world. Our capacity for symbolic culture is as much a part of our heritage as our hand or pelvis, and its evolution is interdependent with all our other attributes. Since the transmission of culture depends upon the facilitation of learning in a social context, such learning, which is typical of the social primates and which in turn maintains social integration, is an essential basis of culture. Our culture itself was a product of human inventive genius, which in turn was the product of intelligence. As we have seen, clear traces of protoculture have arisen in the tightly integrated societies of nonhuman primates. In these groups what has been described by Etkin (1963) as "social sensitivity" is very well developed, for social cues are cues to behavior, and social stimuli control individual response. It is in these tightly knit social groups, in which there is ample opportunity for imitative learning, that the transmission of acquired cultural behavior develops, and more complex behavior only awaits invention.

It is a very important characteristic of material culture that to its creator it is not only an "extrabodily organ" but also part of the environment, a part that affects its bearers closely. When human culture develops, therefore, the human environment changes, and a new adaptation of the organic body will be selected (Fig. 11.1). We therefore find ourselves living in a fast-changing environment and adapting to it with a fast-changing culture. No sooner does the culture adapt than the environment has changed, and further adaptation is selected. In the face of this ever increasing rate of environmental change our physical evolution runs far behind, suggesting that humans are really somewhat ill-adapted genetically to their present environment. In at least one sense that is true: without the protection our material culture forms between us and our environment and the capacity for resource extraction that it provides, we would probably be unable to survive. But it is also open to question whether humans are fully adapted in any sense, for the rate of change in our environment may appear to be outstripping even our cultural adaptations.

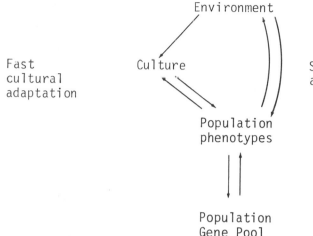

Figure 11.1. Systemic interaction of population, culture, and environment. Note the contrast between slow biological evolution and rapid cultural change.

Huxley (1958) cites "mentifacts" as an aspect of human culture; they are the assumptions, ideas, and values that determine our conscious behavior. Belief in God, belief in the desirability of moral rectitude, belief in the value of the individual are the kinds of ideas that form the basis of modern society and are of extraordinary importance to us. Ideas that constitute the basis of religion and ethics are the most sophisticated aspect of culture, for they involve conscious intelligent thought, a uniquely human capacity. This characteristic of human culture is, we believe, adaptive, and it, too, must evolve, like other cultural traits. In general, mentifacts are by nature static and stabilizing in their effect; paradoxically, they can be lethal if they do not retain some degree of flexibility.

Surprisingly, one aspect of the subject of this section is well documented in geological deposits, and that is the rate of cultural advance. While the remains of early humans are rarely preserved, at least one kind of cultural product almost invariably survives in large numbers: stone tools. The development of stone industries (which have been described as "petrified ideas"), while not necessarily correlated with the development of human culture as a whole, does give at least some guide to our cultural progress. This progress is astonishing, for it is neither constant nor irregular, but exponential: the rate of change increases regularly.

The ever increasing speed of technological progress is evident today. Developments during the 300 years since the beginning of the industrial revolution have altered the face of the earth. Some 6000 years ago the use of metals, and before that the development of agriculture, had already changed the human environment. Even the rate of progress in agriculture and metals, though slow, was rapid compared with the evolution of the stone-using cultures that occupied perhaps three million years. The increase of human cultural resources was at first exceedingly slow, a painful growth of idea and technique. Unlike the evolutionary process, however, the cultural process was both cumulative and progressive, and both the quality and quantity of human cultural assets have grown to where they now surpass the understanding of any single person.

Human culture is not a thing apart. It is clearly dependent on the evolution of both speech and society itself. In this chapter we shall discuss both the internal psychological correlates of culture, such as conceptual thought, memory, intelligence, and speech, and the external correlates, such as social structure. Finally, some of the cultural determinants of human behavior can be summarized.

II. Percepts and Concepts

In our consideration of human evolution and human culture it is necessary to refer to what has been described as a unique human mental* characteristic: conceptual thought.

A *percept* may be considered to be the mental image of the external environment. Just how this mental image comes into existence is not yet known, but it is clearly based on two kinds of information: input from the senses and memory of previous experience. It can be shown experimentally that a given percept is not necessarily a true if limited interpretation of the present environment but that it is strongly influenced by previous experience. In the growth and development of each individual mammal, motor investigation of the environment plays an essential part in building up its appropriate perceptual interpretation. There is no doubt that the content of perception varies beyond our understanding among the different kinds of animals, and we cannot conceive the world of a dog, for example, or a nocturnal prosimian. Our own perception of the outside world has evolved with the higher primates, to which the motor component of exploration is of fundamental importance. Spatial relationships must first be experienced to be later perceived, and we inherit from our primate ancestors a very spatial—a three-dimensional—perception of our environment.

From these considerations follow two conclusions: First, the quality of primate perception will depend on the extent and refinement of motor investigation. Primates more than all other mammals (except bats) live in a three-dimensional world; their eyes are stereoscopic, and their movements are in all three planes of space. They must have precise ideas of spatial relationships, for arboreal locomotion involves a knowledge of space far greater and more detailed than that which may be necessary to ground-living forms. The prosimians as well as the higher primates have no doubt come to perceive the environment as a three-

---

*Mental is the adjective of mind, the functional and subjective aspect of the living brain.

rather than a two-dimensional pattern. The integration of spatial data from the senses to form a composite perception of the environment has clearly advanced further among primates than among other groups of animals. Visual, somatic sensory, and auditory inputs will be analyzed and integrated with the motor and proprioceptive patterns. Such integration appears to take place, at least in part, among the association areas of these different sensory inputs on the parietal lobe of the brain (see Chapter 8, VI).

Second, the memory component of perception is of fundamental importance and must almost inevitably come to include not merely an experiential record of events but some generalizations about spatial relationships. These generalizations are immeasurably expanded and deepened by the information obtained about the environment as a result of the manipulation of objects (Chapter 7, VI). The higher primate can extract an object from the environment, free it, as it were, from spatial implication, and build up a perception of it as a distinct object, not merely as part of a pattern. In time, the primate will come to perceive the environment not only as a three-dimensional pattern but also as an assemblage of objects.

Homo sapiens, though a primate, is no longer an arboreal frugivore, a fact that has played an important part in the evolution of our perception. Kortlandt (1965) records that if food or any other familiar object is given to zoo chimpanzees at a place where they are not accustomed to receive it, the apes are inclined to react to it as though it were something entirely strange, and may even refuse to consume their meal. For apes, things tend to lose their identity when they are displaced. Among carnivores, on the other hand, a displaced food tray or other object usually causes no problem, as every dog or cat owner knows. As Kortlandt puts it, the perception of apes carries a certain quality of "thereness," that is, its position in relation to other things, for among these primarily visual creatures the patterning of the environment is still partly the basis of their perception. Carnivores, on the other hand, identify food primarily by scent, and no strong visual pattern contributes to their perception of it. Visual perception in primates functions to identify static objects that comprise their food; in carnivores olfactory perception functions to identify mobile objects that comprise their prey.

When humans began to hunt, human perception evolved accordingly. Using our primary sense, vision, humans evolved the ability to identify objects on the move without reference to their relationship to the fixed part of the environment; they saw them as totally separate from their environment. Here was a fundamental change in primate perception and something novel among land animals: a carnivore that hunted by sight.

It is clear that manipulation and hunting came in turn to make the human perceptual world different from that of all other animals. Human analytic perception, more than any other factor, opened the door to the development of conceptual thought and eventually to symbolic culture.

Let us now turn to concepts. A *concept*, as generally defined, is an abstraction from the particular to the class. The concept "bird," for example, must be abstracted from the perception of, first, "this flying object" and, later, "that flying object." The concept "bird" does not apply to "many birds" or "all the birds I have seen" but to all birds possible in space and time. Similarly, the concept "food" applies not to "this food" but to all possible and potential food—fruit not yet picked, animals not yet hunted. Consideration of the evolution of perception, such as we have attempted, seems to imply some degree of abstraction from experience, from the particular to the class. The abstraction may not have gone far and may not be complete as a concept, for intermediate stages of abstraction can exist. Nor is such abstraction conscious, for perception is not a conscious activity. But it does seem likely that conceptual thought may perhaps have originated in the classification of experience that was necessary for the sophisticated interpretation of the environment involved in human perception.

It seems likely then, that concepts may form part of the mental activity of animals, but there may still be an important distinction from human mental activity, for thinking is a conscious process, and so, it follows, is conceptual thought. Apes can deal intelligently with objects here and now—objects they can see and the function of which is clear. They can be trained to carry out activities that appear to show foresight, and they can learn appropriate behavior for future needs. For example, chimpanzees will select straws and prepare vines (by stripping of side shoots) even when they are out of sight of any termite hills, and then set out to visit a known but distant group of hills to obtain termites. Apes in zoos will use sticks to reach bananas too high above them to be picked by hand and will pile up boxes to get closer to them; yet they have limited creative and imaginative ability. An experiment was performed by Kohler (1925) to demonstrate the limitations of ape mentality. He loosely constructed a box out of sticks so they looked like boards. When the chimpanzees were presented with bananas out of reach, they were unable to see, within the structure of the box, the sticks they needed to get the fruit. Thus, an ape is unable to make a tool out of a natural object; it is able only to use a natural object as a tool with, perhaps, slight modification. An ape cannot conceive the tool without seeing it; it cannot see the stick in the plank or a hand-axe within a piece of rock. This humans alone can do.

Freedman and Roe (1958) have suggested that frustration may well be one prerequisite in the development of conceptual thought; it is certainly a component of it. If a biological need arises and is not quickly satisfied, the requirement for such satisfaction will appear in the mind, not as a particular object, perhaps, but as a generalized one. Thus, frustration may bring with it imagination, and imagination is the consciousness of sets of concepts, which are the classification of experience. Having in our possession the concept of a small cutting stone, we can look for it in a rock. We can see the possibility of its manufacture in future time, even though we may not have it at present. Abstraction means escape from the present, escape from the immediacy of life. It has been said that what distinguishes humans from animals is the length of time through which human consciousness extends. In animals this dimension is small, stretching a little way into past and future; in humans it grows both qualitatively and quantitatively. The evolution of conceptual thought gives humans greater power to live in the past and in the future by abstraction from the past.

To summarize, we find in mammals evidence for the classification of experience, which might be called unconscious conceptualization. This sort of classification, even at a very low level, seems to be necessary for the development of perception. As perception improves and incorporates more data from sensory investigation, especially in the higher primates, we find the probability that certain associational patterns of such data are crystallizing as concepts that might be equated with "things" and perhaps with relationships. Human conceptual thought appears to be characterized particularly by its conscious nature, but it is no doubt the result of a steady process of evolution from less conscious and indeed unconscious concepts in primates. The human achievement was the fully conscious concept of things not possessed but needed; the recognition of game, weapons, or women as classes brought with it the classification of more and more of the environment and the possibility of foresight of every future need. Leaving behind the narrow limits of present time experienced, humans entered the broad expanse of past memory and future concepts. Foresight, however, depends upon abstraction from the past in a manner that is termed intelligent (see Section IV).

But the conscious concept did not appear unaided. It seems probable that it was finally evoked by the use of symbols, gestures perhaps, but more often words, which make up language. The concept "bird" could be brought into full consciousness only by its identification with the word symbol "BIRD." The word symbol was the twin of the conscious concept, and it seems likely that they were born and grew together.

Human language is not fully programmed in genes, but the capacity for speech and conceptual thought is certainly innate; the word symbols themselves must be learned. In human evolution a totally new structure of complex sounds has evolved to express not only emotions, which animal vocalizations express, but concepts; thought and speech came hand in hand.

What, finally, are the neurological correlates of conceptual thought? We cannot expect to find a concept in the brain, for the brain is a living system, not a hard-wired computer. We can, however, point to one fascinating principle, which may in time become more clearly established than it is now. As we have seen, the brain is a spatially organized and oriented organ in which the two sides are related by contralateral connections to the opposite sides of the body, and in turn by sensory input to the two halves of the spatial environment. Thus, sensory input and motor elaboration within the brain are spatially organized (see Chapter 8, VI). There are, however, some brain functions that are nonspatial, and include all physiological homeostatic control mechanisms (such as body temperature, heartbeat, and breathing controlled by the hypothalamus). These are found in the central infracerebral parts of the brain. In primates we can add perception of depth derived from stereoscopic integration (see Chapter 3, VI), and in humans concept and word memory (see Section III). Spatial relationships involving depth and distance may appear to be predominantly spatial concepts, but they are not *of* space, but are *about* space; in themselves they are spaceless and concerned with pattern rather than place.

We do not know accurately the site in the brain of concept and word memory, but we do know that the so-called "primary speech area" is concerned with both and that it is to be found on one side of the brain only (see Section VI and Fig. 11.8). It was shown by Penfield and Roberts (1959) to occur on the left parietal lobe only, in what has been called the secondary association cortex (see Chapter 8, VI). On the other side of the brain, in the same area, notions of space appear to be elaborated, and damage here often leads to disturbances of form and space recognition, body image movement, and spatial interactions.

We have seen that an object that can be manipulated is in some way freed from complete spatial implication and can be "handled" mentally, apart from its environment, as a concept. Such a concept does not need spatial representation in both hemispheres of the brain; representation on one side is sufficient. Equally, concepts of spatial relationships are not in themselves spatial representations, and they too require representation on only one hemisphere. It appears, therefore, that the neurological correlate of conceptual thought may prove to be unilateral representation in the cerebral cortex. Concepts based on objects and other abstractions that can be symbolized are possibly represented on one parietal lobe; spatial relationships (not symbolized) appear on the other.

III. Memory   Memory is a psychological and neurological attribute of fundamental importance in the evolution of animals. By this term we mean all the information recorded in the brain during the lifespan of an individual. Penfield and Roberts (1959) distinguish three levels of memory in humans, which will form a useful basis for our discussion. In the brain, however, they may not be neurologically distinct from one another.

A. *Experiential Memory.* Experiential memory is a neuronal record of the stream of experience. This remarkable property of the brain appears to preserve *in sequence* all the experiences of which an organism is aware during its life. The developmental extent of this phylogenetically old memory cannot be precisely established in animals, but its existence can be postulated as a necessary basis of learned behavior.

Learning necessitates recording the sequence of events of any action and its accompanying perception, and such a record must be available for comparison with future sequences of experienced events. This memory operates at the subconscious level, and even in humans learning can be and often is wholly unconscious. Thus, when one person meets another, the degree of familiarity of their face will be instantly known, and an appropriate greeting will follow.

Every experience is automatically compared in an instant of time with all other equivalent experiences, every new face with all other known faces. Normally it will not be possible to recall the previous experiences at will; familiarity will not necessarily be accompanied by a full conscious knowledge of the previous circumstances of the meeting. However, it is interesting that ease of recall, both conscious and subconscious, depends on the amount of emotion associated with the original experience. This fascinating phenomenon speeds up the learning process in all animals. Biologically important experience, which is directly concerned with the survival of the individual and the species, is accompanied by emotion—fear, pain, and pleasure. In some way, such emotion (typically involving the release of certain hormones) impresses the neuronal record of the experience more deeply in the brain. Thus, the learning of biologically useful behavior is facilitated at the expense of biologically useless behavior. Humans, however, came to learn facts outside of a context of emotion; they have evolved further levels of memory.

B. *Conceptual Memory.*   Conceptual memory may be considered to be the accumulation of abstract concepts about experience. It is the record of classified experience in no time sequence rather than a full sequence of particular *referents* (that is, individual events and objects). At its simplest level it is little different from experiential memory, for in the recognition of a face the mind will "present" known faces as a class, ready for comparison. In a dog's experience of rat catching, the dog will already have all rat-catching experience, as it were, classified for reference, and will draw on this experience as it tackles its dangerous task (for rats have sharp teeth). If a dog is told "rat," it will look excited and ready to spring, but we may guess that it will experience the emotion and sequence of its last, or its most emotional, rat hunt. The abstract concept "rat" is perhaps of no biological value, but it is interesting to realize that if it was developed, it probably resulted from the dog's master using the linguistic symbol "rat." For a dog, the meaning of the word is learned by hearing it at different rat-catching experiences, and, if an abstraction is to be made, it is made with the use of that symbol and the assistance of a human brain.

Human conceptual memory goes much further. We can record classes of objects and our experience with them (cars, car driving), and we can also record emotions without attached objects (love and hate) and classes of experience, such as emotion itself. Beyond this, we can continue to abstract and classify indefinitely to the point of a concept such as "entropy." Second-order abstractions and others even more remote from direct personal experience are certainly confined to humans, and in view of the importance of symbols in the handling of abstractions (such as $e = mc^2$), it seems unlikely that even the lowest level of abstractions are commonly found in animal psychology.

C. *Word Memory.* Penfield and Roberts distinguish between conceptual memory and word memory on the basis of both clinical and everyday experience. It is not uncommon to recall a concept (for example "stork") but forget the word symbol, so that one is forced to describe the concept ("you know, those white birds, the kind that nest on rooftops, steeples, etc. What are they called?"). It is even possible to learn the concept without ever knowing its word symbol. ("I have often wondered what those large black birds that nest in groups in treetops are called.") Equally, it is easy to forget the meaning of a word that was once known ("I have forgotten the meaning of 'solipsism'."). Words can be learned that have no conceptual basis ("abracadabra"). In clinical studies, electrical stimulation of the cortex may result in the recall of words without concepts or, on the other hand, concepts without their word symbols (described with much difficulty), a condition known as aphasia. Evidence strongly suggests that concept and word memory are closely interconnected and evolved together but are nevertheless distinct.

It appears that concept-word facilitory circuits develop and become fixed as in a conditioned reflex; the nerve cells are quickly habituated to the passage of impulses through a repeated pattern. As Penfield and Roberts (1959) point out:

> A man listening to a speaker may follow the man's words, ignoring the concepts: or he may attend only to the concepts of which the words are symbols, ignoring the words. If this listener is bilingual, he may fail to notice what language the speaker had used. (1959: 234)

Word symbols may be learned with immense speed by a human child; every normal child learns virtually an entire language by the age of five, and in a bilingual household a child can learn two languages by the age of six. A 9-year-old boy has been known to understand, read, and write nine languages fairly fluently; a mature human many more. It is noteworthy that the units of language are words, not letters; words are heard, and words are read. Chinese writing, which, like ancient Egyptian, is based on ideograms—that is, discrete concept symbols—runs closer to brain function than does Indo-European writing with its limited alphabet. The basis of reading is word recognition, not word analysis.

The relationship between these different kinds of memory and the sensory input from the environment, which has been discussed in Chapter 8, is summarized in Fig. 11.2. It will form the basis for our further discussion of brain function.

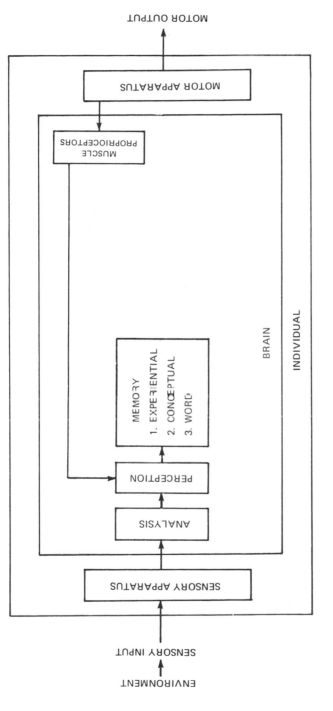

Figure 11.2. Diagrammatic illustration of the process of perception and its relationship to memory. The outer rectangle represents the limits of the individual, surrounded by the environment; the inner rectangle represents the limits of the brain.

A useful discussion of the distinction between
IV. Intelligence   learning and intelligence has been published by
Herrick (1956). We may define learning as a pat-
tern of constructive activity performed unconsciously or consciously; the
act of learning may be a purely empirical process without any end in
view, but its outcome will be a kind of knowledge—an adaptive behavior pattern. The capacity to learn has evolved steadily throughout
the animal kingdom, and Rensch (1956) has shown that it has been
accompanied by an increase in brain size. The processes of learning and
intelligence are closely connected and have something in common, but
there is certainly some distinct feature of human mental activity that is
conveniently defined as "intelligence" and that can be clearly distinguished from a highly developed capacity to learn.

Intelligence has been defined as the relating activity of mind—the
ability to realize the connection between discrete objects and events. It
clearly involves imagination, which may be considered to be the presentation from the experiential memory of memory traces (knowledge) not
obviously directly connected with the immediate experience. It may
involve conscious recall of memory, but the significant feature of imaginative thought is that a much broader range of memory traces is
brought to the interpretation of day-to-day experience than in the process of simple unconscious learning, for only thus could the imaginative
connection between events be realized.

In humans the everyday distinction between intelligence and the
capacity to learn is not always immediately apparent. For example, expert linguists or stenographers have learned complex skills without
necessarily being unusually intelligent, and professors may be famous
for their learning rather than for their native brilliance. In contrast, a
comparatively ill-educated person may be highly intelligent. But intelligence cannot exist alone; it is a capacity that interacts with knowledge, the accumulation of experiential and conceptual memory. Acquired knowledge is the fuel with which intelligence burns and without
which intelligent action is impossible. "Intelligence integrates knowledge and gives it direction"; it is a conscious and "purposively directed
mental process with awareness of means and ends" (Herrick, 1956).

The neurological seat of intelligence cannot be defined, for it is not a
specific activity but a generalized facilitory system. It is a method of
handling knowledge as a whole. Its evolution is correlated with the
evolution of the cerebral cortex, but that does not mean that the cortex is
necessarily the seat of intelligence. On the other hand, a functioning
cortex is necessary for intelligent activity because of its essential role as
input analyzer and memory bank.

In the accompanying diagram an attempt has been made to develop
our diagram of memory structure (Fig. 11.2), to indicate very simply the
role played by mediating mechanisms in mental activity, in particular
the part played by intelligence (Fig. 11.3). Mental activity means, broad-

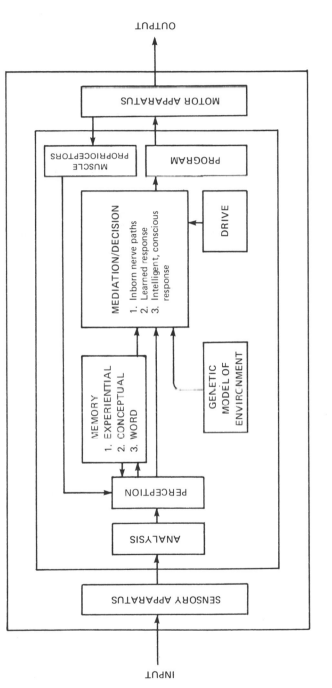

Figure 11.3. An expansion of Fig. 11.2 showing the relationship of perception and memory to mediating mechanisms and motor output. Note that mental activity such as mediation is triggered by a physiological "drive" such as hunger or fear. The output of the mediating mechanisms will be passed to that part of the cortex where motor output is programmed. The box labeled "genetic model of the environment" signifies an innate record of appropriate environmental parameters, deviations from which will elicit behavioral responses.

ly, the neuronal mechanisms that operate between stimulus and response. It effects response according to the stimulus offered by the environment, and such response may be behavioral or physiological. Physiological responses are automatic and inborn (adjustment of body temperature, salivation, etc.) and do not concern us here (see Chapter 1, V). Behavioral responses are often learned and may even follow intelligent thought.

From Fig. 11.3 we see that behavior can be generated in three ways, by three types of mediating mechanisms. Phylogenetically the most ancient is the innate kind of mediation. This is believed to involve the transmission of the impulse between stimulus and response through an inherited pattern of nerve paths (the "hard-wired" mode). The response can be modified only by inhibition.

The second type of mechanism is already found in lowly invertebrates (such as the flatworm), where we have the beginnings of the capacity to learn by the conditioned reflex. In this second kind of mediating mechanism the relevant cerebral facilitory circuits are conditioned by habituation until the appropriate response follows automatically upon the stimulus—the brain is "programmed" (Sperry, 1955).

In the most advanced organisms conscious thought can determine the response on each occasion on the basis of an original assessment of the environmental situation. Unconscious habit (conditioned reflex) can now be superseded by the phylogenetically new and flexible "intelligent" approach. Of course, we do not always act intelligently; a vast amount of our behavior springs from unconscious conditioned reflexes, but, while activity that was originally intelligent may eventually tend to condition our behavior, it is in its origin at least partially free from previous patterns of activity. Truly intelligent thinking (if such a thing is possible) may be considered to be free from the direct influence of any existing cerebral facilitory activity. When, therefore, intelligence mediates our behavior, it carries our response far beyond that of the conditioned reflex into a totally new realm of flexibility and adaptability. On the basis of the refined input analysis that the human brain can provide, a much more precise and delicate assessment of the environment is available; in turn, intelligence makes possible a much more exact and delicately appropriate response. With the considerable intelligence that we possess, the possibilities for novel behavior are practically infinite, and from this remarkable development of the human mind has come our culture and our genius for invention.

The human brain is, as we have said, exceptionally large for our body size. It is generally assumed that this increase in size has come about as a response to the selection for intelligence among humans. It is of interest, in this context, that brain size in primates is closely correlated with the size of the newborn, and it has been suggested by Lynch *et al.* (1983) that our brains might be no more than a product of an allometric shift in the size of the newborn, which preadapted us for intelligent activity. Certainly it is not easy to account for the extraordinary capacities of the human brain if it is seen merely as a product of a hunting and gathering lifestyle.

Whatever the reasons for the size of the brain, cranial capacity has always been broadly related to intelligence in any account of human evolution, and those primates with absolutely larger brains (apes) do seem to be more intelligent than those with smaller brains such as monkeys (Passingham, 1982). Among humans, brain size varies as it does in any other species. All but 5% of men lie between a range of 1180 to 1704 gm, and all but 5% of women lie between a range of 1033 to 1533 gm (Passingham, 1979). Taking body size into account, sex differences are found to be minimal or nonexistent (Gould, 1980).

The question has also been asked as to whether or not variation in brain size among living humans is related to intelligence. While it has generally been indicated that this is not the case by citing famous writers with widely varying brain sizes such as Turgenev (2012 gm) and Anatole France (1017 gm), Passingham (1982) has shown that the evidence is not yet available to resolve the question. Intelligence and brain size may be loosely related (Van Valen, 1974), but the present evidence is equivocal; you cannot measure brain size accurately on living people (to whom you can give intelligence tests), and the brain itself shrinks during old age. (The latter change is accompanied by intellectual decline.) This problem awaits further research.

Sacher (1975, 1976) has demonstrated a close relationship among different species between body size, longevity, and brain size; we know that precocial animals are large and live for a long time. We can look at the evolution of precocity as being associated with an emphasis on brain evolution, and in particular with the evolution of intelligence. K selection favors species in which the females invest much care in individual young. Intelligence will in turn enable those young to survive for a long time, so that during their lifespan they can steadily achieve reproductive success. If the environment were constant or life was very short, innate behavioral responses would perhaps suffice, as they do in many short-lived animals like insects. But long-lived terrestrial animals, which must survive changes in season and fluctuations of climate, and which by their size are enabled to range widely in search of food, need the capability to learn and, through their intelligence, to predict.

V. Nonverbal Communication
Some form of communication is almost invariably found among sexually reproducing animals, both vertebrate and invertebrate, as a means of bringing about the fertilization of the female or her eggs by the male. Signals will be transmitted by the female to attract and trigger a sexual response in the male. In some species, especially birds, "courtship" ceremonies, which are specialized modes of communication, are particularly striking.

But it is among social animals that communication becomes a major evolutionary adaptation; societies depend on communication between individuals for their existence. Social behavior and organization arise from this phenomenon. Communiciation is a means by which one organism can bring about change in one or more others (beyond simple thermodynamic action and reaction); it is the means by which one animal can trigger a response in another. Because communication among social animals is so important, it has been said that the social sciences are the study of the means of communication within social groups. In all animals communication is nonverbal; only among humans has language evolved to supplement nonverbal signals. In this section we shall briefly review the complex subject of nonverbal communication and its evolution among higher primates. In the next section we shall discuss language.

Nonverbal communication can take a wide variety of forms; the signals may pass by a number of different channels. The *output channel* is the means whereby the signal is transmitted. The *output effector*, which generates the signal, may be a specialized organ such as a scent gland or the sexual skin of a female; or it may be an organ partly evolved for communication, such as the facial musculature or the vocal tract; or it may be a totally unspecialized organ, such as the skeletal musculature in general.

The *input channels* are the sense receptors, which can be classified as was shown in Fig. 9.1. Signals transmitted via distant nondirectional ("wide beam") channels are usually received by all members of the group (e.g., calls); signals transmitted by directional or contact channels can be aimed at a particular individual (e.g., tactile or body scent), and these are called "tight-beam" transmissions. Some signals may attract the attention of nonmembers (calls or scentmarking), while others may be limited to reception by members of the group concerned (gesture and touch). Some transmissions are more suitable for a forest environment where vision is restricted (calls), while some are appropriate for communication in open country where predators are a threat (gestures and displays). The channel selected and the evolved transmitting and receiving organs are always appropriate to the environment of the species concerned.

Among the higher primates many channels of communication are exploited, though as we have seen, the emphasis among receptors has changed from olfactory to visual. However, olfactory signals are still important, especially in sexual encounters where the precise condition of the female is made known to the interested male by pheromones. Scent is probably also important in the bonding between mother and newborn infant, but we have no experimental work to confirm this.

Visual receptors receive a vast amount of information in primates, and the optic channel is probably the most important among chimpanzees and terrestrial monkeys. Transmitters vary from the involuntary physiological coloration of the sexual skin of females, which announces estrus, to the facial expressions, gestures, and displays that are so well known among baboons and chimpanzees. Facial expression, clearly visible in a relatively hair-free face, is highly evolved among chimpanzees (see Fig. 8.19) and has remained important throughout human evolution. It is interesting that men who allow their beards to grow freely probably use this channel less than chimpanzees. Its importance is related to the evolution of the facial muscles. Bodily gesture is also important to all the higher primates, including ourselves, but its transmission and reception are often unconscious. Displays, which are active stereotyped motor sequences usually denoting anger or threat, are more common among chimpanzees than other higher primates, though stamping the foot and tearing the hair, as well as scratching the head, can be placed in this category.

An interesting comparison can be made here with dogs, which have a highly evolved gestural system of communication that is centered around the face and the tail; indeed, the whole body is used to communicate in a way humans understand very well (Fig. 11.4). It is not surprising that dogs and wolves, which are social carnivores, have a great ability to signal in this way but express a limited number of vocalizations. Social hunting, as we have seen (see Chapter 9, VIII), requires cooperation, endurance, intelligence, foresight, and a system of communication. As among dogs and wolves, human communication came to serve the complex organization of the hunt as well as the technology associated with it. It is no coincidence that dogs and humans understand each other so well today; until recently they both led the life of the social carnivore. As we shall see, an evolved communication system is not the only psychological attribute they have in common.

Touching is also an important means of communication among higher primates. Because of its tight-beam characteristic, it comes into play in interpersonal bonding and is thus more limited in use than the visual signals described above. In the mother–infant bond and in sexual contacts, touch is vital to the successful outcome of the relationship. Physical contact also plays an important part in the establishment of bonds between adults and in lowering tension levels in a social group. Among chimpanzees we find that the hand-to-hand touch or caress is as reassuring as it is among humans, and in many other species mutual grooming is an important means of bonding members of the troop and reducing tension. Our own culture seems to have neglected the value of touching as a means of communication essential for full human development (Montagu, 1971).

Because our own communication is primarily vocal, research workers have tended to concentrate on the study of vocalizations among higher primates at the expense of other channels. As might be predicted, this channel is exploited most among forest species that have only limited visual opportunites to communicate. It can also be used when an individual is occupied in some other way; for example, to transmit an

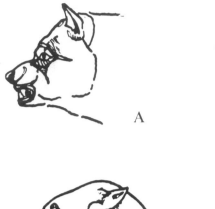

Figure 11.4. Examples of visual communication in dogs. *Above*, some facial expressions: (*A*) confident threat, (*B*) threat with some uncertainty, (*C*) very weak threat, (*D*) uncertainty and suspicion. *Below*, some caudal expressions: (*E*) self-confidence in social group, (*F*) uncertain threat, (*G*) normal carriage (in a situation without special tension), (*H*) depressed mood, (*I*) and (*J*) complete submission (from Schenkel, 1947).

alarm call while watching a predator or running quickly. Bird song can be noisy, but it is nothing compared to the din produced by some forest primates. Schultz (1961) has written:

> The orgies of noise, indulged in especially by howlers, guerezas, gibbons, siamangs and chimpanzees, seemingly so repetitious and meaningless, are probably at least as informative to the respective species as most after-dinner speaking is to *Homo sapiens*. (1961: 61–62)

This brings us to the function of vocalizations and other signals. Briefly, they can be grouped as follows: (1) signals that bring about bonding between mother and infant or within the social group as a whole; (2) signals related to territorial demarcation and defense; (3) signals announcing food; and (4) alarm calls on sighting predators. The gibbon is the only ape that regularly defends a territory, and whose vocalizations are mainly directed to define its extent and warn rival groups away. Howler monkeys show similar adaptations, with their specially modified hyoid bone that acts as a resonator. Most other primates confine their calls to items 1, 3, and 4. Because of the essential part played by communication in a social group, it is obvious that the majority of transmissions will be those of the first category, which maintain social structure and social bonds between individuals.

Calls in category (4) are of interest to us because among vervet monkeys (*Cercopithecus aethiops*) they do seem to carry some of the characteristics of language; they appear to refer to objects in the environment. The alarm calls vary according to the predator sighted: the call is a high-pitched chutter for snakes, a short tonal call for leopards, and a low-pitched staccato grunt for birds of prey such as the martial eagle (Seyfarth *et al.*, 1980). If tape recordings of each of the calls are played to the monkeys, they respond appropriately, scrambling into the trees on hearing the leopard call and looking up or down in response to the other two.

It would be wrong, however, to attribute to animals the kind of conscious intent that we have in making a verbal communication. The signals that we have been discussing are almost certainly generated or motivated by emotion and find their neurological origin not in the cortex but in the limbic system of the brain (see Chapter 2, V). Nonverbal communication can be described as the expression of emotion. By doing this in stereotyped and overt ways, animals inform each other of their needs and fears and trigger appropriate responses in one another. The expression of emotions can also be seen as the basis of social life; it creates and maintains both bond and structure in primate society, and can, therefore, be seen as an evolutionary adaptation. Beyond such physiological changes as in sexual skin or the tumescent penis, this and this alone is the means of communication among primates.

Are the alarm calls of the vervet an exception to this generalization?

Do their vocalizations actually refer to the objects that threaten them rather than to their own emotions? Or are they merely distinguishing snake fear from leopard fear? Or, alternatively, does one call signify surprise, another aggression, and another fear? We do not yet know the answers, or can we be certain that some primate vocalizations do not refer directly to objects in their environment. Experiments to investigate this have not given us positive results. Chimpanzees have been seen to point toward food or individuals in experimental situations, but have not been observed making even this simple gesture in the wild. The evidence at present suggests that limbic system communication is sufficient for their needs.

The vervet evidence does show us, however, that the monkeys are capable of categorizing features in their environment, and presumably they are using concepts of snake, leopard, or eagle. Their ability to categorize is learned and improves with age (Seyfarth *et al.*, 1980).

But there is more to communication than transmission and sensory reception. The appropriate response to the reception of a signal is a vital ingredient in communication. This is the social sensitivity we have already described, and which became the basis of learning and culture. Awareness of minute changes in the posture or facial expression of different individuals is an essential component of this adaptation, as is the appropriate response, which may be learned or innate.

Thus, nonverbal communication has evolved in two interlocking fields: in the expression of emotions and in the awareness of the behavior of others, which alters and directs the behavior of individuals in adaptive ways.

Though a nonverbal call system and a language are both vocal-auditory means of communication, Hockett and Ascher (1964) have shown that there are significant differences between them, so many, in fact, that it is difficult to trace the possible course of evolution of one into the other. These differences may be summarized as follows:

1. A call system depends on the generation of signals or *signs*, which can be defined as anything that announces the existence of some event, the presence of a thing or person, or a change in a state of affairs. The meanings of signs are naturally apparent. They are innate and do not require social agreement. This definition should be compared with that of the symbol, the basis of language (see Section I, this chapter). The distinction is clear-cut.

2. Calls are mutually exclusive. That is, an animal cannot emit a complex double call, using some features of one call and some of another. Only one call can be emitted at a time. This situation is described as "closed"; language, in contrast, is "open" in that we can freely emit utterances that are entirely novel combinations of sounds.

3. A call is emitted only in the presence of the appropriate stimulus; a call system does not show temporal or spatial displacement. Speech, on the other hand, does; we speak freely of things out of sight, in the past,

future, or even nonexistent. This capacity of speech is based on thought processes discussed in the preceding sections.

4. The differences between the sounds of any two calls are total. Speech, on the other hand, is the elaboration of a limited number of components of sound that in themselves have no meaning but, according to their pattern, may have an infinite number of meanings.

5. The call system is innate (generated by the limbic system), so the relationship between stimulus and response is genetically determined and somewhat inflexible. Language, on the other hand, is learned (generated by the cortex), and its transmission must depend on an extensive capacity to memorize and vocalize. However, this capacity for speech is a genetic endowment; no amount of effort will teach a nonhuman primate to speak.

VI. The Evolution of Language

The act of speaking uses a set of body parts of diverse primary function. They have been taken over for this new secondary function with little modification and without interfering with their previous function. The speech apparatus includes the *larynx*, pharynx, tongue, and lips, and while these organs must work closely together, they are controlled by motor nerve supplies of very different origin in the brain. At the same time, as Spuhler (1959) has pointed out, while such coordinated muscular movement usually requires adjustments from proprioceptors, they are absent from the important laryngeal muscles, and feedback control comes by way of the ear. The only place where the different motor organs and steering apparatus of speech could be connected is in the cerebral cortex, suggesting that the corticocortical connections that make speech possible are phylogenetically recent. Their appearance is probably correlated with the evolution of the human cerebral cortex.

Before further consideration of the cortical speech areas, it is necessary to understand how the larynx, pharynx, tongue, and lips came to effect speech. The few slight evolutionary changes that are believed to have effected the speech apparatus of evolving hominids have been described by DuBrul (1958) and Lieberman (1975) and can be summarized as follows:

1. As we have already seen (Chapter 9, V), the chin has been molded in such a way as to avoid any further decrease in the space that forms the floor of the mouth. This floor is effectively sealed (Fig. 11.5) by a hammocklike diaphragm of the *mylohyoid muscle*, which together with the deeper *geniohyoid* lies beneath the tongue and can raise and lower it, aided by the tongue musculature.

2. The relative shortening and broadening of the jaw has brought with it a shortening of the tongue. The genial (anterior) and hyoid insertions of the tongue are brought closer together, and the base of the

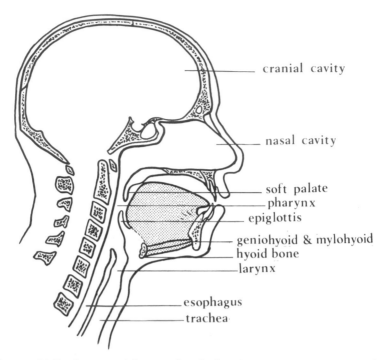

Figure 11.5. Section of human head showing structures concerned with speech production. The tongue and its musculature are shaded.

tongue is thrown back into the pharyngeal area; the whole tongue is thick and bulbous.

3. The hyoid bone and larynx have also retreated a short distance down the neck, so the *epiglottis* and *soft palate* are separated by a small space called the *pharynx*.

4. The *digastric muscles* have been slightly rearranged so they disturb the stability of the hyoid bone as little as possible, and have moved apart to allow up-and-down movement of the floor of the mouth (Fig. 11.6).

The result of these small changes is that the nasal cavity can be closed off completely. Thus, expelled air can be shunted through the mouth, and the tongue and lips can move freely to interrupt this flow of air.

The actual sounds of speech are formed by the vocal cords in the larynx. Here vibrations are set up in the expelled air and modified in the mouth. The shape of the tube formed by the larynx, pharynx, and mouth determines the vowel sounds upon which the consonants are imposed by the tongue and lips.

It seems that the really important human physical characteristic is the possession of the pharynx. The anterior portion of this tube is formed by

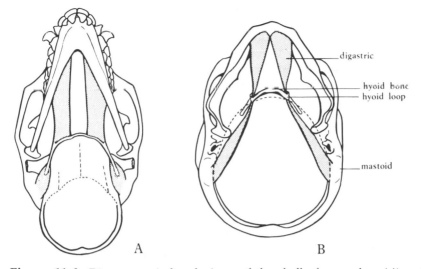

Figure 11.6. Diagrammatic basal views of the skull of a monkey (*A*) and human (*B*). In the monkey, the diagastric muscles are attached to the hyoid bone; in humans, two fibrous loops are formed on the bone through which the muscles pass. The posterior parts of the two muscles are also farther apart in humans to make room for the neck (which has moved forward in relation to the skull) and the tongue. The dotted line shows the approximate extent of the ventral margin of the neck in each animal (redrawn after DuBrul, 1958).

the highly flexible base of the tongue. Easily altered in shape, it generates the whole range of vowels and makes possible resonant human *phonation* (articulation of sounds). The distinction between humans and apes lies in the kinds of sounds humans make and the remarkable way in which we can interrupt resonant notes with the consonants of speech. Movements of the mouth concerned with speech are not greatly different from those concerned with sucking or swallowing. Just as the sucking reflex concerns the lips and tongue, so the consonants are formed: *p*, *b*, *m* at the lips; *d*, *l*, *n*, *t* at the tip of the tongue; *g*, *k* at the tongue base. The proximity of the tongue to the larynx and sides of the mouth may also form gutturals and clicks in certain languages.

It is significant that devoted teaching by the Hayeses could not induce their chimpanzee baby to learn to speak more than three or four poorly pronounced words (Hayes, 1951). Other experiments resulted in similar failures. Neither apes nor monkeys have the anatomical equipment to generate speech sounds of the kind and variety produced by humans.

As Andrew has demonstrated (1965), the baboons show us how human speech may have arisen. In baboons a common form of communication is lip-smacking, a display derived from movements of the lips and tongue that they use in grooming. Lip-smacking is not an uncommon greeting among higher primates, but the baboons alone emit a grunt at the same moment. These grunts, modulated by the tongue and lip movements, sound very much like human vowels and are produced in exactly the same way. The plains-dwelling baboons and the plains-dwelling human ancestors shared a way of life that depended on group cohesion and structure. It is perhaps not a coincidence, therefore, that baboons share with us the ability to make vowel sounds modulated by tongue and lip as an aid to within-group communication.

Speech, however, is more than complex sound. It is the act of codifying thought in sets of controlled and connected sounds, and such codification occurs in the cerebral cortex. The primary motor area for speech, which controls movements of the lips, tongue, etc., lies in the *motor gyrus*, as was shown in Fig. 8.15. In speech and writing, impulses must pass from these areas through the different nerves to the speech apparatus or hand. The resulting movements of the speech apparatus musculature in vocalization and the precise movement of the hand in writing are conditioned reflexes. But the motor gyrus itself is not, of course, concerned with associating words and concepts.

Evidence from cortical excision and electrical stimulation experiments has shown (Penfield and Roberts, 1959) that the parts of the cortex concerned with the source of speech symbols and concepts lie (as we have seen) for the most part only on one side of the brain, the so-called "dominant" hemisphere, usually the left. (Damage to that hemisphere in early life will result in displacement of the speech areas to the right hemisphere.) These speech areas are shown in Fig. 11.7. They include *Wernicke's area*, which appears to be the most important and partly coincides with the uniquely human inferior posterior parietal region (or *angular gyrus*). Damage here results in loss not of verbal fluency, but of sense. The patient's speech seems to be lacking in content; the appropriate words are frequently replaced by others, sometimes related in meaning and sometimes not; and in some patients the sound components of words are incorrectly integrated.

Information passes from Wernicke's area to *Broca's area* on the frontal lobe; damage here results in loss of words and difficulties with grammar rather than failure of sense. These two areas of the cortex create the program for speech (Geschwind, 1972), which is synthesized by a third area, in the supplementary motor cortex where the appropriate muscle patterns are planned. These plans are then fed into the adjacent motor gyrus where muscle contractions are effected. [The parietal area on the right (nondominant) side is concerned with awareness of body scheme and spatial relationships; see Chapter 11, II.] It is of great interest that Wernicke's area overlaps the parietal association area (see Chapter 8, VI), which is found only in humans, as well as the interpretive cortex of the temporal lobe.

In humans there appears to be a high negative correlation between the side of the dominant hemisphere and left- and right-handedness. The majority of people are right-handed and have left cerebral domi-

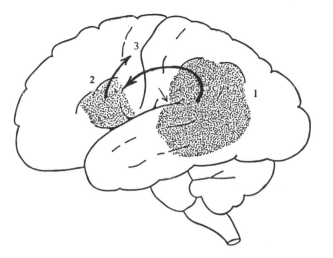

Figure 11.7. The three speech areas of the cerebral cortex. The angular gyrus and Wernicke's area (*1*) is the most important; the frontal Broca's area (*2*) is the next most important but has been found to be dispensable in some patients; the superior frontal area (*3*) is dispensable and lies in the supplementary motor area. The first two are found in the dominant hemisphere only; the third is duplicated on both sides. The small arrow indicates the planum temporale of Wernicke's area (redrawn from Penfield and Roberts, 1959).

nance, while at least half the left-handed people have right cerebral dominance. Though some degree of handedness is found in monkeys, this hemispheric dominance with preferential use of limbs is a characteristic found only in humans. It is also interesting that it is not found at birth but emerges in the course of growth, coinciding with the development of speech from about two to twelve years of age. The brain often becomes asymmetrical in that part of Wernicke's area where it extends into the *sylvian sulcus*. This part of the cortex, the *planum temporale*, is one-third larger on the left side of the brain in 65% of individuals (Geschwind, 1972; see Fig. 11.7). We do not yet have the evidence, however, to demonstrate a probable correlation between this factor and cerebral dominance (Passingham, 1982).

What is the function of cerebral dominance? Passingham (1982) has pointed out that not only do the organs of speech occur in the midline (and, therefore, do not effectively require bilateral motor control), but to achieve high precision in muscular contractions of the mouth and tongue it is better to have one central program generated in a single hemisphere. This would be more efficient than the generation of two programs which, to ensure coordination, would have to utilize the relatively long pathways through the connecting *corpus callosum*—the nerve tracts that connect the two cerebral hemispheres. It is significant that bird song, which is often very intricate, is also generated by a brain with somewhat similar asymmetry in function.

Sperry (1966) has done some remarkable experiments with the help of patients in whom the corpus callosum has been surgically severed. They show that in the right-handed patients only the activities of the right side of the body (with nerve supplies to the dominant left cerebral hemisphere) are reported to be conscious; the left side of the body can be shown to behave quite independently, and the patient is not able to report such activity as conscious experience. We also have confirmation from the experiments that speech and writing are unique correlates of the dominant hemisphere, and it follows that speech is necessary to express consciousness. There is, however, no reason to deduce that consciousness is not also characteristic of the nonverbal hemisphere even though it cannot be verbalized. We are certainly vividly aware of ourselves in relation to our environment without the assistance of symbolic thought.

During the last 20 years enormous efforts have been made to teach language skills to apes. This work was triggered by the Gardners, who realized that if apes were incapable of producing the sounds of speech, they might nevertheless be able to communicate by other symbolic means, such as gestures. In their now-famous teaching experiments, the Gardners (1969) taught the chimpanzee Washoe the American Sign Language for the Deaf (Ameslan), which depends on gestures of hand and arm. Soon after, Premack (1976) set out to teach his chimpanzee Sarah to read and write using plastic magnetic symbols for words, which could be placed and moved on a vertical board. Rumbaugh (1977) used a keyboard with symbols that were projected onto a screen to teach his chimpanzee Lana.

These experiments and many others, which have included gorillas and an orangutan, surprised everybody, because the apes proved to be able to communicate symbolically with considerable success about the various matters that concerned them, such as, food, drink, and company. But the experimental results have been difficult to interpret. None of the reports suggests that the apes are fully capable of using language as we know it, but they do show that apes have some capability to learn and to use symbols in a linguistic manner. They also cite evidence that apes appreciate a simple rule of word order. In a recent review of the work, Passingham (1982) concludes that their competence is that of a very young child. The whole question has been investigated in detail by Terrace (1979) and Desmond (1979).

It only needs to be added that apes have not been reported to use symbols under natural conditions. It is possible that they do so, but observations of symbol use have not yet been made. Their use of language is always a product of intensive training by humans. In contrast, children will learn language skills without being taught, but merely by exposure to speech.

The fossil evidence is of limited value in our study of the evolution of language. Attempts have been made to reconstruct the vocal tracts of fossil hominids by a study of the base of the skull and its supposed relation to the hyoid bone (which is never preserved). From this work it has been shown that the relatively long bent pharynx peculiar to adult humans is not present in apes or newborn humans and could not have been present in *Australopithecus* or *Homo habilis*, which had an upper respiratory system like that of living apes (Laitman and Heimbuch, 1982). [It is relevant that individuals with Down's syndrome (mongolism) retain the infant morphology and are unable to speak.] The functional human vocal tract develops during the second year of life. A short apelike pharynx limits the variety of vowel sounds and the consonants that can be imposed upon them, and alters the possibility of speech in many other ways (Lieberman, 1975). It is claimed that the fully human pharynx makes its first appearance in the Steinheim skull (ca. 200,000 BP) (Laitman *et al.*, 1979), but this does not imply that earlier humans were not capable of an extensive linguistic reportoire that was adequate for their needs. The critical changes in this structure probably began in the Middle Pleistocene, though definite evidence is still lacking.

We have stated that there is evidence of asymmetry in the human brain, which is possibly related to cerebral dominance, handedness, and even to language. It is interesting, therefore, that Holloway and de la Coste-Lareymondie (1982) have measured asymmetries in ape and hominid endocasts and found that while all species show some assymetries, all the hominids, including *Australopithecus*, are grouped together in possessing the peculiar human pattern of cerebral asymmetry. Holloway has also pointed out (1981) that the configurations of the gyri and sulci of *Australopithecus* were of the human pattern, and that Broca's area can be identified in some fossils (Holloway, 1974). Altogether this evidence at least suggests that there were important changes in brain structure, organization, and function during the evolution of *Australopithecus*, which could well be significant in the evolution of language.

An equally important correlate of linguistic ability is the size of the brain in hominid evolution. As we have seen, the endocranial capacity expanded from a range of 400 to 800 cc about 2 mya to a range of 1000 to 2000 cc today. This represents a brain three times the size of what would be expected for a higher primate with a body size similar to that of humans. This size increase was sudden in evolutionary terms and came relatively recently in hominid history. The peak of brain expansion came one to one-half million years ago with the evolution of well-organized social hunting and the technology associated with North Temperate adaptations. It is also probable that the most rapid deceleration in postnatal growth occurred at this time. This cerebral expansion, which is perhaps correlated with the expansion of linguistic ability, was the product of a much older transformation—the cerebral reorganization that language implies.

While the large-brained elephant and whale have no language, human "bird-headed" dwarfs, whose brains are in absolute size no bigger than those of gorillas (400–600 cc), can talk on the level of a 5-year-old child (Holloway, 1968). Evidently they carry the anatomical basis for language, and they show that a large brain alone is not an essential correlate of linguistic ability. In contrast, we have seen in the previous section that apes are not capable of human language. Even though their cranial capacity may reach 700 cc, their brains are not constructed for speech.

It seems reasonable, therefore, to conclude that language evolved over a long period. The hominid and ape lineages have been independent for at least four million years and probably more, and this seems a reasonable length of time for the evolution of the extensive corticocortical connections that make language possible. No doubt auditory perception and analysis evolved step-by-step with vocalization. The evolution of the modern vocal tract would have followed, responding to the advantages of improved phonation.

One thing is certain: language does not effectively replace the primate call system, but is a new and supplementary means of communication. While the call system, the expression of emotions, holds pride of place in the most profound aspects of human relationships, words have evolved to symbolize environmental processes and the abstract concepts arising from the new complexities of social structure. Language has made possible the vast development of human culture and brought us the unique human consciousness of ourselves and others.

Language must have constituted a powerful new adaptation, and we can summarize its adaptive role as follows:

1. Language constitutes a method of *environmental reference* through symbols. It allows communication about objects out in the environment, rather than about the inner state of individuals.

2. Symbols allow the *storage* of far more information. Thus, the record of the environment, stored in the memory, can be immensely enriched and further enlarged through the increasing development of exploratory behavior.

3. The symbolic nature of a language permits *classification* of such referents, and indeed of all memory data. A classified data bank allows far more efficient retrieval than the unclassified sequence of experiential memory. Classification of data also allows storage of cultural information and norms of behavior and provides better based criteria leading to more adaptive decisions. The extended data bank is especially valuable where there is environmental unpredictability with rare events, as occur in semiarid and cold regions.

4. Language permits the *transmission* of such symbolic data. Thus it comes about that each individual carries the pooled experience of the social group, past and present, as knowledge is passed on and accumulated. Language also allows the transmission of information between bands or groups, which may be intermittently extended through space

and time: it maintains information sharing through the division of labor. It is an essential factor for successful cooperation among extended groups, permitting broader dispersion and wider interdependence.

5. Language, therefore, implies *displacement,* so that reference can be made not only to the present, but to the past and future: this fosters the development of instruction as well as mere observational or social learning.

6. Language permits *discussion,* bargaining, planning, and democratic processes generally. Individuals could, for the first time, share their thoughts as well as their feelings. It allows more flexibility in social behavior and more originality and provides new grounds for both antagonism and cooperation.

7. The articulation of the demands of *reciprocity* was probably essential for the development of complex webs of social obligations in time and space, which bond extended social groups.

8. Language permits *prescription,* or instruction, as well as proscription, or prohibition, which alone is well developed among social non-human primates.

9. Language permits the appearance of more *complex social organizations,* with different, named, individual roles.

Language is a genetically determined characteristic of modern *Homo sapiens,* but our various languages, the actual words we use, are part of our culture. This remarkable, priceless, and uniquely human kind of behavior rests on our peculiar endowment of a large, somewhat reorganized and slowly maturing brain and the precisely controlled musculature of our vocal tract.

VII. Social Stability and Social Institutions

The evolution and survival of a cultural society, as well as a population without culture, depends on factors of variability and stability. Just as a population of animals evolves by the stabilizing action of natural selection upon a random source of genetic variability to produce a controlled and more or less steady evolution, so customs and technology "evolve" by a process of variation channeled by stabilizing factors. The source of variation lies in the infinite variety of human behavior and invention, and the factors that stabilize it are just as important to the survival of society as are those effecting it. The individualistic and imaginative nature of humans makes possible an immense variety of behavior patterns, and it is characteristic of social institutions among both higher primates (e.g., the dominance hierarchy) and humans (e.g., one-male grouping) that stability in these patterns is promoted; in effect, they protect society from anarchy. Thus, the potential plasticity and flexibility of individual behavior, although precious sources of novelty, are controlled by customs, social sanctions, and eventually by religious and legal institutions. Such institutions protect and transmit both sanctions and knowledge, which are necessary to integrate and maintain social structure for the survival of the social group.

Some of the most important ways in which social stability may be maintained will be briefly considered, for, although the variety of human customs and beliefs is endless, social behaviors and situations are basically the same everywhere. Various customs and institutions have been selected because they have helped to ensure the survival of human societies.

First, the preservation of a body of knowledge essential to the social group is an important factor in maintaining continuity in social life and is the basis of tradition, which helps to promote regularity in behavior patterns. Knowledge about how to behave as well as more abstract knowledge about the supposed bases of human existence are preserved in preliterate societies in the form of ritual (a verbal and behavioral statement) or myth (a verbal statement). Both ritual and myth are means of condensing and encoding knowledge; both are, characteristically, regularly enacted in public as a means of maintaining their power, meaning, and integrity.

Ritual has been described as typical of both humans and animals. Although the two forms of ritual have superficial characteristics in common, their function is different. Both human and animal rituals are stereotyped behavior patterns used as means of communication; well-known examples among animals are the dance of the bee, which communicates the direction and distance of a source of honey, and the courtship dances of birds. The complexity of these animal rituals has only one aim: to transmit an unambiguous communication signal relating, in these instances, to the whereabouts of food and the prospect of copulation. The signal is related to immediate biological needs. Human rituals may be distinguished insofar as they rarely relate to immediate biological needs, are concerned with social rather than individual communication, and are far more than unambiguous signals; in fact, they might better be described as complex statements.

Of course, humans also have predictable courtship patterns that can be described as ritualistic and relate to the prospect of copulation (drinks, dinner, dancing, kissing, petting, etc.), but there is a flexibility here not found in birds. In a discussion of human rituals we are concerned with customary behavior patterns normally associated with set words, patterns that gain their power by being carried out in an exact form and that are consciously enacted. Human ritual serves to perpetuate knowledge necessary for survival and, by combining speech and action, constitutes a much more compact form of social memory than words alone (Leach, 1966). Compared with speech, the language of ritual is enormously condensed; many different meanings are implicit in the same category sets (also an attribute of mathematics). Ritual is a special form of language with a very high information content, but, unlike language, the meaning of ritual depends upon its social context.

A ritual act is a social act; it is society's meditation of traditional knowledge and behavior.

In its remarkable way ritual records knowledge about social origins, about social structure, kinship, and obligations. It records behavior patterns of a fundamental nature, such as hunting, toolmaking, and food preparation. Ritual is, as it were, the DNA of society, the encoded informational basis of culture; it is the memory core of human social achievement.

Myth is the verbal derivative of ritual, and, with the evolution of writing, the informational content of ritual has to some extent taken a verbal form. The written early Sumerian myths are probably only the verbal part of a ritual and as such are only half the language. As literate peoples, we feel more at home with the creation myth of the Book of Genesis than with complex religious rituals, but the character of this myth, as of others, is to have many meanings.

It is impossible to speculate about the origins of social behavior based on a study of primitive society today; nevertheless, it seems a reasonable postulate that ritual behavior was an early and effective means of recording cultural achievement in the absence of a developed language. With the evolution of language, the trend may have been from ritual to myth, from a more to a less compact means of encoding knowledge necessary for the survival of society. Such knowledge will contribute to the continuity of social norms and the stability of behavior, with its recognition by individuals as a tradition worthy of respect.

Second, claim has been made that the word *religion* is derived from the Latin verb *religare* (to bind). It this is correct, it is not merely a coincidence that an important function of religious institutions is to bind society together, even if it is done by binding humans to God at the same time. In primitive societies a personal religion, such as some varieties of Western Christianity, is very weakly developed, and public ritual forms the basis of religious activity. In fact, all rituals may be described as religious, for not only do they bind individuals to a core of social knowledge, but, by performing them, individuals are bound to each other in a common activity often requiring much effort and skill.

The binding function of religious ritual is, then, its second important function. Not only are traditions ritualized, but many important social activities appear to take on a sacred and ritual quality. An element in the common life of a society may come to have a special significance and in time become an act of religious observance. Especially important in this respect are *rites de passage* (birth rites, puberty rites, marriage rites, death rites), which consolidate social roles and social structure as well as bind members of the society together. However, religion does more than that, for it directs social sentiments toward one stable and symbolic center.

In many primitive societies, this further binding property of religion takes the form of ancestor worship, which creates as it were, a continuum between the living and the dead. The transition from contemporary to ancestor is believed to be a gradual one; the ancestor's identification with the next world is only slowly completed. In time, the sacredness of the deceased increases until in due course great reverence is paid to one common ancestor, worship of whom binds the members of descent groups together and gives them a common center for their sentiments. Archaeologists have uncovered evidence of burial customs of prehistoric humans dated some 70,000 years ago. Skeletal remains of large-jawed Neandertal people from western Europe and Asia have been discovered with patterned arrangements of rocks, bones, and stone artifacts on and around the bodies. There are also remains of cave bears (especially skulls) arranged ceremonially in sites that apparently were not inhabited (Fig. 11.8). These people left evidence of religious ritual, buried their dead with great ceremony, and probably believed in their survival after death.

In clans the totem may fulfill the same function as the ancestor, and in tribes or nations the ancestor/god may be personified by the living king or queen. Like the ancestor, the king or queen, although alive, is invested with a sacred quality, which enhances the social effectiveness of the institution. Divine kings are known in Africa, and the divinity of the monarch was still recognized in England until the end of Stuart times (1688). The English monarch today, though not explicitly divine, fulfills a religious function and acts as a symbolic center, especially in times of national danger. Society needs both ritual and symbol to give it form and continuity.

Third, another device that appears to stabilize and maintain social structure is economic interdependence. We have already seen (Chapter 10, X) that human society is bonded by economics and kinship. What we have to note in this section is that the bonding is sometimes strengthened and channeled by ritualized economic exchange. In societies with a subsistence economy in which an economic superstructure does not "naturally" exist, we often find a network of economic bonds superimposed in the form of the bride-price or, more generally, in the form of "tribute" or "royalties." The goods thus accumulated by a king or chief priest are in practice redistributed to the members of the society who supplied them; nevertheless, the ritual economic transaction serves to bind the society together in a common act and at the same time enhances the role of the king or priest. The modern system of taxes can be seen as a redistribution of goods for the common good and is, in fact, one of very few transactions in which the individual is directly and personally associated with the state.

These different means of binding and stabilizing society are clearly factors of overriding importance in human evolution. It remains to point out that religious and legal institutions help control human behavior at an individual level for the benefit of the society. What "is done" and "is not done" is a powerful determinant of human behavior; it is dictated by tradition, religion, and law, through each individual's peers and elders. Religious and legal sanctions control human behavior at its most fun-

Figure 11.8. The discovery of cave bear skulls in a pit suggests that either they were hunting trophies or they formed the center of some religious cult. Such structures date from the time of Neandertal man. This imaginative reconstruction is based on a discovery in France.

damental level, and the two kinds of sanctions are often so closely allied as to be indistinguishable, as in the Ten Commandments. Among these vital functions of religious and legal institutions falls control of the way people express their biological needs, by means of sanctions against theft, aggression, adultery and so on. From the need for social control of human behavior arises the mechanisms whereby an individual is enabled to direct his or her own behavior toward social ends—that is, the evolution of ethics.

VIII. The Evolution of Ethics

Humankind has been described as the ethical animal, which is undeniably one striking characteristic that distinguishes us from the rest of the animal kingdom. Its early recognition is recorded in the Book of Genesis, where the story of Adam and Eve describes our acquisition of the knowledge of good and evil. When Adam and Eve ate the forbidden fruit of the tree of knowledge, "the eyes of them both were opened, and they knew that they were naked; and they sewed fig leaves together . . . " (Genesis 3:7). Possession of this consciousness resulted in the expulsion of humankind from the paradise of ignorance into the bittersweet world of knowledge.

This ancient myth embodies one of the most important events in human evolution—the evolution of ethics, of the "knowledge of good and evil." Ethics arose as a direct result of the appearance of self-awareness in the growing human consciousness. Cortical control of motor activity and the establishment of connections to the limbic system from the frontal lobes of the cortex slowly provided evolving humans with consciousness of their actions and motivations, and beyond that, consciousness of themselves as individual beings. As it has often been written: the animal knows, but only humans know that they know.

How did this come about? What was the adaptive value of self-consciousness? Humphrey (1983) has suggested that the advantage to its possessor lay in the insights self-consciousness gave to an individual about the behavior of others in the social group. Such insights would be of great value in social relationships. As individuals became aware of the workings of their own minds, they gained insight into the workings of the minds of others. Thus, they were able to predict the behavior of others and plan their own strategies accordingly. If self-consciousness was advantageous in this way, then such an evolutionary adaptation would soon spread through the populations of evolving humans.

From this step the problem of ethical behavior followed inexorably. Humans could see, as a result of their self-consciousness, how many of their activities were directed to satisfy their own needs, their basic requirements for life; but they could also see that certain of their actions satisfied only social needs and led not to personal satisfaction but perhaps to frustration.

It is significant that, among animals, socially oriented, learned behavior is possibly most highly developed in dogs and wolves, not monkeys. When a pack of dogs hunts, it performs a cooperative activity that requires considerable discipline. In rounding up their prey, the members of the pack may have to show remarkable self-control in the interest of the group. As social carnivores, dogs have evolved the possibility of self-discipline, which is learned from and imposed by their peers and elders. As a result, as Kortlandt (1965) has pointed out, dogs are the only creatures besides humans that appear to experience guilt, the only creatures that seem to have evolved a recognizable superego, or conscience. Domesticated dogs can easily be taught not to touch meat left on the kitchen table. They can generally be relied upon in the master's absence, and they can certainly feel guilt if they disobey, as any dog owner knows. Human and dog understand self-discipline, and that is one reason why they understand each other.

If dogs have a social conscience as members of a social group, they are certainly not far from being ethical animals, but, as we have seen, ethics is more than this, for it is the result of social conscience *and* self-awareness. Without the latter the dog will not consciously make an ethical decision. This is not to deny altogether the possibility of self-awareness in dogs; a certain degree of self-awareness may exist in them, as it may in monkeys and apes. It seems likely that it does, for, as we know from our own experience, consciousness is not an all-or-none phenomenon; it varies by degree. However, it seems unlikely that a dog would experience an agony of indecision as to whether or not to succumb to strong temptation. Because guilt follows the act, we need not assume that foresight of guilt necessarily precedes it.

Ethics as we know it may have arisen, then, from the full development of self-consciousness in a demanding social context. One important key to human nature lies in the development of social hunting; like the dog, *Homo* was a social carnivore. In such circumstances the actions of every individual directly affected others in the social group, and what is important is that it was the social group (as part of the population) that was evolving, not the individual. In gaining awareness of our own needs and activities we also became aware of the needs of the society in which we lived. It is the conflict between the two needs, those of the individual and those of the society, that makes necessary some device for directing a choice between possible actions.

As we saw in Chapter 1, VII, natural selection operates on the expression—the phenotype—of an individual's genes, and by extension, on the genes of his or her relatives. This concept of kin selection has been used to account for the kinds of altruism (bioaltruism) that we find in small social groups of closely related kin. To account for altruism between individuals sharing a social group but not closely related, Trivers (1971) has developed the concept of *reciprocal altruism*. This refers to behavior that is likely to be reciprocated during an individual's lifetime and where the risk to the altruist is low in relation to the risk to the victim being rescued. If conditions are reversed at some future time, the altruist may benefit greatly from his original act by having his own life saved. Members of social groups will tend to encourage such behavior for their mutual benefit, so that reciprocal altruism will become part of the culture of the society.

To account for altruism on a larger scale, it has been proposed that selection operates not just on individuals but on social groups, even tribes (Darwin, 1871). This concept of *group selection* has gained some support, especially where warfare develops, and is discussed by Wilson (1975) and Grant (1977). If a band had the capacity to understand the significance of adjacent social groups and their home range or territory, they might realize that to kill them all (or only the males) and appropriate the territory would be of benefit to them by increasing the resources available for them and their children. Genocidal aggression would surely pay, and bands that showed the appropriate behaviors would benefit. Thus, we can imagine how some of the "noblest" traits of humans might have evolved, including the team spirit, cooperation, true altruism, bravery in battle, patriotism, and so on. Territorial expansion, which depends on group consciousness, has been a major factor in human history. Selection will, in these circumstances, operate on groups as well as individuals (Wilson, 1975). Altruism would be strongly selected within bands. It seems quite possible that in this way a society could develop an ethical system that would strongly channel human behavior for social ends.

It should be added that this important development in the making of humankind was possible because of our extraordinary flexibility of behavior, our educability, and our facility to internalize social rules. It is not so much a case of the selection of simple egotism or altruism, but the possibility of creating a superego that will almost unconsciously direct our behavior. The value system created in human history is toward the survival and protection of the society; this surely is the direct product of the operation of group selection.

Ethics, therefore, arose when humans found that they had to make conscious choices in a social context. The agony of the human condition arises from our self-consciousness and our knowledge of the price we pay as individuals for our membership in society. Wilson's genetic leash (Chapter 1, VII) is long enough to give humans the room to make choices; self-awareness makes it possible for us to recognize the whisperings from within, from our biology, and to turn against them, to give prior place to needs that are socially determined and can even be individually destructive. The evolution of the human condition is descibed as "the fall" because the authors of Genesis describe the coming of human self-awareness as alienation from a state of holiness (wholeness) and the entry of evil into humankind. Clearly, the evolution of self-awareness brought with it a new and terrible situation: human activity was to be directed, not by the straightforward operations of an unreflecting brain created as a tool of the genes, but by the conscious functioning of the human mind, which can foresee the pleasant and unpleasant, "good" and "evil" results of its actions, and choose how to act with freedom. Self-consciousness and foresight, therefore, brought discord to the human mind.

Freedman and Roe (1958) have accordingly described the emergence of *Homo sapiens* as the beginning of the age of anxiety. The need for values can be seen to arise from the fact of our possessing an "archaic neurological and endocrinological system partially but not completely under cortical control." On the one hand, the innate and unconscious homeostatic system of needs and satisfactions affects the motor system and physiological structures in a broad and relatively imprecise manner. On the other hand, the cortical analysis of sensory data responds relatively accurately to distinctions and social cues of no obvious physiological significance. In practice, there is a dislocation and delay between biological needs and their satisfaction that is due to social demands. Conscious frustration of biological needs, then, becomes the human condition and leads to internal conflict and anxiety.

> Only in man is there simultaneously such a rigidity of social channeling, and such a degree of potential plasticity and flexibility for the individual. Incompatible aims and choices which are desirable but mutually exclusive are inevitable conditions of human development. This discrepancy between possibility and restriction, stimulation and interdiction, range and constriction, underlies that quantitatively unique characteristic of the human being: conflict (Freedman and Roe, 1958:461).

Internal conflict and frustration may, however, be among the most important stimuli of cultural progress, of the development of adaptive behavior and technology. The development of ethical values is certainly adaptive, and different ethical systems reflect suitable adaptations for different social groups, which can sometimes be related fairly directly to features of their environments (Harris, 1974). Ethical systems are nearly always incorporated into religious dogma, ritual, myth, and belief, and in this way are stabilized and strengthened with the greatest authority.

Is there fossil or archaeological evidence of self-awareness? Curiously enough, there is. In Section VII we noted that the Neandertal people buried their dead as long as 70,000 years ago. Their bodies were not treated as rubbish or like dead animals, but were laid to rest with ornaments, flowers, and even food supplies for the next world. They foresaw a future. This awareness of the significance of death implies self-awareness and the concept of a value system. It is most likely that these phenomena are still older, but by 70,000 years BP the human mind had reached a level of consciousness that was perhaps little different from our own.

We can see, then, that with the evolution of self-awareness we stand in possession of conscious knowledge, purpose, choice, and values, all of which involve responsibility. Responsibility is humankind's special burden, and though society supports the individual at every turn, it is the individual who must make the final choice and upon whom the social group always depends. This, however, was not the only way in which the individual assumed a new and magified importance in human society.

IX. The Rise of the Individual

Although it is the breeding population that evolves and is the characteristic and permanent feature of organic life, our knowledge of the theory of evolution makes it clear that, until late in human evolution, when group selection may have played a part, the individual organism is the functional unit in terms of adaptation and survival. One of the striking features of humankind is the variety seen in individual personality and behavior, which seems to enhance the role and value of the individual in society. Such individuality is not, however, unique to humans, but has evolved slowly, like other characteristics.

Most social animals recognize other members of their social groups. The members of a troop of chimpanzees, to take an extreme example, are almost as different anatomically and as variable in personality and appearance as human beings are. Nevertheless, variations in personality and behavior are most striking among humans. These growing differences between individuals, and the importance of particular individuals to each other, appear to have evolved hand in hand.

The division of labor between the sexes, which we presume must have developed among our ancestors, is the beginning of the process of enhanced individualization. The fact that particular males and females are interdependent means that one individual is very important to another, if only at an economic level. With the appearance of one-male groups, the interdependence became sexual, expanding still further with the evolution of the family. Men and women recognized each other as economic and sexual partners. Role names such as "wife," "husband," "father," "mother," appeared and reflected the division of labor within the family. At this level, individuals came to depend on each other more and more; that interdependence within the family was the basis of love, which began as the expression of need. Individualization appears to have grown through the recognition of an individual's dependence on other particular individuals both within and beyond the family.

At a later stage in human evolution we find a diversity of occupations stretching beyond the family. The important function of the elders, the priest, or the king, gave these individuals immense importance in the society that recognized them and their vital role. In Neolithic times individuals made different contributions to social needs; some made tools and weapons and some produced food. Different goods and services were produced by different people in the more tightly knit urban communities that evolved. More people came to depend on more individuals to satisfy their different needs, and individualization developed with the increased complexity of society.

Besides the division of labor there was possibly a more fundamental reason for the individualization of *Homo sapiens*. The more behavior depends on learning and the more complex it becomes, the more it will vary among individuals. Variation in competence in performing tasks is commonplace in the animal world and is striking among the Japanese monkeys (Itani, 1965). It is a very important factor in human evolution because, the more behavior patterns vary, the more important individual action—individual achievement—is to the population as a whole. Through their individual behavior, people can determine the fate of the social group of which they are a part, for better or for worse. Society comes to care about the behavior of individuals, especially when they perform their more socially oriented activities, which, as we have seen, society attempts to control.

What is especially important in this thesis is that human behavior is to some extent directed by human intelligence, and the use of intelligence is a characteristic of the individual, not of the population. Intelligence itself is variable in its development and to a considerable extent genetically determined; genes for intelligence are selected as are any other genes. Intelligence, however, as we have seen, is a generalized faculty, and its varied application in problem solving (and problem setting) is the property of the individual. The evolution of intelligence will, therefore, by its nature enhance the value of the individual in the social group. The evolution of culture is due not only to natural selection for its potential within a varying gene pool but also to the selection of a series of imaginative inventions that arise from the intelligent activities of individuals. Thus, the individual, as the source of invention, is ultimately the source of culture.

It is relevant that intelligence and intelligent activity are not directly correlated with ability to learn (Section IV) or with good citizenship, sociability, or a strict adherence to social norms. As a result, it would appear that society should attempt to support and value intelligent individuals, even if they are not socially well-adapted, for the survival of society depends upon individual people. In a cultural society the contribution of each individual to the population is no longer only via the gene pool but may be direct and immediate, as a result of intelligent behavior. Cultural evolution finds its sources of variability, its "mutations," in individual behavior, which can spread through the population within one generation if proved of value.

X. Human Behavior

We have attempted briefly to describe and understand the origin and function of culture. Culture appears to comprise behavior and ideas that are the property of a society rather than an individual and are learned by social animals from each other. Human culture has developed from some basic primate characteristics that have been rapidly evolved. In particular, primate visual three-dimensional color perception characterizes the human adaptation, and to it have been added conceptual thought, self-consciousness, and a remarkable level of intelligence. As noted earlier, the social behavior of the Arnhem chimpanzees foreshadows human behavior to a remarkable degree.

Among primates, protoculture evolved in tightly knit social groups, which, though of modified structure, remained the necessary basis of human culture and made possible the extensive transmission of this kind of learned behavior. The modifications involved the evolution of the human family and band, and the appearance of exogamy, which, respectively, improved cultural transmission within the social group and improved cultural diffusion between groups.

Culture must have come to contain its own means of transmission at an early stage. Observation was reinforced by instruction, which was made possible by the evolution of language. Humans had something to communicate that could not be transmitted by gesture, facial expressions, or simple vocalizations; the evolution of human culture depended on the development of a symbolic language that was adequate to transmit it in all its complexity.

The newly evolved flexibility of human behavior that arose from the cortical control of behavior patterns might have tended to weaken social bonding, and it seems clear that one of the functions of religious and other social institutions was to stabilize individual behavior for social ends. Social bonding was maintained by these institutions in various ways so that the newly evolved versatility of human behavior would not disrupt the social group and endanger its survival, yet would remain available as a source of new cultural invention. Tradition and convention thus served to control human behavior, especially in the biologically important realms of food resources, reproductive strategies, and defense; humans became conscious of the different motives for their behavior, and the human condition of internal conflict and anxiety developed. Group selection may have operated in some areas to reinforce altruism and other "noble" values.

Culture now controls every aspect of human behavior. It is clear that even the simplest most basic biological activities (such as suckling or defecation) are in part culturally controlled, and that the socially important activities, like fighting and copulating, are under very strict cultural control. In our work we fulfill roles determined by society for the benefit of society; at the same time, society aims at arresting the activities of unethical and antisocial individuals such as thieves and gangsters. Only a very stable and secure society can afford much deviation from the social norm, valuable though it may be. Humans are social animals, so their survival as a species depends on the maintenance of their society; as a result, individual behavior, though varied and culturally enriching, must be controlled and directed by social needs. While humans invented symbolic culture, it has in turn created human society.

We are cultural animals, therefore, not because we alone have culture but because culture has come to condition our every act. We owe our nature to an evolutionary trend in which cultural adaptations have been constantly selected. Culture has made possible the unprecedented success of a species of primate that perceived so clearly the nature of its environment that it learned to change it to its own advantage. Animals adapt to their environment by changing their genes; humans have come to change their environment to preserve their genes.

Suggestions for Further Reading

A zoologist's view of culture similar to that expressed here is described by H. Kummer, *Primate societies: Group techniques of ecological adaptation* (Chicago: Aldine, 1971). For a cultural anthropologist's view of culture, see R. Linton, *The tree of culture* (New York: Knopf, 1955). For a discussion of perception and consciousness, see R. E. Ornstein, *The psychology of consciousness* (San Francisco, Calif.: Freeman 1972).

Our understanding of animal communication and its evolution has been discussed in R. Hinde (Ed.), *Non-verbal communication* (Cambridge: Cambridge Univ. Press, 1972). The biological basis of language is reviewed by E. H. Lenneberg, *Biological foundations of language* (New York: Wiley, 1967). For a review of experimental work with apes see A. Desmond, *The ape's reflexion* (London: Blond and Briggs, 1979). The evolutionary and adaptive nature of cultural change is clarified by D. T. Campbell, "Variation and selective retention in sociocultural evolution," in H. R. Barringer, G. I. Blanksten and R. W. Mack (Eds.), *Social change in developing areas* (Cambridge: Schenkman, 1965).

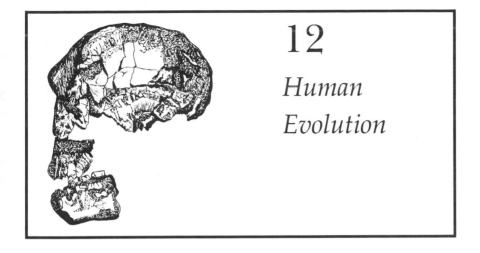

12

Human Evolution

Nothing in biology makes sense except in the light of evolution.

Th. Dobzhansky, 1973

I. The Longest Journey

Although life on earth stretches back some 3000 million years, only humans have ever known the meaning of time or considered the immense progression of living species. For almost all this vast period of time there was no one to gaze in awe at the earth and its seething organic life; no one to wonder and speculate on its origin. Only in the last millenia has a species appeared on earth capable of seeing the extraordinary nature of its environment and the miraculous complexity of the organic realm. Only in the last 100 years has humankind developed the vision to see so far into the past and understand for the first time—albeit at a superficial level—the real nature of life and the creative process. Since the work of Charles Darwin we have suddenly become able to see who we are and where we come from. It is an astonishing event in human history.

Our understanding is very new, and we are still beginners in the science of anthropology. Nevertheless, at this point in the earth's history we do have some idea of the detailed path of the process of evolution and a good idea of the main steps in our own creation. We have recorded some 400,000 species of plants and over 1,500,000 species of animals. That the process of evolution is also responsible in an immediate sense for our presence here is no longer in doubt, but questions about the ultimate meaning of life are not yet a subject for scientific research.

Half a century of research reviewed in a recent book by Ciochon and

Corruccini (1983) makes it clear that fossil primates from the Fayum desert of northern Egypt can be considered members of the basic stock of apes and Old World monkeys, the Catarrhini. This group of fossils (family: Propliopithecidae), which date from the Oligocene, are usually classified with the Hominoidea because they carry a dentition similar to that of later apes, though being so primitive, they predictably do not share with them any "derived"characteristics (Fleagle and Kay, 1983). Fleagle (1983) concludes on the basis of the known forelimb elements (and some fragments of the hindlimbs) that *Aegyptopithecus*, the best known genus of the group, was a small but heavily muscled, slow-moving arboreal climber (Fig. 12.1). The Old World monkeys, on the other hand, as Andrews (1983) points out, are comparatively specialized, with their bilophodont dentition, folivorous diet, habitat preferences, and peculiar locomotive and postural adaptations. With all these derived and specialized characteristics, the Cercopithecidae can be seen to be a group that diverged from the basic hominoid stock which led to the apes, and that probably separated from the ape lineage at least 26 mya (Boas, 1983).

Following the Fayum primates we have a sample of apes from the middle Miocene of East Africa (the *Proconsul* species are the best known), which have received some consideration in this text. They show us that, while the dentition and the cranium were by now recognizably apelike (Fig. 12.2), their locomotor adaptations were still rather generalized. Fleagle (1983) concludes that *Proconsul africanus* was an arboreal quadruped with some suspensory and climbing abilities but with no obvious adaptations for either brachiation (as in gibbons) or knucklewalking (as in *Pan*). Neither these specializations nor the full climbing adaptations of later apes were yet fully developed. There is no evidence suggesting that one of the *Proconsul* species was not a possible ancestor to the living African apes and hominids.

It seems clear today that our own part of this immense journey began some 5 to 6 mya in Africa, as Darwin surmised. At this time our ancestors parted company with those of the African apes and formed an independent lineage with new adaptations, both behavorial and anatomical. There are no fossil data (yet) that relate directly to this fork in the primate evolutionary tree, but there is no doubt that it occurred. The anatomical evidence we have reviewed in this book makes it perfectly clear that we have far more in common with the chimpanzee and gorilla than with any other primate. This close link is also made clear by a consideration of biochemical and genetic evidence.

Living animals are astonishingly similar in chemical composition. The same compounds, nucleic acids and proteins, are found everywhere; the same energy carriers and enzymes (with identical functions) are present in the most diverse organisms. Proteins are composed of some 20 amino acids, usually the same, and all are left optical isomers. Thus, all living creatures are closely related, and we, too, are an integral part of the

Figure 12.1 This reconstruction of *Aegyptopithecus* was made under the direction of Elwyn Simons, who discovered it. This primitive hominoid has an apelike dentition but retains a tail and a generalized limb structure (courtesy Elwyn Simons).

living world. The differences that separate us from other organisms at the chemical level are very slight.

We have already seen that we share with the chimpanzee some 98.5% of our DNA, and as we move away from the apes, across the less closely related primates to the other mammals, this percentage is reduced. As may be expected, this pattern is also reflected in the biochemistry of these groups, as we pass from the more closely to the more distantly related forms. The protein hemoglobin, so important as an oxygen carrier in our blood, shows this point clearly, since more and more amino acid differences are seen as we move away from ourselves (Table 12.1). Nevertheless, it should be noted that chickens and even fish bear far more similarities with humans than differences in their hemoglobins, and these are due to our common ancestry; they cannot be a result of chance. Similar work has revealed similar progressions of difference in other proteins, such as cytochrome *c*, myoglobins, and fibrinopeptides. The evidence always places us very close to the chimpanzee and gorilla. Many biochemical techniques have been used to measure molecular differences, including amino acid immunodiffusion (Table 12.2), microcomplement fixation, radioimmunoassay, electrophoresis, together with nucleotide sequence and restriction endonuclease mapping. All these tell the same story, as does the study of cytogenetics (the form of the chromosomes).

When all these data are gathered for the higher primates, they show a consistent pattern of relationship, which was plotted earlier in Fig. 1.2. The closeness of humans and the African apes is always clear. They

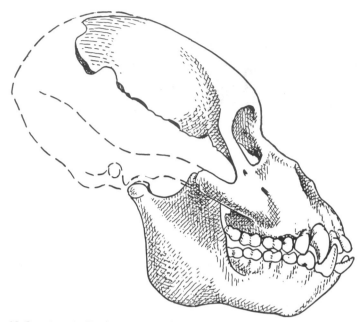

Figure 12.2. The skull of *Proconsul africanus* is a recognizable ape skull with ape dentition. The locomotor anatomy, however, was still generalized. [Courtesy the Trustees of the British Museum (Natural History).]

share not only most of their biochemistry, but certain particular molecular changes which must have occurred during the period of time that they shared a common ancestor exclusive of any other living species; that is, since the orangutan lineage split away about 10 mya. These shared derived molecular features are listed in Table 12.3, and show the distinctness of the orangutan, which forms a sister lineage to the *Homo/ Pan* lineage, as indicated in Fig. 1.2.

Thus, it is well established that humans are closely related to the chimpanzee and gorilla and more distantly related to all other primates. When did the split occur between humans and African apes? As we have said, we have no fossil evidence relating to this question beyond that of the earliest known single fossil believed to be hominid: a jaw fragment that dates from about 5 mya and was found at Lothagam in Kenya. The earliest fossil evidence that occurs in any abundance (and so is entirely convincing) is the collection of fossil jaws from Laetoli in Tanzania, dated from ca. 3.5 million years (Fig. 12.3). We know that by this time the Hominidae were an established group, adapted to bipedalism.

The biochemists can tell us more. The molecular data can be used to elucidate the timing of evolutionary events. It is believed that because the molecules appear to change regularly during the process of evolution, they may serve as an evolutionary clock to date the points of divergence in the evolutionary tree (Sarich and Cronin, 1977). This is not

TABLE 12.1. Amino Acid Substitutions in the Hemoglobin Chains of Various Animals Compared to Humans[a]

Alpha chains		Beta chains	
Gorilla	1	Gorilla	1
Macaque	5	Macaque	10
Mouse	19	Mouse	31
Sheep	26	Sheep	33
Pig	20	Pig	28
Horse	22	Horse	30
Rabbit	28	Rabbit	16
Chicken	45	Kangaroo	54
Carp	93		
Lamprey	113		

[a]From Dobzhansky et al., 1977.

TABLE 12.2. Immunological Distances between Albumins of Various Old World Primates[a]

Species tested	Antiserum to		
	Homo	Pan troglodytes	Hylobates
Homo sapiens	0	3.7	11.1
Pan troglodytes	5.7	0	14.6
Pan paniscus	5.7	0	14.6
Pan gorilla	3.7	6.8	11.7
Pongo pygmaeus	8.6	9.3	11.1
Symphalangus syndactylus (siamang)	11.4	9.7	2.9
Hylobates lar	10.7	9.7	0
Old World monkeys (average of six species)	38.6	34.6	36.0

[a]Calculated from data in Sarich and Wilson, 1967.

the place to explain the method used or to argue its validity, but the data obtained by this means do resemble the paleontological data in their broadest terms. They are summarized in the legend to Fig. 1.2: according to these data the hominid lineage diverged from that leading to the African apes at 5 ± 1.5 mya.

What were the early hominids like? Because we are closely related to

TABLE 12.3 Molecular Comparisons among the Great Apes and *Homo*[a]

Molecular comparisons	Homo-Pongo	Pan/Gorilla-Pongo	Homo-Pan/Gorilla
Fibrinopeptide sequence (no. of amino acid differences)	2	2	0
DNA hybridization (% base pair mismatch)	2.4	2.0	1.1
DNA hybridization (% base pair mismatch)	—	4.5	2.4
Immunological distance (ID units)	37	39	14
Immunodiffusion distance (antigenic distance)	1.6	1.6	0.8
Electrophoresis			
Plasma proteins	>>2	>>2	~1.6
Cladistic analysis based on number of electrophoretic charge changes at 23 loci	8	9	7
Genetic distance value based on 23 loci	0.349	0.300	0.367
Mitochondrial DNA comparisons using endonuclease restriction enzyme maps of site shared	1*	2†	7‡

[a]From Andrews and Cronin, 1982. *Homo shares a specific site with *Pongo* but others do not; †*Pan* and *Gorilla* share a site in common with *Pongo* but *Homo* does not; ‡*Homo* shares this number of sites with at least one of the species of *Pan* or *Gorilla* but not with *Pongo* (or *Hylobates*).

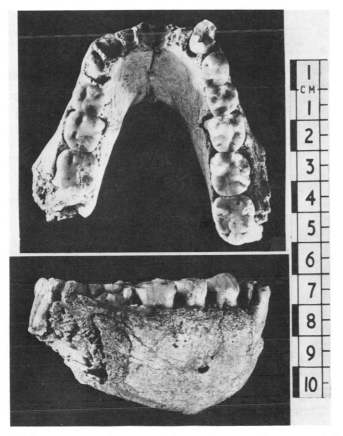

Figure 12.3. This jaw of *Australopithecus* from Laetoli is one of the oldest known, dated from 3.5 mya. The upper photograph shows the relatively large but damaged canine tooth (*top, right*). The lower photograph shows the body of the mandible from the right hand side. Note its thickness and depth (courtesy Dr. Mary Leakey).

the chimpanzees, we need not assume that they were chimpanzeelike; on the contrary, the chimpanzee ancestor might well have been hominidlike. While we can find no evidence of bipedalism in fossil or living chimpanzees, we can find clear evidence of climbing adaptations in both fossil and living hominids, and based on this we believe that the common ancestor *was* probably more chimpanzee than hominidlike. However, that ancestor was not like a modern chimpanzee, and we can find no evidence in fossil or living humans of a knuckle-walking ancestry. We also know that knuckle-walking is an ancient adaptation, since it is shared by the chimpanzee and gorilla; the biochemical evidence tells us that these two species split from each other at about the same time the Hominid lineage split away. Indeed, if it is not an extraordinary case of parallel evolution, the knuckle-walking adaptation must have evolved during the relatively short interval that must have occurred between the separation of the hominid lineage and the divergence of the two ape lineages (Fig. 12.4).

Thus, our early ancestors were evidently small, apelike creatures traditionally adapted for tree climbing (as children climb trees today), yet becoming efficient at bipedal locomotion on the ground. They retained the ability to cope in both forest/woodland and the more open savanna with its scattered trees and wooded riverbeds. The country of the Gombe Stream Reserve, where Jane Goodall has long studied the chimpanzee, is exactly the kind of border country between the two biomes that would have given maximum opportunities to an evolving hominid (Fig. 12.5). Here the protohominids could take advantage of both zones, harvesting the forest fruits in their season, and the berries, tubers, and game of the savanna at other times. These earliest hominids can be considered the first members of the genus *Australopithecus*.

II. The Protocultural Phase: *Australopithecus*

The protocultural phase of human evolution, which is documented by the remains of *Australopithecus*, is now well known from numerous fossil specimens. Ranging from Taung in South Africa to Hadar in northern Ethiopia, *Australopithecus* appears to be associated with the vast areas of savanna grasslands that stretch from northern to southern Africa down the eastern side of the continent. Fossils are particularly well preserved along the Gregory rift and in the dolomite caves of the high veldt (Fig. 12.6). More discoveries from these areas may be expected.

However the taxonomy may develop, the broad picture of the history of this genus is becoming clear. We have a gracile ancestral group, here referred to as *A. afarensis*, leading, on one hand, to a very robust and specialized lineage terminating in *Australopithecus boisei* and *A.robustus*, and, on the other hand, a generalized and rapidly evolving gracile lineage leading to humankind (Fig 12.7). The exact placing of the South African species (*A. africanus* and *robustus*) is unclear, and it now seems more probable than it did previously that they both may represent a side branch, relatively isolated from the main center of hominid evolution, which appears to be further north.

Of the gracile lineage the anatomy is clear. Even the earliest known of these creatures, from Laetoli, were evolved bipeds, and were almost certainly capable of an efficient and effective walk and run. The footprints preserved at this site confirm the anatomical deductions based on the slightly later Hadar specimens. However, among the early forms from Hadar we also find clear indications of adaptations to arboreal climbing in both fore- and hindlimbs as well as in fingers and toes. This seems to imply that, although bipedal, these protohominids were also effective arboreal climbers and probably spent some time in arboreal

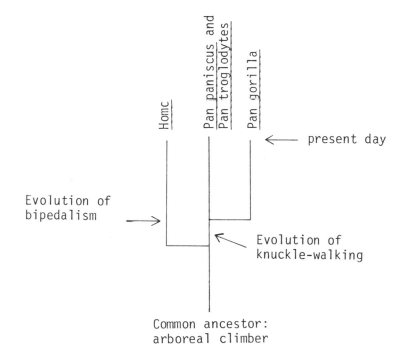

Figure 12.4. This diagrammatic branching diagram shows the successive locomotor adaptations in the African Hominoidea.

feeding and other arboreal activities. The cranial capacity is still small—a little bigger (relative to body size) than among the great apes—but the skull and dentition do show some evolution in the hominid direction: in particular, the canines are reduced and the lower first premolars carry an incipient second cusp.

Archaeological evidence does not tell us much about the lifestyle of *Australopithecus*. The remains from South Africa are from stratified deposits from caves and sink holes (Fig. 12.8), which (at these levels) carry no artifacts and do not represent living floors. The discoveries at Laetoli and Hadar are from lacustrine and aeolian deposits, respectively; they, too, have no archaeological associations. We can only surmise the lifestyle of *Australopithecus* on the basis of its anatomy and taxonomic relationships. We can suppose that the gracile species were omnivorous and probably ate a variety of fruit, seeds, and tubers, together with the meat of small animals. They were almost certainly scavengers as well as hunters.

Because its successor was a tool user and toolmaker, it seems probable that *Australopithecus* used tools and possibly made them to a limited extent. They may well have used both sharp stones for cutting and bone

Figure 12.5. Photograph of woodland and grassland savanna on the hills along the eastern shore of Lake Tanzania, occupied today by the chimpanzees studied by Jane Goodall. This kind of forest fringe country is probably similar to that occupied by the earliest hominids. (Photo courtesy Harold Bauer.)

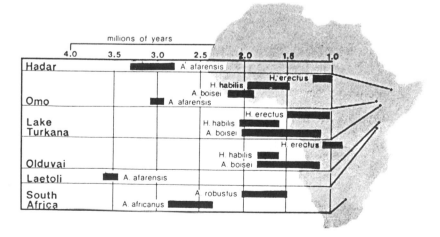

Figure 12.6. African sites that have yielded remains of early hominids are all found on the low-rainfall, eastern side of the continent along the savanna belt.

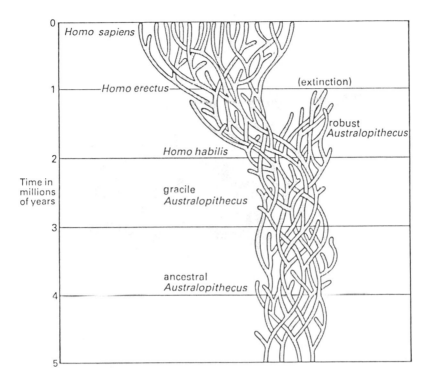

Figure 12.7. This evolutionary "tree" diagram of the hominid lineage is designed to emphasize the complexity of the evolutionary process. Each evolving species comprises unnumbered subspecies and races. The diagram indicates the succession and relationship of the various hominid species represented in the fossil record.

and wood as levers and digging sticks. It is quite unrealistic to draw a hard line between a tool-using and a toolmaking hominid. The evolution of a material culture was gradual and certainly very slow.

Australopithecus is a good candidate for the role of "missing link" between ape and human: a link no longer missing. After Raymond Dart discovered the first specimen in 1924, he wrote:

> It is obvious, meanwhile, that it represents a fossil group distinctly advanced beyond living anthropoids in those two dominantly human characters of facial and dental recession on one hand, and improved quality of the brain on the other. . . . At the same time, it is equally evident that a creature with anthropoid brain capacity, and lacking the distinctive, localised temporal expansions which appear to be concomitant with and necessary to articulate man, is no true man. It is therefore logically regarded as a man-like ape. I propose tentatively, then, that . . . the first known species of the group be designated *Australopithecus africanus*, in commemoration, first, of

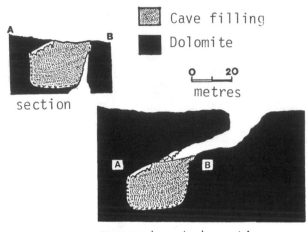

Figure 12.8. Reconstructed section of the Sterkfontein cave (*right*) as it was at the time the fossils were deposited, and as it remains today (*left*). (After C. K. Brain, 1981).

> the extreme southern and unexpected horizon of its discovery, and secondly, of the continent in which so many new and important discoveries connected with the early history of man have recently been made, thus vindicating the Darwinian claim that Africa would prove to be the cradle of mankind (Dart, 1925: 199).

This is the importance of *Australopithecus*: it links us anatomically and behaviorally with our cousins, the African apes, in a clear and unquestionable manner. After 50 years of research, investigation, and argument, this much is certain.

The successive generations of *Australopithecus* may have constituted a stable element in the African fauna for as much as three million years. The rate of evolutionary change was probably slow once bipedalism evolved. During this time the robust forms evolved and diverged, eventually becoming extinct. The gracile forms came to use stones, bones, and wooden tools with increasing frequency. The rate of change may have accelerated a little; tool use began to pay, and cultural evolution accelerated. Silently, and without the sound of trumpets, the first humans found themselves here on earth.

Let us end this section with a further quotation from Raymond Dart, who is 90 years old:

> In anticipating the discovery of the true links between the apes and man in tropical countries, there has been a tendency to overlook the fact that, in

the luxuriant forests of the tropical belts, Nature was supplying with prof-
ligate and lavish hand an easy and sluggish solution, by adaptive specialization, of the problem of existence in creatures so well equipped mentally as
living anthropoids are. For the production of man a different apprenticeship
was needed to sharpen the wits and quicken the higher manifestations of
intellect—a more open veldt country where competition was keener between swiftness and stealth, and where adroitness of thinking and movement played a preponderating role in the preservation of the species. Darwin has said, 'no country in the world abounds in a greater degree with
dangerous beasts than South Africa', and, in my opinion, Southern Africa,
by providing a vast open country with occasional wooded belts and a relative scarcity of water, together with a fierce and bitter mammalian competition, furnished a laboratory such as was essential to this penultimate phase
of human evolution (Dart, 1925: 199).

III. The
First
Humans:
Homo habilis

The earliest hominid fossils with known archaeological associations are those of *Homo habilis*. We
have seen (Chapter 4, II) that this species can be
considered to range in time from about 2.3 to 1.5
mya and that it was confined to Africa. The earliest
securely dated occurrences are found at Omo
in southern Ethiopia; here both skull fragments and teeth date from
about 2.2 mya.

The oldest known evidence for tool manufacture comes (as we have
seen in Chapter 7, VI) from Hadar, where an old land surface with an
elephant molar, bone fragments, and basalt tools of the Oldowan type
has been discovered. The site overlies a deposit of volcanic ash dated by
potassium argon and fission track methods at 2.7–2.8 mya; as a result,
the tools have been dated conservatively at about 2.5 mya. The dating,
however, needs further confirmation. As we know from many sites at
Olduvai, hunting or scavenging (or both) was a fairly regular occupation
of the early hominids after this time. Game varied from small reptiles,
frogs, birds, and rodents to middle-sized antelopes and pigs to giraffe,
buffalo, and even large pachyderms. It seems improbable that the latter
were killed or even moved at this early date, so some of these remains
may have been scavenged. Some sites at Olduvai look like living sites
(see Fig. 9.18), some like butchery sites (see Fig. 9.19), and some like
workshop sites. The evidence of so much butchery, however, should
not cause us to forget that hominids have almost certainly always eaten
a preponderance of vegetable food.

Reliance on meat would have developed slowly. The proportion of
meat in the diet would have increased in the dry season or when vegetable foods were scarce. That meat would ever have composed more
than 50% of the diet of early hominids seems extremely improbable; a far
smaller percentage is more likely. Today the San Bushmen of the Central
Kalahari Desert survive in much more arid regions, and when game is
absent they survive for long periods on vegetable foods. (Diets high in

meat are a recent adaptation to winter arctic desert conditions in areas where there is no vegetation.) The dentition of the hominids at all stages is oriented toward chewing and grinding rather then cutting and slicing, and carries clear evidence of a preponderately vegetable diet.

We can suppose, therefore, that the early hominids came to exploit a range of fauna as a subsidiary food resource, with the adaptations of a bipedal posture and run and, later, the use of the cutting tool. There is no convincing evidence at this early stage of the use of weapons to kill game, but it is quite possible that sticks and bones would have been used as clubs (Wolberg, 1970). Sticks are thrown in agonistic display by apes and would almost certainly have been used as clubs by hominids. Techniques of killing must have been developed that did not involve the use of sharp canine teeth.

There is little doubt that during this period of adaptation the social structure of the hominids underwent some modification. Even among living chimpanzees we see a tighter social structure in savanna woodland than in forest areas, and this pattern was surely followed by early hominids. The relationship between social structure and ecology, clearly demonstrated among primates, is important for our consideration of early hominid adaptations, because the social group sets the scene for social hunting.

Cooperative hunting cannot have arisen from a vacuum, but must have come about gradually from another related cultural adaptation. Just as a physical structure may prove preadaptive for different functions, so cultural structures may serve a novel purpose. From what we know of the behavior of baboons, it is reasonable to suppose that behavioral adaptations that served cooperative defense gave rise to cooperative offense. There is, in fact, considerable evidence of cooperative hunting among baboons. One eyewitness in South Africa repeatedly observed for 20 years:

> apparently organized hunts which often result in the death of the intended victim. The baboons, usually led by a veteran of the troop, surround an unsuspecting three-parts grown Mountain Reedbuck, or Duiker, as the case may be, and on one occasion, a young Reedbuck doe was the victim. It would appear that on a given signal the baboons close in on their quarry, catch it and tear it asunder. . . . In nine cases out of ten, the game animal is devoured limb by limb, and after the affair is all over all that is to be found is the skull and leg bones (Oakley, 1951: 78).

Such behavior may be unusual for baboons, but it has been observed elsewhere (Harding, 1975) and does suggest that the difference between cooperative defense and cooperative offense is not great. Many primates use signals for troop control; what is interesting here is that the baboons show the sort of discipline that we associate with dogs when they act as cooperative hunters. It appears that primates are sufficiently versatile in their behavior to develop this particular pattern when it proves to be adaptive.

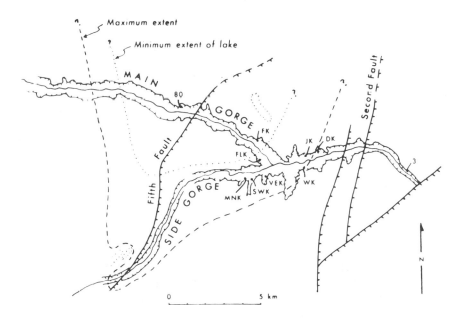

Figure 12.9. Like most primates, early hominids required a regular supply of fresh water. All the hominid-bearing sites at Olduvai Gorge (which carry code letters on this plan) are situated in areas that once constituted the shore of an ancient lake, into which freshwater streams flowed (from Hay, 1970).

By the time of the deposition of Bed I, Olduvai Gorge, above the 1.9 mya larva flow, a stable adaptation to woodland and savanna was almost certainly achieved. Food remains on the living floors at many sites testify to successful hunting as well as scavenging. We know, too, that the Olduvai sites were on a lake shore, and it seems certain that the hominids were dependent on both a constant supply of fresh water and the presence of trees (Fig. 12.9). Arid areas without fresh water but with saline lakes would not have been satisfactory sites for hominid occupation.

Early Pleistocene African sites usually lie along sandy stream channels. These ephemeral stream courses would carry strips of riverine bush as well as trees; shade and fruit would be available as well as essential drinking water. There is also less well documented evidence of settlement in montane forest. At present the data could be made to fit any dietary hypothesis, and it is the environmental diversity that should be stressed. We can certainly predict as well as recognize this characteristic among early Pleistocene hominids. Adaptability in this respect is essential as a preadaptation for expansion into yet more varied environments, including those of the North Temperate regions.

Further adaptations to savanna regions must have appeared by Olduvai times. We know that the stone tool kit was already extensive (M.

Leakey, 1971), and probably included tools for making implements of wood and bone. Large stones suitable for toolmaking were collected from point sources as much as eight miles from the lake shore. Animal products such as skins and ligaments would probably have been used and prepared with stone tools. One Olduvai site carries evidence that has been interpreted as indicating some sort of shelter from rain or wind (see Fig. 9.18).

Modern social carnivores such as lion or hyena use all possible techniques to obtain meat: hunting, scavenging, and chasing other animals from their kill (Schaller and Lowther, 1969). It seems highly probable, therefore, that *Homo habilis* similarly took advantage of any means available to get meat. This is also suggested by the animal bones on occupation floors; there is great variety and a complete absence of regularity in prey species.

Homo habilis is not yet well represented in the fossil and archaeological record, but we have sufficient evidence to know that it forms a link between the protocultural *Australopithecus* and the later (and extraordinarily successful) *Homo erectus*. *Homo habilis* was, however, the first hominid to make tools to a regular pattern, and the species developed material culture to the point that it triggered an extraordinary spurt in the evolution of the Hominidae, a spurt that was most strikingly documented in the evolution of the hominid brain.

IV. The Temperate Adaptation: *Homo erectus*

There is no doubt that throughout the Pliocene the evolving hominids survived as a result of their social structure as much as their intelligence, for social life is one of the primates' most important adaptations. By the early Pleistocene, therefore, we would expect *Homo habilis* to have a stable and well-adapted social system.

It now seems possible that the expansion of the hominid populations into temperate zones was the most significant step in the evolution of *Homo*. It was more than a shift between adjacent tropical biomes such as may have occurred among earlier hominids; it involved a major climatic change and was accompanied by many important new adaptations. Apart from the evolution of the brain, the most striking adaptations were almost certainly cultural and social rather than biological.

Adaptations to temperate climates in particular involved some developments of profound importance. These cooler regions required dependence not only on tools but on *facilities* (see Table 9.2), defined by Wagner (1960) as objects that restrict or prevent motion or energy exchanges (such as dams or insulation), so that anything that retains heat is included (tents, houses, or clothing). Containers of various sorts— skins for carrying food or water, pots or boats, fences or even cords—fall into this category. Temperate adaptation requires far more facilities than tropical adaptation; the most important aspect of facilities was the extent to which they enabled humans to become increasingly independent from certain limiting factors in the environment.

Unlike the tropical biomes, the temperate regions are subject to extensive seasonal fluctuations in temperature: there are usually two or three months of frosty weather in winter when plant growth ceases. The critical climatic factor limiting adaptation to temperate regions is not the mean annual temperature, but the seasonal variation—the lack of climatic equability. On one hand, a long harsh winter will stress temperate species to the utmost, and would have placed a premium upon those human groups who had shelter, fire and stored food, all of which became necessary for survival. A hard winter was the most demanding experience hominids had faced since they spread into the savanna country millions of years earlier. On the other hand, temperate biomes carry an abundant fauna and a richly diverse flora, and the temperate woodland is second only to the tropical rain forest in species diversity. The rainfall is evenly distributed, and in the woodland regions lakes, permanent streams, and rivers are common.

Adaptations to this biome would have opened up extensive food resources to early humans. While geographical (mountain) barriers may well have delayed expansion into this environment, cultural adaptations to winter were undoubtedly the most important factor enabling humans to expand northward.

We have tropical fossil hominids from Java from nearly 1 mya that are broadly similar to the somewhat earlier *Homo erectus* fossils from Kenya (see Appendix). These populations could have entered Southeast Asia via a tropical woodland and savanna corridor at this time if not earlier. Perhaps the earliest archaeological evidence of occupation of a temperate zone is that recorded at the Vallonet Cave in southern France (Alpes-Maritime), which is believed to be of Gunz age (Fig. 12.10). As we have seen (Chapter 9, X), there is evidence of hearths in the Escale Cave in the nearby region of Bouches-du-Rhone; these deposits date from at least the early Mindel period. Neither of these sites, however, is so convincing as the inter-Mindel site at Vértésszöllös, where there are hearths with burned and split bones of a substantial mammalian fauna, particularly rodent bones, but also those of bear, deer, rhinoceros, lion, and dog. There are also several hundred artifacts of chert and quartz, mainly choppers and chopping tools, flake tools, and side scrapers.

The great cave of Choukoutien, mentioned earlier, has deposits (now dated 500,000 to 450,000 BP) containing numerous hearths with food remains of 45 different species, including sheep, zebra, pigs, buffalo, and rhinoceros, together with deer, which form about 70% of the total. The tool kit has much in common with that of Vértésszöllös and indeed with the "developed Oldowan" from Olduvai Gorge.

Dating from the late Mindel we have two other important European sites, and both show the use of fire. At Torralba and Ambrona in Spain (two contemporary sites three kilometers apart) we have evidence interpreted as butchery by bands of *Homo erectus*. Bones of more than 20

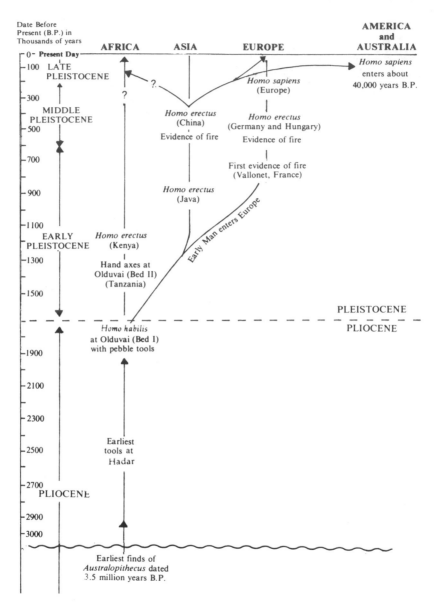

Figure 12.10. This chart indicates approximately some of the dated archaeo-logical discoveries from the Pliocene and Pleistocene and their possible rela-tionships in terms of migrations of hominid populations. From about 1 mya migrations between Africa, Asia, and Europe were probably continuous, and no attempt has been made to show the complexity of these movements.

elephants have been found in a small area together with remains of horses, cervids, aurochs, rhinoceros, and smaller animals. There are shaped and polished tools of wood and bone and there is a rich Early Acheulian industry (Fig. 12.11).

At Terra Amata in present-day Nice we have evidence of seasonal habitations on coastal dunes. There are ovoid arrangements of stones with regularly spaced postholes. Within the shelters these represent, the floors were covered with pebbles or animal hides (imprints are preserved). Hearths occur in holes or on stone slabs sheltered by low stone walls. Food residues include elephant, deer, boar, ibex, rhinoceros, small mammals, and marine shells and fish. The industry is of Early Acheulian type and includes a few bone artifacts.

Although humans probably entered this latitude during the preceding warm interglacial, present evidence indicates that humans were well established in North Temperate zones by the time of the Mindel glaciation (variously dated between 700,000 and 400,000 BP), when highly efficient and productive hunting techniques were employed; it appears that no animal was too large or too dangerous to be killed by hunting bands (Fig. 12.12). The product of the hunt would have served to support an increased population through an improved food supply and the use of other animal products, such as skins, that were needed in an advancing technology.

Because at most times of the year vegetable foods in the temperate regions would have been fairly plentiful, systematic hunting was primarily a means of supplementing a diverse vegetable diet. Mammal meat would have become a primary food resource only during late winter and early spring. Berries and seeds would have been eaten by this time, and the new succulent vegetation would not yet have grown. Even the game would have dispersed to find food. Because this limiting period of the temperate year would have kept hunting populations fairly sparse, both gene flow and the transmission of cultural traits would have been restricted. New developments in human evolution were to be tied to adaptations in the future that increased the extraction rate by the exploitation of additional food resources and by food storage techniques.

There is very limited evidence of biological adaptation to cold in modern humans, even among Eskimo. Biologically adapted to the tropics as *Homo erectus* was (and as we still are), its survival through the cold winters of the temperate zone required extensive cultural adaptations unmatched by the evidence from Africa. Archaeological evidence strongly supports the notion that fire was used in temperate biomes for a very long period in human prehistory.

Cold winters also necessitated considerable development in social behavior. It seems inescapable that there would have been a fairly complete division of labor by this time: the men hunting and the women minding the babies and gathering vegetable foods, water (in skins), and fuel. Perhaps for the first time babies were put down and left in the

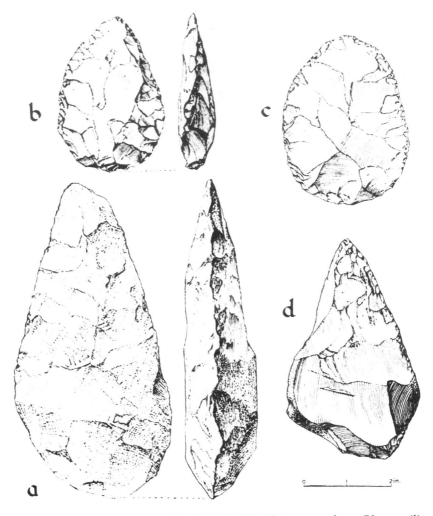

Figure 12.11. Acheulian tools of the Middle Pleistocene, from Olorgesailie, Kenya (*a*), St. Acheul, France (*b*), Wady Sidr, Palestine (*c*), and Hoxne, England (*d*). [By permission of the trustees of the British Museum (Natural History).]

charge of siblings or aunts at the base camp. The division of labor and separation of the sexes must have increased the need to communicate abstract ideas with language, and vocabulary no doubt expanded. Perhaps the expression of emotions (which language could replace) was first inhibited in a closely knit cave-dwelling band, and this emotional inhibition was to become increasingly important. It may prove to have been one of the most fundamental social developments in the shaping of human psychology.

Figure 12.12. Imaginative reconstruction of the life of *Homo erectus* from Peking. [By permission of the trustees of the British Museum (Natural History).]

From the skeletal evidence at Choukoutien we can deduce that more than 50% of the population died before they reached the reproductive age of 14. This suggests that it would have been necessary to produce four to five children per female simply to maintain the population level, and it is probable that many more than four children were born to each woman. The impossibility of hunter-gatherers carrying and nursing more than one child at a time indicates that a cultural adaptation such as infanticide was possibly common, as we shall see.

The cold made demands on human ingenuity to devise protective facilities such as clothing and tents. It was surely an important factor in the evolution of human intelligence. All these adaptations had their anatomical correlates in the brain, the skeleton, and soft parts of the body. In this respect the people of Choukoutien show a great advance over *Homo habilis*. Their mean endocranial capacity is twice that of *H. habilis* and just falls into the range of modern humans. At Choukoutien the cranial capacity varies between 915 and 1225 cc, while at the Vertésszöllös site in Hungary (which is at least half a million years old) the cranial capacity of one skull has been estimated to be about 1400 cc. (This is above the mean of 1330 cc for modern humans.) The people at Choukoutien were also anatomically more modern than *Homo habilis*; they were of greater stature and more sturdily built, with a larger and more balanced skull. Yet they still carried a heavily built masticatory apparatus that clearly distinguished them from ourselves.

The cultural adaptations that *Homo erectus* must have developed to survive in temperate biomes were an extraordinary achievement; northern winters were undoubtedly a factor of great importance in the evolution of *Homo sapiens*.

V. The Rise of Modern Humans

None of the sites mentioned so far carries evidence of permanent cold, only of seasonal frost. The fauna and pollen data suggest a cool climate becoming either colder (as at Vertésszöllös) or warmer (as at Torralba). The climate at Terra Amata near the sea was certainly mild, and that at Choukoutien was of warm interglacial type.

At this time humans were able to survive cold temperate winters, but there is no evidence that they had yet adapted to arctic conditions. Following the Mindel period we have more northerly fossils of late Riss-Wurm interglacial date (Swanscombe and Steinheim), which represent the expansion of human populations northward during warm temperate spells (Fig. 12.13). In view of the extreme difficulty of survival in northern coniferous forest and arctic tundra, it is not to be expected that humans would have entered these zones at a very early date. Today it is hard to see how they could possibly have adapted to arctic conditions without domesticated animals such as reindeer or dogs. The presence of Neandertal people during the first Wurm glaciation of western Europe is surprising. It suggests an advanced use of both tools and facilities.

Though the present evidence suggests that the Neandertal people did not survive throughout the first major advance of the Wurm glaciation, it does demonstrate that they could survive extreme cold and must have lived some thousands of years under arctic conditions. This Wurm advance of the northern ice sheets brought a cold moist climate characterized by animals such as mammoth, woolly rhinoceros, reindeer, musk-ox, ibex, blue fox, and marmot. All these were hunted, together with the formidable cave bear.

The Neandertalers were well established in southern and central Europe before the colder weather descended, and they survived the cold to a great extent by using caves and rock shelters. Judging by the extent of their cultural remains the Neandertal people adapted successfully to the climate and were able to exploit the huge herds of reindeer and other animals. In some areas such as the Dordogne in France, the local topography must have offset the extreme conditions. The Dordogne River and its tributaries dissect deeply into a limestone plateau and offer a number of sheltered valleys. Possibly the vast herds of animals that must have undertaken regular seasonal migrations used these valleys as migratory routes. It seems possible that here, as in southwestern Asia, people came to rely on harvesting migratory animals. Many temperate animals, particularly arctic species, migrate regularly in the spring and autumn between coastal plain and mountain pasture. These people,

Figure 12.13. Imaginative reconstruction of the life of early *Homo sapiens* from Swanscombe. [By permission of the trustees of the British Museum (Natural History).]

then, were able not only to harvest "earned" resources (which gain their food within the local habitat where they live), but to tap "unearned" resources [animals that pass through or spend some portion of their annual life cycle in one biome and yet gain most of their food (energy) in another biome]. To settle along the migration routes of herd mammals such as reindeer, musk-ox, or ibex and intercept them between their summer and winter feeding grounds is a sophisticated adaptation that we can fairly safely attribute to the later Neandertalers. It was a simple step to allow autumn-killed meat to dry and freeze for use in winter, as many Eskimo do today. We can also deduce that the game was sufficient for their needs, for they must have relied to a great extent on meat during the winter.

An interesting clue is provided by the teeth of the male skeleton from La Ferrassie. They show a particular type of extreme wear also found today among the Eskimo and some other hunters caused by chewing animal skins to soften them for clothing. It is indeed highly probable that the Neandertal people had exploited the whole range of animal products, and especially skins, to develop a well-differentiated material culture. They could well have made the kind of clothing that we find among the Eskimo, though the ready-made shelter of rocks and caves would probably have stood in place of the warm and intimate family igloo.

Probably the most difficult problem facing these arctic people was transport. Without dogs or sleds they would have been confined to the valleys in which they lived during the winter months. This restriction shows how successfully they had been able to exploit the local food resources of the region.

The successors of the Neandertal people in western Europe, the Aurignacian and Magdelenian peoples, have left us a far more detailed picture of their adaptations than their predecessors. Migratory herds of reindeer were harvested in large numbers, and often formed 85–90% of the faunal assemblage. Other mammals hunted include mammoth, bison, and horse. When climatic conditions became severe again, as they did toward the end of this period, the later Magdelenians (14,000–12,000 BP) began the systematic hunting of new kinds of unearned resources: migratory birds, aquatic mammals, and fish. The significance of these additions to their food supply can scarcely be overstated. Migratory fish and fowl appear in early spring, a time of maximum food shortage, and make possible the survival of a much larger population throughout the year. At the same time, it is probably important that fish oils (unlike terrestrial animal fats) contain vitamin D, and this may have been an essential vitamin to people living in areas where insolation was minimal and where the wearing of clothing was always essential. Evidence of rickets is present among Neandertal skeletal remains; as far as we know, these people were unable to catch fish and complement their own low vitamin D production. Shortage of this vitamin may indeed have been one of the factors that acted against their survival.

This extensive exploitation of migratory animals, which must have arisen slowly through the later stages of human evolution, was to have a profound effect on the evolution of social life. The most obvious result was that it allowed a more sedentary lifestyle to develop. Home bases could be occupied for longer periods of time during animal migrations and the ensuing winter, and there was no longer such a premium on the mobility of hunting bands. During the early phases of hunting and gathering, possessions and infants were limited by the need for mobility. A mother with a baby or a small child who must break camp frequently and transport her baby as well as her household gear will not welcome a second infant to care for and carry, or a mass of material possessions. She will have few compunctions about taking any means necessary to limit family size. Today, hunter-gatherers practice infanticide, abortion, and other means of birth control to retain their essential mobility.

Sedentism changed all this. As soon as a band could remain more or less permanently in one place, an increase in possessions and population densities was possible. The limitation on numbers was removed, and population could now expand to a level supportable by the increased food supply. When we compare the sites of the earlier Magdelenian to those of the later period (since 14,000 BP), we find that the living sites are larger, more numerous, and more often situated low on riverbanks, frequently at places where the river narrows. Many of these sites have yielded evidence that they were inhabited throughout the year. Thus, we find these permanent settlements associated with an increase in density and group size. It is also clear that these developments may have required a much more complex social structure compared with the essentially egalitarian local bands that characterize most hunter-gatherer groups. It opened the way to the developments that evolved in the Neolithic period.

In adapting as hunter-gatherers to the particular conditions of each biome and each region, our ancestors remained part of a more or less stable natural ecosystem. Because food supplies were not the limiting factor in population growth, except perhaps at certain critical periods, we do not find evidence of overkill or any serious instability following the appearance of humans in a region. As hunters they were competing with carnivores for herbivores, but the ecosystem is characterized by a functional dynamism that allows it to equilibrate in the face of climatic and other changes, especially if it is diverse in species.

Farming is the protection of food plants and animals at the expense of wild forms of less nutritive value. It also involves domestication, which is the selective breeding of certain species for their tameness and their value as food. The overall effect of the practice is a reduction in the diversity of organisms in an area, which is balanced by an increase in the domestic species. A larger proportion of the solar energy in the area is channeled into human food, either as plants or meat. Pastoralism can be a surprisingly effective adaptation in this respect, especially when more than one animal species is herded in a single area. Where field agriculture results in whole areas of the ground being covered in one or two food species, the conversion rate of solar energy into human food is even higher. Although we cultivate only 10% of the earth's land surface today, it has been estimated that the human population has increased from a potential maximum of about 10 million hunter-gatherers to its present size: some 4.4 billion (many of whom suffer from malnutrition and starvation). But the ecological cost of the introduction of pastoralism and agriculture is high; it implies the destruction of the natural ecosystem and the diversity of species that assure its stability. The Neolithic was the start of such destruction, and as the rate of population increase grew, the rate of destruction increased.

In the past 5000 years humankind has altered the ecosystem in many parts of the world and destroyed the natural balance. Pastoralism itself has been one of the most destructive forces; it is clear that wherever it has been carried out in semi-arid regions, whether in Australia, Asia, Africa, the Americas, or limited areas in the Mediterranean regions of Europe, there has been degradation of the grasslands and the threat or reality of soil erosion. The local fauna has been destroyed and the existing ecosystem degraded beyond the point where it can naturally equilibrate. Where soils are eroded the loss is irrevocable. The displacement of game and the destruction of their environment has done far more damage to natural life than all the hunting of the Pleistocene. In the same way, agriculture and deforestation for timber have involved the destruction of vast areas of forest (areas of naturally high rainfall), and we have lost both the forest with its associated flora and the forest animals (which often have a very limited distribution). All these developments, though they may eventually prove to endanger human survival, have enabled us to increase the extraction rate from our environment and place ourselves at the top of the energy food chain of the biosphere.

VI. The Human Species: *Homo sapiens*
The final scenes in this story of human evolution show the replacement of the large-jawed forms such as the Neandertal people by their small-jawed successors—ourselves. In the continuously evolving organic world, what we see around us today is, as it were, one frame in the continuing motion picture of human evolution; "mankind evolving" is the subject of Dobzhansky's important book (1962). We find ourselves as we are today, living in the places we occupy, doing the things we are doing. Our genes and our environment together have created us; our physical nature and our behavior are so determined. Our society and culture have brought about the situation in which we read this book now; it is all part of our complex process of adaptation to each other and to the planet on which we live.

But much has happened in these most recent phases of human evolution. Since the appearance of people indistinguishable from ourselves, a development recorded in the fossil record in France, Israel, South Africa, and southeastern Asia between 60,000 and 30,000 years ago, our ancestors have spread throughout the land masses of the Old and New World, and we now occupy a vast range of environments. Stable populations are adapted for life in the arctic and the desert. Our adaptations are both physical and cultural; small variations in body form and body chemistry are adaptations to the environment, as are variations in culture.

We are not yet able to account for all such variations in terms of their functions, but we can understand some, especially the most obvious ones, such as the ratio of surface area to weight (Fig. 12.14), which is correlated with temperature and relative humidity; or skin and hair color, which is correlated with solar radiation (see Chapter 8, IX); or nose shape, which is correlated with absolute humidity (see Chapter 8, IV). These kinds of physical adaptations are clear enough to make recognizable the sorts of differences that zoologists would classify as subspecific, and, as a result, some anthropologists recognize between six and nine different subspecies that have arisen as human adaptations to different ecological situations (Garn, 1971) (Table 12.4 and Fig. 12.15). Their isolation, however, is incomplete, and gene flow continues between them; otherwise they would be recognized as distinct species. Humankind is a single widely dispersed polytypic species showing adaptation to a wide range of environments.

One of the difficulties in understanding the adaptations that distinguish the different subspecies of modern humans lies in our nomadic

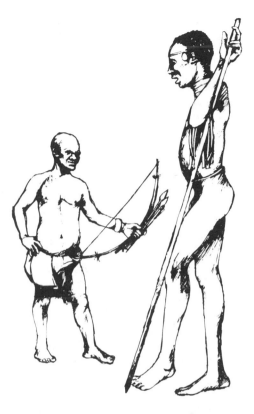

Figure 12.14. Two neighboring tribes in East Africa show very different physical adaptations. On the left is one of the pygmies from the northern rain forests of the Congo Basin; on the right, one of the Dinka, a tribe that lives in the desert regions of the upper Nile. They represent the tallest and shortest people in the world. [Courtesy British Museum (Natural History).]

nature. Evolution is a slow process, but geographical expansion can be rapid; as a result, populations originally genetically adapted to climate A are now, as a result of the high adaptive value of culture, living in climate B. Thus, as we noted earlier, human organic adaptation lags far behind environmental change. The problems introduced by migration and expansion in understanding racial differentiation and characterization are great. Understanding the relationship between the present pattern of *Homo sapiens* populations and their environment must involve, in practice, a consideration of their past history and cultural adaptations as well as their anatomical and physiological characteristics.

The recent phases in human evolution are striking at both an organic and a cultural level. As Washburn pointed out (1963a), there were perhaps from three to five times as many San Bushmen as there were Europeans only 15,000 years ago, when the ice sheets of the last glaciation reduced the habitable area of Europe to one-half that available to the San in eastern and southern Africa. Since that time, differential population growth has changed the whole pattern of the races of *Homo sapiens* on the face of the earth; as a result, although the majority of racial

TABLE 12.4. Subspecies or Geographical Races of Modern Humans[a]

Race	Notes
1. Amerindian	
2. Polynesian	Polynesians, Melanesians, and Micronesians are
3. Melanesian	combined by some authors as a single group
4. Australian	
5. Asiatic	
6. Indian	Indians and Europeans are combined by some
7. European	authors as Caucasians
8. African	
9. Micronesian	

[a]After Garn, 1961.

Figure 12.15. Distribution of nine surviving geographical subspecies of *Homo sapiens*. Each is bounded by natural barriers in the form of sea, ice, mountains, or desert (from Garn, 1971).

elements have probably survived, their proportions have changed out of all recognition.

As we all know, one after another population has undergone the so-called "cultural revolutions" that marked the beginnings of the ages of agriculture, metallurgy, and industrialization. Each revolution has made available greater resources and has resulted in population growth. Today we find industrialized countries supporting an immensely larger population per square mile in comparison with those at an earlier stage of cultural development. This increased population has been made possible by a whole range of adaptations based on a greater food supply and, recently, advances in medical science. Life expectancy has vastly improved. In South Africa, a white woman can still expect nearly 25 more years of life than a black woman can (Washburn, 1963a). Changes in life expectancy are readily subject to cultural adjustment; they can rapidly change the distribution pattern of *Homo sapiens*, and will probably do so in the future as they have in the past.

Probably the most important product of the cultural developments of the last 100 years is the world population explosion. Space travel and nuclear bombs make news and terrify us all, but human population (though it may not directly affect the readers of this book) presents a gnawing and devastating problem of pain and misery that will not be easily resolved.

Angel (1975) has estimated the average age of death of upper Paleolithic peoples to be 28.7 years for females and 33.3 years for males. The life expectancy for San Bushmen today is 32.5 years (Howell, N., 1976); that for the population of the United States in 1900 was 47.3. Today, the figures for the United States are: 75 for females and 67 for males. Clearly, if people live twice as long, the population will be doubled. But more people are surviving into adult life as well; biological constraints have been overcome.

Animal populations are known to be adapted to the level of their food resources or other limiting factors by a variety of biological mechanisms rather then merely by starvation. These mechanisms probably come into effect through the animal's perception of incipient overcrowding well in advance of resource shortages. The perceptions trigger mechanisms that either reduce fertility or the survival rate of fetuses or neonates. There is reason to believe the a number of these mechanisms have survived in human evolution (Stott, 1969), though the linkage between cause and effect is difficult to demonstrate. Miscarriage and infant mortality have been shown in a number of instances to be a product of stress upon the mother during or possibly prior to pregnancy. The stress will be the product of either overcrowding, food shortages, or social disharmony, all of which may be due to problems of resource availability. The most unfavorable conditions induce sterility or stillbirth; less harsh conditions result in impairment of the young, which will reduce their chance of survival and in the past would have meant their early death. More moderate stress due to malnutrition, crowding, or war appear to affect the intelligence, vigor, and motivation of the young.

In the past these factors may have helped to keep the human population at a level that imposed little stress on the adult members of a group due to resource shortages. The short lifespan of most females would also have cut down the effective potential for population growth, though if women started to have babies at the age of 16, they still had the potential to produce six or seven children by the time they were 30. In practice, however, as we saw in Chapter 10 (see Fig. 10.7), the birth spacing would probably have been closer to four years, as it is in the San, and there is no doubt that lactation under these circumstances acts as an effective limit to fertility (Short, 1976a). We also know that in most tribal peoples there are taboos and other rules against sexual intercourse that keep the birthrate low. There is much evidence that where, in spite of this, the children came too fast, infanticide was widely practiced in Christian and non-Christian countries alike. In the Ellice Islands of the Pacific, for example, infanticide was ordered by law; only two children were allowed to a family, as the islanders feared the scarcity of food (Wilkinson, 1973). Even today, infanticide is treated in English law as manslaughter—not as murder.

What we have seen in the last 100 years is a weakening of the cultural controls on sexual intercourse accompanied by advances in medical knowledge that have made it possible for many fetuses and neonates, which would previously have died, to survive. We can lay the responsibility for the population explosion at the door of humankind's understandable wish to create life and save it. A healthy woman today can produce 20 healthy children, and as we saw in Fig. 10.7, the average family size among the Huttites is 11. Science has provided as a solution a range of contraceptive devises that offer hope to the overpopulated countries of the world. With a finite resource base, the world can only carry a limited population without endless suffering and needless death. Since 500 million people today suffer gross malnutrition, that limit has surely already been passed.

Estimates of past world populations are liable to error, but they do give us some idea of what might have been. Certainly the expansion of the population is well attested (Table 12.5).

As we find the human species today, *Homo sapiens* is in a state of rapid change. This change is due, not so much perhaps to an increase in rate of evolution (such a rate cannot at present be satisfactorily measured) as to an increase in rate of cultural change, which has brought with it greater mobility and more rapid adaptation. These developments place us in an unusual situation, which requires some further elucidation.

VII. Tools, Technology, and Culture

From what has been said in the preceding pages, it seems reasonable to trace the origin of our material culture to two of humankind's special mental attributes: perception and intelligence.

We have discussed the importance of the evolution of human perception at some length (see Chapter 11, II), and its

TABLE 12.5. Estimates of Past Human Populations[a]

Date	Cultural period	Population in millions
From 1 million	Palaeolithic hunter-gatherers	2-5
12,000–2,000 BP	Agricultural age rising to	200
At 300 BP	Literate age	500
At 100 BP	Industrial age	1000
At 40 BP	Nuclear age begins	2300
Present day		4400

[a]Partly after Deevey, 1960 and Westing, 1981.

significance as a correlate and determinant of culture cannot be over-rated. Humans came to see their environment in a unique manner, one that carried a greater informational content about its nature than any other. We know that the real structure of our environment is beyond our perception, for it is an emptiness sparsely occupied by electrical charges moving at great speed. There is no world corresponding to the world of our common experience. But the way the human organism came to perceive the average effect of these electrical phenomena has fashioned our world. Our senses have enabled us to form in our minds the percepts of objects, solid, colored, and textured; these objects lie in our minds, not in the environment, and what other animals make of these electric charges, which constitute the environment, we cannot tell.

One special characteristic of human perception was, however, that since it depended so much on manipulation it came in turn to facilitate manipulation; because it evolved to a great extent from the motor investigation of texture and form through exploration, it came in time to increase the motor investigation of the objective world as it was perceived.

The evolution of manipulation for the manufacture of tools (Fig. 12.16) and eventually for the advance of modern technology has come to depend, appropriately, on the development of perception. Humans have used technology to increase their sensory awareness and to reveal still further the nature of matter and organic life. By the beginning of the 17th century we had invented the telescope, which enabled us to see more clearly the nature of the universe, and the microscope, which enabled us to perceive the minute structure of minerals and organic matter. Today, powerful telescopes and electron microscopes have vastly increased our visual perception, just as audio amplifiers, thermometers, and chemical analyses have effectively increased the range and sensitivity of our other senses.

Modern technology appears to be based on this expanding knowledge of our environment, which springs from our increased powers of

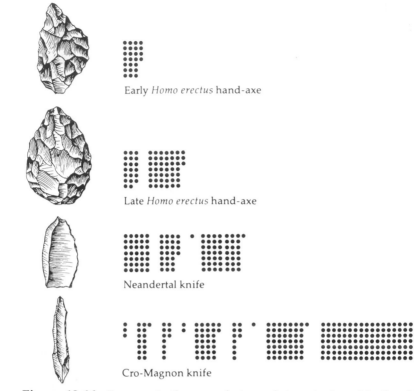

Early *Homo erectus* hand-axe

Late *Homo erectus* hand-axe

Neandertal knife

Cro-Magnon knife

Figure 12.16. Progress in the manufacture of stone tools and in the skill of the toolmaker is shown in this diagram. Increasing numbers of blows (dots) and distinct steps in manufacture (clusters of dots) have led to finer tools and more efficient utilization of material. The first two are Acheulian tools made by *Homo erectus*; they were roughhewn from single pieces of flint. The third was made in Neandertal times by chipping a flake from a prepared core and then modifying the flake. The bottom tool—a knife so sharp that one edge had to be dulled so that it could be grasped—was shaped by upper Palaeolithic people of modern type.

perception. There is a limit to this increase, however, because we find that at the microcosmic level we are dealing with electrical phenomena that can no longer be termed objects, and at the cosmic level we are limited by the distance from which light has reached the earth. It follows that, while there is much to be done in developing aids to perception, technological advances are beginning to receive impetus from the mechanization of other mental processes that characterize the brain: in particular, the processes of computation and prediction. The computer, another superorganic extension of the human brain, is already beginning to modify our relationship with the environment and alter our society in significant ways.

Another factor that is complementary to our highly evolved perception is, of course, our intelligence: our ability to make *conscious* deductions on the basis of our experience and our power to reason. The hypothesis springs to mind that among other factors it was also the adaptive value of reason that brought the evolution of self-awareness (see Chapter 11, VIII). Whether or not this was so, it seems that human perception and human reason, interacting under the competitive conditions of organic life, are the twin bases from which we may suppose the growth of human material culture to have developed.

We must also recall that human culture has one essential characteristic: it embodies a complex means of communication. Just as sexual reproduction allows a more rapid spread of novel genetic information within the gene pool than would be possible by asexual heredity, so the aspects of cultural behavior that form systems of communication (speech and writing) allow cultural diffusion within an "idea pool" at a fast rate. Writing is, at the same time, culture's own memory, another attribute previously known only in organic life. Today, machines can sense, analyze, and record; memory of percepts and experience are commonplace, stored in books or on videotape. We have learned to project into a cultural form most of the activities of our brain.

The human achievement was, therefore, first the extension of motor function, and second the extension of sensory and neural functions, all by means of a material culture (Fig. 12.17).

The "brain-like" nature of culture, so obvious in its store of knowledge and ability to compute probabilities, is derived from the abilities not of an individual human but of a whole society. No one without education could build a computer or write an encyclopedia. Culture is the long-accumulated experience and knowledge of the whole society. The basis of the idea of *involution* is that cooperation replaces evolutionary radiation and by cooperation we can create machines of a superhuman kind. Culture is in one sense the group mind of *Homo sapiens*.

The advantages of cooperation have, of course, resulted in the evolution of society not only among humans but among many different kinds of animals. As we have seen, these societies have in turn evolved sophisticated means of communication, yet only among humans have the material products of cooperation been preserved as an external record of experience. It is the development of human society's high-speed communication systems, its sensory and motor machines, and its memories and computers that differentiates it from any other. Soon, as a result of the development of multiple access to computer facilities, we shall be able to transmit complex problems to the state-run computer by telephone. When we can do that, our function will no longer be to solve problems but to invent them. As Herrick has written, "the most significant characteristic of intelligence is the ability to invent problems. This capacity for imaginative or creative thinking marks the highest level of integration in the organic realm" (1956, p.359).

We have tried to understand the adaptive nature of culture. Yet when

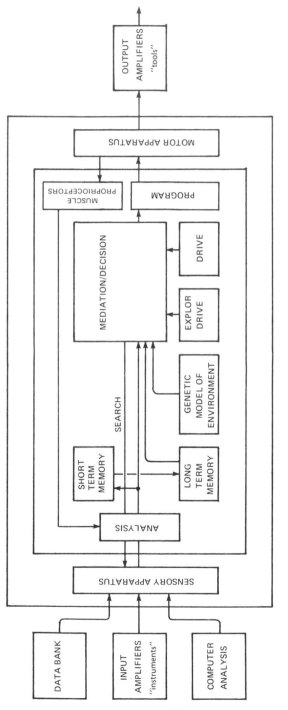

Figure 12.17. Building upon the diagrams in Figs. 11.2 and 11.3, we can add boxes indicating the way that external cultural tools and instruments can be used to enhance the capability of an individual by increasing the effectiveness of the motor output and the five senses. We can also add data banks (books, magnetic tape, disks, etc.) and computers for analysis and prediction, all of which reflect different aspects of brain function.

we turn to consider the present state of humankind, it looks as though culture is failing as an adaptive mechanism, as though it may lead to destruction rather than preservation of the species. In this connection there are three features of modern culture that may throw some light on the complexity of our cultural adaptation and help us to understand it.

First, modern culture makes an effective barrier between ourselves and the external environment. The result is that under the shield of culture humans are able to survive and breed with a genetic endowment that, in the absence of culture, would not allow them to survive (the survival of diabetics is an obvious example). The dangers of this situation have been repeatedly stressed and have been discussed by Dobzhansky (1962). They will not, therefore, be pursued here; instead, let us consider another aspect of this cultural barrier.

The present acceleration of cultural development may be due not to a primary failure in adaptation but to an imprecision in cultural adaptation. Thus, a particular novel cultural trait that arises in response to one environmental condition may alter a second environmental condition as a by-product. While such imprecision may also occur in organic adaptation, the speed and novelty of cultural change renders it more significant. For example, in North Temperate regions clothes were developed to cut heat loss from the body, and, although they are most effective, they also have far-reaching effects on skin, humidity, solar radiation, abrasion, external parasites, and many characteristics of a sematic kind, especially epigamic features. The original equilibrium is adjusted only to result in a new imbalance; yet, while this imbalance may seem to increase continually, the power to adjust that is allowed us by a developing culture can also increase. Human evolution is characterized by a great increase both in immediate environmental variability (through culture) and in cultural adjustment to it. Every development may be an attempt at adaptation, yet it may in turn bring with it new factors requiring further adaptation.

Second, it is clear that one of the central problems of the modern world arises from the fact that cultural adaptation has gone further in some social groups than in others, and in any case is of a different nature in different places. Nations that have steadily evolved cultural adaptations find themselves today (in the absence of war) in a relatively stable environment. Nations with different and more limited cultural adaptations may, as a result of modern communications and transportation receive inappropriate culture "secondhand" from more "advanced" nations, which may cause a gross change in the environment and in turn cause a biologically unstable period requiring considerable further cultural adjustment on the part of the native population. To give a simple example: in England the increase in the means of production that resulted in an increased food supply occurred at the same time as (if not before) the developments in medical science that raised the life expectancy from about 40 to about 65 years of age. There was, in fact, enough food available for the vastly increasing population.

Today in many countries we see medical science contributing to an

improvement in life expectancy, without an accompanying and necessary improvement in food supply; the result is intermittent famine. Culture contact, by its nature, introduces instability into the environment of
human populations insofar as such populations absorb behavior patterns that for them are inappropriate and nonadaptive. They must then
adapt to these patterns. In fact, it is a remarkable sign of the adaptive
efficiency of culture that contact is not always disastrous. On occasions it
has been fatal (the Tasmanians could not survive it), and it has often
been nearly fatal (as for the North American Indians), but in most instances adaptation rapidly occurs.

Third, we must recognize that cultural adaptations are not always
directed toward the impact of the external environment. They also serve
to protect society against the individual, for human survival depends on
the integrity of human society, and, as we have seen (Chapter 11, VII),
cultural behavior is directed toward maintaining that integrity. The
problem is particularly acute in the large social groups that characterize
the modern world. We may come to see that national enterprises, such
as war and space exploration, are among other things adaptations to
maintain the bonding of the societies that undertake them. Modern
advances in communication are certainly essential to the successful integration of the vast social conglomerates that are modern nation-states.

Today those of us who live in the West are no longer occupied merely
with our own basic subsistence. We are concerned with increasing our
economic resources and with the search for something we seem unable
to find, which we call "happiness." When we are not "at work" (a
concept that is by no means universal), we can ponder our condition
and look into our own minds with intelligent self-awareness; and we
discover there something more intractable than the external environment. As Eiseley has written (1958), "ancestral man entered his own
head, and he has been adapting ever since to what he finds there."
There is no doubt that a great deal of human behavior that appears to be
nonadaptive, often termed religious and frequently (though not necessarily) ritualized, arises in response to some form of projection into the
environment of mental experience that is not rationally understood. In
particular, the *regular* repression of emotion (which seems to be a
species-specific characteristic of *Homo sapiens*) can result in inappropriate
symbolic projections that are associated with maladaptive (neurotic) behavior. With our increase in leisure, we turn from society's god to a
personal god, as well as to our psychiatrist; we still must come to terms
with the phenomenon of our own consciousness.

We see then that human culture springs in some very general sense
from the projection into the environment of human neural attributes
and mental experience. We extend our means of perception, our means
of computation and memory, into the outside world as adaptations to
the environment. In understanding our needs, in using our reason, we
finally come to look into ourselves, and we must in turn adapt to what
we find in our own consciousness. We have to live not only with our

environment and with our fellows but also with ourselves, and our behavior is an adaptation to all three.

VIII.
The Dreamer

The supreme problem for the biologist lies in the evolution of consciousness and its relationship to matter. Because consciousness is often believed to be a particularly human characteristic and because of the difficulties of any attempt to study a phenomenon so subjective, this mysterious problem has received scant attention from biologists. Hinshelwood (1959, 445) is one of the few distinguished scientists who have drawn attention to it and discussed the form of the problem: "The question of the relation of the internal and external worlds cannot and should not be ignored by men of science."

Both he and Thorpe (1965) argue that human behavior is not machinelike, and it is clear that the mind-matter relationship should not be ignored in any text on human evolution. The difficulty for the scientist arises from the fact that he is used to dealing with observable data, and the existence of consciousness in others can only be inferred. But, Hinshelwood points out that somewhat the same comment could apply to the existence of the atomic nucleus; in the absence of direct observation, we can regard its existence only on the basis of inference as being in the highest degree probable. As was said in the Introduction to this book, all that science is really ever in a position to offer is a coherent body of evidence that has emerged from a large number of varied tests and observations.

The evidence for human consciousness arises ultimately from our social intercourse; therefore, it does not allow us to postulate with any certainty the existence of consciousness in animals, though evidence from those who have lived with animals intimately, whether with birds, dogs, lions, or primates, suggests strongly that some degree of consciousness exists, at least in the warm-blooded vertebrates (Thorpe, 1965). It is probable that consciousness, like any other human characteristic, evolved slowly, and it may indeed appear in its simplest form at much lower levels of life than has been supposed. If consciousness is a function of brain, and the two are certainly concomitant, then we can postulate its evolution to coincide at least with the evolution of the vertebrate central nervous system. This is, in fact, the simplest and therfore the most acceptable hypothesis, though at present it cannot be tested.

If, like Hinshelwood (1959), we accept the existence of consciousness—and few are likely to deny it today—we can account for it in two ways;

First, the potential for life and consciousness exists in every atom (of carbon, hydrogen, oxygen, nitrogen, etc.), and these properties are revealed to us as the atoms are combined into organic molecules of increasing complexity (see, for example, Rensch, 1960). In other words, the fundamental particles have a mental component that is apparent only when they constitute appropriate structures. Thorpe (1965, p.26) maintains, however, that this suggestion will not stand because it means that an unbelievable degree of potential organization was in fact present in the randomly moving atoms of the gases of the primordial nebulae, a potential for both organic and mental synthesis of vast complexity.

Second, an alternative explanation lies in the idea of emergence, the idea that at some stage in the process of evolution a completely new and, in principle, unpredictable quality may appear. Since physics and chemistry know nothing of mind in matter, we must confront this possibility. The only alternative, it seems, is to accept the first hypothesis and allow biology to modify physics.

Humankind, however, has something more than mere consciousness; we also have self-awareness, for, while the animal knows, only we know that we know. Here we are on even more difficult ground, and indeed we have no evidence and no clue to the history of this heightened level of consciousness. For the present, we must stick to our interpretation of the "fall of man" (see Chapter 11, VIII). With that in mind, we may see, with Dobzhansky (1962) and others, that self-awareness is humankind's distinctive mental characteristic, a characteristic that, linked with imagination, has raised us to the status of Lord of creation.

Only speech, in fact, can reveal to an individual this self-objectification that is characteristically human; speech is necessary for self-awareness and self-awareness is necessary for the moral order of human society. While some self-awareness may exist in apparently moral animals like dogs, its full development is surely confined to humans.

With the evolution of humankind, therefore, a new loneliness comes into the stream of life; our separate nature is revealed to us by the evolution of new mental processes. As Eiseley (1958: 125) has put it, "for the first time in four billion years a living creature has contemplated himself and heard with a sudden unaccountable loneliness, the whisper of the wind in the night reeds." We have looked at our environment and known ourselves to be no part of it; we are alone, and divided from the source of our being.

IX. Human Evolution: Past and Future

In the first chapter of this book evolution was described as a homeostatic adjustment in a living system in response to envirommental change. Evolution is the product of organic homeostasis. We have seen throughout our account of humankind's adaptations to a changing environment that this homeostatic adjustment, whether at a physiological, anatomical, or social level, is made by shifts in the mean of variable characteristics.

We have seen how the primitive mammal—already a remarkable and immensely complex evolutionary product—became adapted to arboreal life, and observed that the beginnings of erect posture occurred in the forest with concomitant changes in limb structure, hand, and foot. We have seen how some primates were preadapted for terrestrial life by their diet and posture either as quadrupeds or as bipeds, and we have tried to elucidate how hominids further adapted once they entered their new environment.

Perhaps the most essential humanizing adaptation was the increasing importance of society and culture as mechanisms for survival; society and culture have saved us from extinction and made us human, just as we in turn have created our society. In the words that Gordon Child chose as the title for his famous book, man makes himself. And that is the curious feature of human evolution; we ourselves seem to be involved in the creative process, up to the present unconsciously, perhaps in the future consciously.

But what we have recounted is no more than a rough sketch of our subject. We know so little that we have only been able to glimpse a process of great duration and immense complexity. We might be forgiven for doubting that anything less than a miracle was needed to produce, by a process of continuous adjustment of some strings of amino acids, a Mozart or an Einstein. We might also be forgiven for doubting if natural selection could really produce the kind of altruism we call selfless love. Eros we can understand, and maternal love lies deep in our genes; both clearly contribute to the survival of the genotype and to its inclusive fitness. But the selfless love of Mother Theresa, or the altruism of the blood donor who gives blood free for someone he will never meet,* does seem to have transcended the iron law of the selfish genes. This question is surely central to a real definition of humankind, a vital key to the nature of humans. Have we or have we not the possibility of moving beyond our genetic heritage: are we or are we not still attached to Wilson's genetic leash?

In Chapter 1, VII we concluded that we had the power to escape from Wilson's leash because we can and often do choose to bear no children. How did we gain this power? We gained it through the rational use of our intellect in our investigations of our environment in this case its physical aspects. We gained the power because we brought intelligence and reason to bear on our problems. Intelligence and reason imply detachment from the limbic system, and it is this detachment that gives us

---

*The example of Mother Theresa is given by Wilson (1975), who discounts her selflessness by reference to her belief in Christianity and the rewards she can expect in the next life. The example of the blood donors is given by Singer (1981). This is a particularly good example because in the United Kingdom all blood is collected by voluntary contribution. The donor receives no reward other than a cup of tea. The blood is stored in central blood banks for the use of anyone who needs it (and who receives it free of charge). Similar systems operate successfully in Australia, Holland, and some other countries.

freedom from our leash: that is why intelligence and reason have given us the possibility of moving outside genetic control.

Another way of stating this is to say that reasoning humans are not bound to restrict their behavior to what makes evolutionary sense. Nonreciprocal altruism directed toward complete strangers *does* occur. What humans can do and do do is to take evolutionarily sensible behavior, like bioaltruism, and apply it to a larger group, and eventually, perhaps, to the entire world. This we do regularly when we give to UNICEF or famine relief. We also give time and money to World Wildlife. It is impossible to argue that such behavior is selfish, unless we use this term is a sense so broad that it includes all human behavior ("everything you do is what you want to do, and is therefore selfish"). Such use of the word selfish is meaningless.

Socrates was put to death in 399 BC on the charge of corrupting the youth of Athens. He subjected traditions and custom to the fire of pure reason and in so doing threatened the traditions of the society. As one of the first and greatest philosophers and as one who unhesitatingly espoused reason, he showed us once and for all that it is possible to break with tribal lore, traditions, and the cultural luggage that we have brought into the world and look at ourselves anew, in the light of reason. Socrates, more than any other teacher, has shown us that we need not be slaves to the promptings within, the whispers from the limbic lobes. That is where our genes speak, where they hold our hearts: reason alone can free us from their ancient leash.

It may be appropriate, therefore, to define the human species as the *reasonable animal*. It is also clear that few people ever use pure reason unless they are writing a computer program. In our daily lives we move as in a trance, directed by feeling and tradition. Socrates said that "the unexamined life is not worth living." It may feel good to follow our heart, but it looks very much as if our salvation lies in reason; it may well be the only way to freedom and ultimate security.

Finally, although we can now escape from the dictates of natural selection if we wish, and although we have the power of reason, we have not found any serious inadequacies in the hypothesis that we are a product of organic evolution. It is clear that humans show a particular kind of adaptation to the terrestrial environment through a degree of independence from it, and that humankind and its power to reason are indeed the outcome of the process of natural selection.

Yet there are two broad problems to be acknowledged before we close. One concerns the limits of science; the other concerns the limits of our understanding. There are some special difficulties that are encountered when we attempt to apply the scientific method to human problems (Campbell, 1964), difficulties arising from the special relationship between the observer and the observed. We know that perception is far more greatly influenced by previous experience than by the nature of the object that is perceived. Posthypnotic suggestion can completely destroy the reliability of our perceptions; at a more everyday level, it can easily be demonstrated that people tend to perceive what they expect to perceive rather than what is before them. The frailty of human perception is notorious, and it is particularly subject to aberration when the subject and object already have some relationship, real or imagined. For that reason, our interpretation of the evidence relating to human evolution may be subject to prejudice of which we know nothing. The dispassionate intellect exists, but an approach to it is made increasingly difficult as the subject and object are more closely involved with each other. The difficulties are particularly intense when we attempt to study our own behavior and culture; they are well demonstrated by the different schools of psychoanalysis. A perfectly objective analysis of human behavior could not be made by any human being.

In this context it is also well to recall that we are in the power of the *zeitgeist*, the contemporary way of looking at the world and at ourselves—another important kind of aberration that destroys the clear focus of our science. This book, like almost every other, is a product of the times, with an emphasis that is typically slanted. The distortions in human studies are inherent and unavoidable. It is clear, however, that any approach to the truth can result only from the accumulation and record of human experience of every kind.

Finally, we must recognize some of the limits of our understanding, if only to indicate the most exciting directions for future research. The origin of life and the evolution of consciousness each presents a problem that biologists have not yet solved. We cannot foresee the final solutions to these problems, though they may come somewhat nearer in our lifetime. But let us not suppose that the ultimate truth is at all simple. A simplification of any hypothesis or data always results in a loss of truth, and we are guilty of simplification at every turn. In closing, we would do well to recall a comment by the philosopher Alfred North Whitehead (1920, p.163):

> The aim of science is to seek the simplest explanation of complex facts. We are apt to fall into the error of thinking that the facts are simple because simplicity is the goal of our quest. The guiding motto in the life of every natural philosopher should be, seek simplicity and distrust it.

Suggestions for Further Reading

The most recent and most useful compilation of research papers on human ancestry is R.L. Ciochon and R.S. Corruccini (Eds.), *New interpretations of ape and human ancestry* (New York and London: Plenum, 1983). For a review of the state of our knowledge of paleoanthropology, see B.G. Campbell, *Humankind emerging* (Boston: Little, Brown, 1985). For a cultural anthropologist's views on the origins and nature of human cultures, see M. Harris, *Cannibals and kings: The origins of culture* (New York: Random House, 1977) and *Cows, pigs, wars and witches: The riddles of culture* (New York: Random House, 1974).

For a brief account of the history of human ecology, see B.G. Campbell,

Human ecology: The story of our place in nature from prehistory to the present (New York: Aldine, 1985). For an account of the biology of human races, see J.S. Weiner, *The natural history of man* (New York: Universe Books, 1971), and S.M. Garn, *Human races* (Springfield, Ill.: Charles C. Thomas, 1961).

For a philosopher's discussion of the relationship between sociobiology and ethics, see P. Singer, *The expanding circle* (London and New York: Oxford Univ. Press, 1981), and for a collection of research papers on the problem of the relationship of mind and body, see J.C. Eccles (Ed.), *Brain and conscious experience* (Heidelberg: Springer-Verlag, 1966).

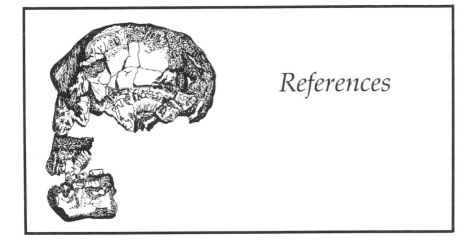

References

Abbie, A. A.
  1969 *The original australians.* Sydney: Rigby.
Aiello, L. C.
  1981a The allometry of primate body proportions, *Symp. Zool. Soc. London* **48**, 331–358.
  1981b Locomotion in the Miocene Hominoidea, *in* C. B. Stringer (Ed.), *Aspects of human evolution.* London: Taylor and Francis Ltd.
Aiello, L. C. and Day, M. H.
  1982 The evolution of locomotion in the early Hominidae, *in* R. J. Harrison and V. Navaratnam (Eds.), *Progress in anatomy*, Vol. 2, pp. 81–97. Cambridge; Cambridge University Press.
Andrew, R. J.
  1965 The origins of facial expressions. *Sci. Amer.* **213**, 88–94.
Andrews, P.
  1983 The Natural History of Sivapithecus, *in* R. L. Ciochon and R. S. Corrucini (Eds.), *New interpretations of ape and human ancestry.* New York: Plenum.
Andrews, P. and Aiello, L. C.
  1984 An evolutionary model for feeding and positional behaviour, *in* D. Chivers (Ed.), *Food acquisition and processing*, pp. 429–466. New York: Plenum.
Andrews, P. and Cronin, J. E.
  1982 The relationships of *Sivapithecus* and *Ramapithecus* and the evolution of the Orang-utan, *Nature (London)* **297**, 541–546.
Andrews, P., Lord, J. M., and Evans, E. M. N.
  1979 Patterns of ecological diversity in fossil and modern mammalian faunas. *Biol. J. Linn. Soc.* **11**, 177–205.
Angel, J. L.
  1975 Paleoecology, paleodemography and health, *in* S. Polgar (Ed.), *Population, ecology and social evolution.* The Hague: Mouton.

Arambourg, C. and Hoffstetter, R.
    1963   Le gisement de ternifine, Vol. 1, No. 32. Mem. Inst. Paleontol. Hum.,
           Paris.
Ashton, E. H. and Zuckerman, S.
    1951   Some cranial indices of Plesianthropus and other primates. Amer. J.
           Phys. Anthropol. 9, 283–96.
Auffenberg, W.
    1972   Komodo dragons. Nat. Hist. 81, 52–59.
Avis, V.
    1978   Brachiation: the crucial issue for man's ancestry. Southwest. J. An-
           thropol. 18, 119–148.
Basmajian, J. V.
    1978   Muscles alive: Functions revealed by electromyography. Baltimore: Wil-
           liams and Wilkins. 1962.
Beach, F. A.
    1947   Evolutionary changes in the physiological control of mating behavior
           in mammals. Psychol. Rev. 54, 197–315.
Beall, C. M. and Goldstein, M. C.
    1981   Tibetan fraternal polyandry—a test case of sociobiological theory,
           Amer. Anthropol. 83, 5–13.
Benedek, T.
    1952   Psychosexual functions in women. New York: Ronald Press.
Biegert, J.
    1963   The evaluation of characteristics of the skull, hands, and feet for
           primate taxonomy, in A. L. Washburn (Ed.), Classification and human
           evolution, pp. 116–45. (Viking Fund Publs. Anthrop., No. 37). Chi-
           cago: Aldine.
Binford, S.
    1968   A structural comparison of disposal of the dead in the Mousterian
           and Upper Paleolithic. Southwest. J. Anthropol. 24, 139–154.
Birch, L. C.
    1957   The meaning of competition. Amer. Natural. 91, 5–18.
Black, D. R., De Chardin, T., Young, C. C., and Pei, W. C.
    1933   Fossil Man in China: The Choukoutien Cave Deposits, with a Synopsis of
           Our Present Knowledge. Mem. Geol. Surv. China, Ser. A, No. 11.
Boas, N. T.
    1983   Morphological trends and phylogenetic relationships from Middle
           Miocene hominoids to Late Pliocene hominids, in R. L. Ciochon and
           R. S. Corruccini (Eds.), New interpretations of ape and human ancestry.
           New York and London: Plenum.
Bonin, G. Von
    1963   The evolution of the human brain. Chicago: University of Chicago
           Press.
Bonin, G. Von and Bailey, P.
    1961   Pattern of the cerebral isocortex. Primatologia 2, Part 2, Chap. 10.
Bowlby, J.
    1969   Attachment and loss, Vol. 1. London: Hogarth Press.
Broom, R. and Robinson, J. T.
    1952   Swartkrans ape-man, Paranthropus crassidens. Transvaal Mus. Mem.,
           No. 6. Pretoria.
Broom, R., Robinson, J. T., and Schepers, G. W. H.
    1950   Sterkfontein ape-man, Plesianthropus. Transvaal Mus. mem., No. 4.
           Pretoria.

Broom, R. and Schepers, G.
  1946 The South African fossil ape-men, The Australopithecinae, Transvaal Mus.
        Mem., No. 2. Pretoria.
Brues, A. M.
  1977 People and races. New York: Macmillan.
Burton, M.
  1962 Systematic dictionary of mammals of the world. London: Museum Press
        Ltd.
Cachel, S.
  1975 A new view of speciation in Australopithecus, in R. H. Tuttle (Ed.),
        Paleoanthropology, morphology and paleoecology. The Hague: Mou-
        ton.
Campbell, B. G.
  1963 Quantitative taxonomy and human evolution, in S. L. Washburn
        (Ed.), Classification and human evolution, pp. 50–74. (Viking Fund
        Publs. Anthrop. No. 37.) Chicago: Aldine.
  1964 Science and human evolution. Nature (London) 203, 448–51.
  1972a Conceptual progress in physical anthropology: Fossil man. Annu.
        Rev. Anthropol. 1, 27–54.
  1972b Sexual selection and the descent of man, 1871–1971. Chicago: Aldine.
  1985 Humankind emerging. Boston: Little, Brown.
Campbell, B. G. and Bernor, R. L.
  1976 The origin of the Hominidae: Africa or Asia? J. Human Evol. 5, 445–
        458.
Cannon, W. B.
  1932 The wisdom of the body. London: Kegan Paul.
Carpenter, C. R.
  1940 A field study in Siam of the behavior and social relations of the
        gibbon. Comp. Psychol. Monogr., No. 16. Republished 1964, in C. R.
        Carpenter, Naturalistic behavior of nonhuman primates. University Park,
        Pennsylvania: Pennsylvania State Univ. Press.
Cartmill, M.
  1974 Rethinking primate origins. Science 184, 436–443.
Cartmill, M. and Milton, K.
  1977 The Lorisiform wrist joint and the evolution of "brachiating" adapta-
        tions in the Hominoidea. Amer. J. Phys. Anthropol. 47, 249–273.
Chance, M. R. A.
  1961 The nature and special features of the instinctive social bond of pri-
        mates. in S. L. Washburn (Ed.), The social life of early man, pp. 17–33.
        (Viking Fund Publs. Anthropol., No. 31.) Chicago: Aldine.
Ciochon, R. L. and Chiarelli, A. B. (Eds.)
  1980 Evolutionary biology of the new world monkeys and continental drift. New
        York and London: Plenum.
Ciochon, R. L. and Corruccini, R. S. (Eds.)
  1983 New interpretations of ape and human ancestry. New York and London:
        Plenum.
Clark, J. D.
  1970 The prehistory of Africa. New York: Praeger.
Clark, W. E. Le Gros
  1947 Observations on the anatomy of the fossil Australopithecinae. J.
        Anat. (London) 81, 300–33.
  1950 New palaeontological evidence bearing on the evolution of the Homi-
        noidea. Quart. J. Geol. Soc. London 105, 225–64.

1955 The os innominatum of the recent Pongidae with special reference to that of the Australopithecinae. *Amer. J. Phys. Anthropol.* **13**, 19–27.

1971 *The Antecedents of Man,* 3rd ed. Chicago: Quadrangle Books.

1978 *Fossil evidence for human evolution,* 3rd ed. (with B. G. Campbell), Chicago: Univ. Chicago Press.

Clark, W. E. Le Gros and Leakey, L. S. B.

1951 *The Miocene Hominoidea of East Africa.* London: British Museum (Nat. Hist.)

Clutton-Brock, T. H. and Harvey, P. H.

1977 Primate ecology and social organization. *J. Zool. (London)* **183**, 1–39.

1979 Home range size, population density and phylogeny in primates, *in* I. S. Bernstein and E. O. Smith, *Primate ecology and human origins.* New York and London: Garland STPM Press.

Colbert, E. H.

1955 *Evolution of the vertebrates.* New York; Wiley.

Conroy, G. C. and Fleagle, J. G.

1972 Locomotor behavior in living and fossil pongids. *Nature (London)* **237**, 103–104.

Corruccini, R. S. and Ciochon, R. L.

1978 Morphoclinal variation in the anthropoid shoulder, *Amer. J. Phys. Anthropol.* **48**, 539–542.

Cronin, J. E., Boaz, N. T., Stringer, C. B., and Rak, Y.

1981 Tempo and mode in hominid evolution, *Nature (London)* **292**, 113–122.

Crook, J. H.

1972 Sexual selection, dimorphism, and social organization in the primates, *in* B. G. Campbell (Ed.), *Sexual selection and the descent of man, 1871–1971,* pp. 231–281. Chicago: Aldine.

Dart, R. A.

1925 *Australopithecus africanus:* the man-ape of South Africa, *Nature (London)* **115**, 195–199.

1949 Innominate fragments of *Australopithecus prometheus. Amer. J. Phys. Anthropol.* **7**, 301–338.

1957 *The osteodontokeratic culture of Australopithecus prometheus.* Transvaal Mus. Mem., No. 10. Pretoria.

1958 A further adolescent australopithecine ilium from Makapansgat. *Amer. J. Phys. Anthropol.* **16**, 473–479.

Darwin, C. R.

1859 *On the origin of species by means of natural selection.* London: Murray.

1871 *The descent of man, and selection in relation to sex.* London: Murray.

Davis, P. R.

1964 Hominid fossils from Bed I Olduvai Gorge: a tibia and fibula. *Nature (London)* **201**, 967–968.

Davis, P. R. and Napier, J.

1963 A reconstruction of the skull of *Proconsul africanus* (R. S. 51). *Folia Primatol.* **1**, 20–28.

Dawkins, R.

1976 *The selfish gene.* New York and Oxford: Oxford Univ. Press.

Day, M. H.

1969 Femoral fragment of a robust australopithecine from Olduvai Gorge, Tanzania. *Nature (London)* **221**, 230.

1971 Postcranial remains of *Homo erectus* from Bed IV Olduvai Gorge, Tanzania. *Nature (London)* **232**, 383–387.
1973 Locomotor features of the lower limb in hominids, *Symp. Zool. Soc.* **33**, 29–51.
1976a Hominid post-cranial material from Bed I, Olduvai Gorge, *in* G. L. Isaac and E. R. McCown (Eds.), *Human origins.* Menlo Park: Benjamin.
1976b Hominid post-cranial remains from the East Rudolf succession, *in* Y. Coppens, F. C. Howell, G. L. Isaac, and R. E. F. Leakey (Eds.), *Earliest man and environments in the Lake Rudolf Basin: Stratigraphy, palaeoecology and evolution*, pp. 507–521. Chicago: Chicago University Press.
1982 The *Homo erectus* pelvis: punctuation or gradualism? *Proc. 1st Congr. Intern. Paleontol. Hum.* **1**, 411–421.

Day, M. H. and Mollison, T. I.
1973 The Trinil femora, *in* M. H. Day (Ed.), *Human evolution.* London: Taylor and Francis.

Day, M. H. and Wickens, E. H.
1980 Laetoli Pliocene hominid footprints and bipedalism, *Nature (London)* **286**, 385–387.

Delson, E.
1977 Catarrhine phylogeny and classification: principles, methods and comments. *J. Human Evol.* **6**, 433–459.

Desmond, A.
1979 *The ape's reflexion.* London: Blond and Briggs.

Devore, I.
1965 *Primate behavior: field studies of monkeys and apes.* New York: Holt, Rinehart and Winston.

Dickemann, M.
1978 Female infanticide and the reproductive strategies of stratified human societies: a preliminary model, *in* N. A. Chagnon and W. Irons (Eds.), *Evolutionary biology and human social behavior: an anthropological perspective.* North Scituate, Mass.: Duxbury Press.

Dobzhansky, T.
1962 *Mankind evolving.* New Haven, Conn. and London: Yale Univ. Press.
1963 Cultural direction of human evolution: a summation. *Hum. Biol.* **35**, 311–16.
1970 *Genetics of the evolutionary process.* New York: Columbia Univ. Press.
1973 Nothing in biology makes sense except in the light of evolution, *Amer. Biol. Teacher* **35**, 125–129.

Dobzhansky, T. and Montagu, M. F. A.
1947 Natural selection and the mental capacity of mankind. *Science* **105**, 587–90.

Dobzhansky, T., Ayala, F. J., Stebbins, C. L., and Valentine, J. W.
1977 *Evolution.* San Francisco: Freeman.

Donaldson, J. F.
1983 The human menopause and the savanna. *Maturitas* **5**, 47–48.

Dubrul, E. L.
1958 *Evolution of the speech apparatus.* Springfield, Ill.: Charles C. Thomas.

Dubrul, E. L. and Sicher, H.
1954 *The adaptive chin.* Springfield, Ill.: Charles C. Thomas.

Dunbar, R.
   1980   Determinants and evolutionary consequences of dominance among
          female Gelada baboons, *Behav. Ecol. Sociobiol.* **7**, 253–265.
Eckstein, P. and Zuckerman, S.
   1956   The oestrous cycle in the mammalia, *in* A. S. Parks (Ed.), *Marshall's
          physiology of reproduction*, Vol. 1, Part 1, pp. 226–396. 3rd ed. London
          and New York: Longmans, Green.
Eiseley, L.
   1958   *The immense journey*. London: Victor Gollancz.
Eisenberg, J. F., Muckenhirn, N. A., and Rudran, R.
   1972   The relationship between ecology and social structure in primates.
          Science **176**, 863–874.
Endo, B.
   1966   Experimental studies on the mechanical significance of the form of
          the human facial skeleton. *J. Fac. Sci. Univ. Tokyo* **3** (Sec. 5), 1–106.
Erickson, G. E.
   1963   Brachiation in New World monkeys and in anthropoid apes. *Symp.
          Zool. Soc. London* **10**, 135–164.
Etkin, W.
   1963   Social behavioral factors in the emergence of man. *Hum. Biol.* **35**,
          299–310.
Evernden, J. F. and Curtis, G. H.
   1965   The potassium-argon dating of late cenozoic rocks in East Africa and
          Italy. *Curr. Anthropol.* **6**, 343–85.
Feldesman, M.
   1982   Morphometric analysis of the distal humerus of some cenozoic catar-
          rhines: the late divergence hypothesis revisited. *Amer. J. Phys. An-
          thropol.* **59**, 73–95.
Fleagle, J. G.
   1976   Locomotion and posture of the Malayan siamang and implications
          for hominoid evolution. *Folia Primatol.* **26**, 245–269.
   1977   Locomotor behavior and muscular anatomy of sympatric Malaysian
          leaf-monkeys, *Amer. J. Phys. Anthropol.* **46**, 297–307.
   1978   Locomotion, posture, and habitat use in two leaf-monkeys in West
          Malaysia, *in* D. J. Chivers and J. Herbert (Eds.), *Advances in primatolo-
          gy*, Vol. 1. London: Academic Press.
   1983   Locomotor adaptations of Oligocene and Miocene hominoids and
          their phyletic implications, *in* R. L. Ciochon and R. S. Corruccini
          (Eds.), *New interpretations of ape and human ancestry.* New York and
          London: Plenum, pp. 301–324.
Fleagle, J. G. and Kay, R. F.
   1983   New interpretations of the phyletic position of Oligocene hominoids,
          in R. L. Ciochon and R. S. Corucini (Eds.), *New interpretations of ape
          and human ancestry*, pp. 181–210. New York and London: Plenum.
Fleagle, J. G. and Mittermeier, R. A.
   1980   Locomotor behavior, body size, and comparative ecology of seven
          Surinam monkeys. *Amer. J. Phys. Anthropol.* **52**, 301–314.
Fleagle, J. G., Stern, J. T., Jungers, W. L., Susman, R. L., Vangor, A. K.,
and Wells, J. P.
   1981   Climbing: a biomechanical link with brachiation and with bipedal-
          ism. *Symp. Zool. Soc. London* **48**, 359–375.

Ford, C. S. and Beach, F. A.
    1951 *Patterns of sexual behavior.* New York: Harper & Bros. (Reprinted 1965;
          London: Methuen.)
Freedman, L. Z. and Roe, A.
    1958 Evolution and human behavior, *in* A. Roe and G. G. Simpson (Eds.),
          Behavior and evolution, pp. 455–79. New Haven, Conn.: Yale Univ.
          Press.
Frisch, K. Von
    1950 *Bees, their vision, chemical senses and language.* Ithaca, N.Y.: Cornell
          Univ. Press.
Gardner, R. A. and Gardner, B. T.
    1969 Teaching sign language to a chimpanzee. *Science* **165**, 664–672.
Garn, S. M.
    1971 *Human races,* 3rd ed. Springfield, Il.: Charles C. Thomas.
Garn, S. M. and Lewis, A. B.
    1963 Phylogenetic and intraspecific variations in tooth sequence poly-
          morphism. *Symp. Soc. Study Human Biol.* **5**, 53–73.
Geschwind, N.
    1964 The development of the brain and the evolution of language, *in* C. I.
          J. M. Stuart (Ed.), *Report of the 15th annual R.T.M. on linguistic and
          language studies,* pp. 155–169. Monograph Series on Languages and
          Linguistics, No. 17.
    1972 Language and the brain. *Sci. Amer.* **226** (4), 76–83.
Goodall, J., Van Lawick
    1968 The behaviour of free-living chimpanzees in the Gombe Stream Re-
          serve. *Animal Behav. Monogr.* **1**, 161–311.
    1971 *In the shadow of man.* Boston: Houghton-Mifflin.
Goodhart, C. B.
    1960 The evolutionary significance of human hair patterns and skin col-
          ouring. *Advan. Sci.* **17**, 53–59.
Gould, S. J.
    1977 *Ontogeny and phylogeny.* Cambridge and London: Harvard Univ.
          Press.
    1980 *The panda's thumb.* pp. 127–132. New York: Norton.
    1982 The meaning of punctuated equilibrium and its role in validating
          a hierarchical approach to macro-evolution, *in* R. Milkman (Ed.),
          Perspectives on evolution. Sunderland, Mass: Sinauer Associa-
          tion.
Gould, S. J. and Eldredge, N.
    1977 Punctuated equilibria: the tempo and mode of evolution recon-
          sidered. *Paleobiology* **3**, 115–151.
Grant, V.
    1977 *Organismic evolution.* San Francisco: Freeman.
    1983 The synthetic theory strikes back, *Biol. Zentrabl.* **102,** 149–158.
Gregory, R. I.
    1966 *Eye and brain.* London: Weidenfeld and Nicholson.
Haldane, J. B. S.
    1956 The argument from animals to men: an examination of its validity for
          anthropology. *J. Roy. Anthropol. Inst.* **86**, 1–14.
Hamburg, D. A.
    1963 Emotion in the perspective of human evolution, *in* P. Knapp (Ed.),

Expression of the emotions in man, pp. 300–317. New York: International Univ. Press.

Hamilton, W. D.
   1963   The evolution of altruistic behavior, *Amer. Natural.* **97**, 354–6.

Harcourt, A.
   1979   Contrasts between male relationships in wild gorilla groups, *Behav. Ecol. Sociobiol.* **5**, 39–49.

Harcourt, A. H., Harvey, P. H., Larson, S. G., and Short, R. V.
   1981   Testis weight, body weight and breeding system in primates. *Nature (London)* **293**, 55–57.

Harding, R. S. O.
   1975   Meat-eating and hunting in baboons, *in* R. H. Tuttle (Ed.), *Socioecology and psychology of primates*. The Hague: Mouton.

Hardy, M.
   1934   Observations on the innervation of the macula sacculi in man. *Anat. Rec.* **59**, 403–418.

Harlow, H. F.
   1959   The development of learning in the rhesus monkey, *Amer. Sci.* **47**, 450–479.

Harris, M.
   1974   *Cows, pigs, wars and witches.* New York: Random House.

Hay, R. L.
   1970   Silicate reactions in three lithofacies of a semi-arid basin, Olduvai Gorge, Tanzania. *Mineral. Soc. Amer. Spec. Pap.* **3**, 237–155.

Hayes, C.
   1951   *The ape in our house.* New York: Harper.

Herrick, C. J.
   1946   Progressive evolution. *Science* **104**, 469.
   1956   *The evolution of human nature.* New York: Harper & Bros.

Hiatt, L.
   1981   Polyandry in Sri Lanka: a test case for parental investment theory. *Man* **15**, 583–602.

Hinde, R. A. and Fisher, J.
   1951   Further observations on the opening of milk bottles by birds, *Brit. Birds* **44**, 393–396.

Hinshelwood, C.
   1959   The internal and external worlds. *Proc. Roy. Soc.* **A253**, 442–49.

Hockett, C. F. and Ascher, R.
   1964   The human revolution. *Curr. Anthropol.* **5**, 135–68.

Holloway, R. L.
   1968   The evolution of the primate brain: some aspects of quantitative relations, *Brain Res.* **7**, 121–172.
   1974   The casts of fossil hominid brains, *Sci. Amer.* **231**(7), 106–115.
   1981   Revisiting the South African Taung australopithecine endocast. *Amer. J. Phys. Anthropol.* **56**, 43–58.

Holloway, R. L. and De La Coste-Lareymondie, M. C.
   1982   Brain endocast asymmetry in pongids and hominids: some preliminary findings on the paleontology of cerebral dominance, *Amer. J. Phys. Anthropol.* **58**, 101–110.

Holt, A. B., Cheek, D. B., Mellits, E. D., and Hill, D. E.
   1975   Brain size and the relation of the primate to the non-primate, *in* D. B. Cheek (Ed.), *Fetal and post-natal cellular growth: hormones and nutrition*. New York: Wiley.

Howell, F. Clark
    1960   European and Northwest African Middle Pleistocene hominids. *Curr. Anthropol.* **1**, 195–232.
    1967   Recent advances in human evolutionary studies. *Quart. Rev. Biol.* **42**, 471–513.
Howell, N.
    1976   The population of the Dobe area !Kung, *in* R. B. Lee and I. Devore (Eds.), *Kalahari hunter-gatherers: Studies of the !Kung San and their neighbours.* Cambridge, Mass.: Harvard Univ. Press.
Hrdy, S. B.
    1981   *The woman that never evolved.* Cambridge and London: Harvard Univ. Press.
Hubel, D. H.
    1963   The visual cortex of the brain. *Sci. Amer.* **209**, 54–62.
Humphrey, N.
    1983   *Consciousness regained.* Oxford: Oxford University Press.
Huxley, J. S.
    1958   Cultural process and evolution, *in* A. Roe and G. G. Simpson (Eds.), *Behavior and evolution,* pp. 437–54. New Haven, Conn.: Yale Univ. Press.
Imanischi, K.
    1960   Social organization of sub-human primates in their natural habitat. *Curr. Anthropol.* **1**, 393–407.
Inge, W. R.
    1922   *Outspoken essays.* 2nd ser. London and New York: Longmans, Green.
Isaac, G. L.
    1978   The food-sharing behavior of proto-human hominids. *Sci. Amer.* **238**(4), 90–108.
Isaac, G. L., Leakey, R. E. F., and Behrensmeyer, A. M.
    1971   Archeological traces of early hominid activities, east of Lake Rudolf, Kenya. *Science* **173**, 1129–1123.
Itani, J. S.
    1965   Social organization of Japanese monkeys. *Animals* **5**, 410–17.
Janzen, D. H.
    1978   Complications in interpreting the chemical defence of trees against tropical arboreal plant-eating vertebrates, *in* G. G. Montgomery (Ed.), *The ecology of arboreal herbivores.* Washington, D.C.: Smithsonian Institute.
Jerison, H. J.
    1973   *Evolution of the brain and intelligence.* New York: Academic Press.
Johanson, D. C. and Edey, M. A.
    1981   *Lucy: The beginnings of humankind.* New York: Simon and Schuster.
Johanson, D. C. and White, T. D.
    1979   A systematic assessment of early African hominids. *Science* **202**, 321–330.
Johanson, D. C., Taieb, M., and Coppens, Y.
    1982   Pliocene Hominids from the Hadar Formation, Ethiopia (1973–1977): Stratigraphic, chronologic, and paleoenvironmental contexts with notes on hominid morphology and systematics. *Amer. J. Phys. Anthropol.* **57**, 373–402.
Jones, F. Wood
    1916   *Arboreal man.* London: Arnold.

Jungers, W. L. and Stern, J. T.
    1983 Body proportions, skeletal allometry and locomotion in the Hadar
          hominids: a reply to Wolpoff. *J. Human Evol.* **12**, 673–684.
Kawai, M.
    1965 Japanese monkeys and the origin of culture. *Animals* **5**, 450–55.
Kay, R. F.
    1977 The evolution of molar occlusion in the Cercopithecidae and early
          catarrhines, *Amer. J. Phys. Anthropol.* **46**, 327–352.
Keesing, F. M.
    1958 *Cultural anthropology.* New York: Rinehart.
Keith, A.
    1923 Man's posture: Its evolution and disorders. *Brit. Med. J.* **1**, 451–
          672.
Kimbel, W. H., White, T. D., and Johanson, D.C.
    1984 Cranial morphology of *Australopithecus afarensis:* A comparative study
          based on a composite reconstruction of the adult skull. *Amer. J. Phys.
          Anthropol.* **64**, 337–388.
Kimura, T., Okada, M., and Ishida, H.
    1979 Kinesiological characteristics of primate walking: its significance in
          human walking, *in* M. E. Morbeck, H. Preuschoft, and N. Gomburg
          (Eds.), *Dynamic interactions in primates: environment, behavior and
          morphology,* pp. 297–312. New York: Gustav Fischer.
Kohler, W.
    1925 *The mentality of apes.* New York and London: Kegan Paul.
Kohts, N.
    1935 *Infant ape and human child.* Moscow, Sci. Mem. Mus. Darwin.
Kortlandt, A.
    1965 Comment on the essential morphological basis for human culture.
          Curr. Anthropol. **5**, 320–25.
Krantz, G. S.
    1963 The functional significance of the mastoid process in man. *Amer. J.
          Phys. Anthropol.* **21**, 591–93.
Kroeber, A. L. and Kluckhohn, C.
    1952 *Culture: A critical review of concepts and definitions.* Pap. Peabody Mus.
          No. 47. Cambridge, Mass.
Kummer, H.
    1971 *Primate societies.* Chicago: Aldine.
Kurten, B.
    1972 *Not from the apes.* New York: Pantheon. Reprinted 1983.
Kuypers, H. G. J. M.
    1964 The descending pathways to the spinal cord, their anatomy and func-
          tion, *in* J. C. Eccles and J. P. Schade (Eds.), *Progress in brain research,*
          Vol. 2, pp. 178–202. New York, Amsterdam: Elsevier.
Kuypers, H. G. J. M., Szwarcbart, M. K., Mischkin, M., and Rosvold, H. E.
    1965 Occipitotemporal corticocortical connection in the rhesus monkey.
          Exp. Neurol. **11**, 245–262.
Laitman, J. T. and Heimbuch, R. C.
    1982 The basicranium of Plio-Pleistocene hominids as an indicator of their
          upper respiratory systems, *Amer. J. Phys. Anthropol.* **59**, 323–343.
Laitman, J. T., Heimbuch, R. C., and Crelin, E. S.
    1979 The basicranium of fossil hominids as an indicator of their upper
          respiratory systems, *Amer. J. Phys. Anthropol.* **51**, 15–34.

Lancaster, J. B. and Lee, R. B.
  1965 The annual reproductive cycle in monkeys and apes, *in* I. Devore (Ed.), *Primate behavior,* pp. 486–513. New York: Holt, Rinehart and Winston.
Leach, E.
  1966 Ritualization in man in relation to conceptual and social develop ment. *Phil. Trans. Roy. Soc. London* **B251**, 403–408.
Leakey, M. D.
  1970 Early artifacts from the Koobi Fora area. *Nature (London)* **226**, 228–230.
  1971 *Olduvai Gorge, Vol. 3: Excavations in beds I and II, 1960–1963.* Cambridge: Cambridge Univ. Press.
Leakey, M. G. and Leakey R. E.
  1978 *Koobi Fora research project,* Vol. 1: *The fossil hominids and an introduction to their context, 1968–1974.* Oxford: Oxford Univ. Press.
Lee, R. B.
  1968 What hunters do for a living, or, how to make out on scarce resources, *in* R. B. Lee and I. Devore (Eds.), *Man the hunter,* pp. 30–48. Chicago: Aldine.
Lee, R. B. and Devore I. (Eds.)
  1976 *Kalahari hunter-gatherers: Studies of the !Kung San and their neighbours.* Cambridge, Mass.: Harvard Univ. Press.
Lerner, I. M.
  1954 *Genetic homeostasis.* New York: Wiley.
Lewis, O. J.
  1973 The hominoid *Os Capitatum,* with special reference to the fossil bones from Sterkfontein and Olduvai Gorge. *J. Human Evol.* **2**, 1–11.
  1981 Functional morphology of the joints of the evolving foot. *Symp. Zool. Soc. London* **46**, 169–188.
Lieberman, P.
  1975 *On the origins of language.* New York: Macmillan.
Lovejoy, C. O.
  1973 The gait of *Australopithecus. Yearb. Phys. Anthropol.* **17**, 147–161.
  1981 The origin of man. *Science* **211**, 341–350.
Lovejoy, C. O. and Heiple, K. G.
  1970 A reconstruction of the femur of *Australopithecus africanus. Amer. J. Phys. Anthropol.* **32**, 33–40.
Lovejoy, C.O., Heiple, K. G., and Burstein, A. H.
  1973 The Gait of *Australopithecus. Amer. J. Phys. Anthropol.* **38**, 757–780.
Lovejoy, C. O., Johanson, D. C., and Coppens, Y.
  1982a Hominid lower limb bones recovered from the Hadar formation: 1974–77 collections. *Amer. J. Phys. Anthropol.* **57**, 679–700.
  1982b Hominid upper limb bones recovered from the Hadar formation: 1974–77 collections. *Amer. J. Phys. Anthropol.* **57**, 637–649.
Lowe, C. Van Riet
  1954 The cave hearths. *S. Afr. Arch. Bull.* **33**, 25–29.
Luckett, W. P.
  1975 Ontogeny of fetal membranes and placenta: Their bearing on primate phylogeny, *in* W. P. Luckett and F. S. Szalay (Eds.), *Phylogeny of the primates.* New York: Plenum.
Lynch, G., Hechtel, S., and Jacobs, D.
  1983 Neonate size and evolution of brain size in the anthropoid primates. *J. Human Evol.* **12**, 519–522.

MacArthur, R. and Wilson, E. O.
  1967 *The theory of island biogeography.* Princeton, N.J.: Princeton Univ.
        Press.
McClean, P.
  1978 The evolution of three mentalities, *in* S. L. Washburn and E. R.
        McCown (Eds.), *Human evolution; biosocial perspectives.* Menlo Park,
        Calif: Benjamin/Cummings.
McHenry, H.
  1973 Early hominid humerus from East Rudolf, Kenya. *Science* **180**, 379–
        741.
McHenry, H. and Corruccini, R. S.
  1975 Distal humerus in hominoid evolution, *Folia Primatol.* **23**, 227–244.
  1976 Analysis of an early hominid ulna from the Omo basin, Ethiopia.
        Amer. J. Phys. Anthropol. **44,** 295–304.
Magoon, H. W.
  1960 Concepts of brain function, *in* S. Tax (Ed.), *The evolution of man*, pp.
        187–209. Chicago: Univ. of Chicago Press.
Malthus, T. R.
  1798 *An Essay on the principle of population.* London: Johnson.
Mark, V. H. and Ervin, F. R.
  1970 *Violence and the brain.* New York: Harper and Row.
Martin, C. J.
  1902 Thermal adjustment and respiratory exchange in monotremes and
        marsupials—a study in the development of homeothermism. *Phil.
        Trans. Roy. Soc. London* **B195**, 1–37.
Martin, R. D.
  1975 Strategies of reproduction, *Natur. History* (NY) **84**, 48–57.
  1979 Phylogenetic aspects of prosimian behavior, *in* C. A. Doyle and R. D.
        Martin (Eds.), *The study of prosimian behavior.* London and New York:
        Academic Press.
  1981 Field studies of primate behaviour. *Symp. Zool. Soc. London.* **46,** 287–
        336.
Martin, R. D., Chivers, D. J., Maclarnon, A. M., and Hladik, C. M.
  1984 Gastrointestinal allometry in primates and other mammals, *in* W. L.
        Jungers (Ed.), *Size and scaling in primate biology.* New York:
        Plenum.
Mason, A. S.
  1960 *Health and hormones.* London: Penguin.
Mayr, E.
  1963 *Animal species and evolution.* Cambridge, Mass.: Harvard Univ. Press;
        London: Oxford Univ. Press.
Mednick, L. W.
  1955 The evolution of the human ilium. *Amer. J. Phys. Anthropol.* **13**, 203–
        16.
Mettler, F. A.
  1956 *Culture and the structural evolution of the neural system.* New York:
        Amer. Mus. Natural History. Also *in* M. F. Ashley Montagu (Ed.),
        Culture and the evolution of man. pp. 155–201. New York: Oxford Univ.
        Press, 1962.
Miller, G. S.
  1931 The primate basis of human sexual behavior. *Quart. Rev. Biol.* **6**,
        379–410.

Mills, J. R. E.
    1963   Occlusion and malocclusion of the teeth of primates. *Symp. Soc. Study Human Biol.* **5**, 29–51.
Montagna, W.
    1965   The skin. *Sci. Amer.* **212**, 56–66.
Montagu, M. F. Ashley
    1962   Time, morphology and neotony in the evolution of man. *in* M. F. Ashley Montagu (Ed.), *Culture and the evolution of man,* pp. 324–42. New York: Oxford Univ. Press.
    1971   *Touching: The human significance of the skin.* New York: Columbia Univ. Press.
Morbeck, M. E.
    1975   *Dryopithecus africanus* forelimb. *J. Human Evol.* **4**, 39–46.
Morton, D. J.
    1927   Human origin. *Amer. J. Phys. Anthropol.* **10**, 173–203.
    1964   *The human foot.* New York: Haffner.
Moss, M. L. and Young R. W.
    1960   A functional approach to craniology. *Amer. J. Phys. Anthropol.* **18**, 281–92.
Musgrave, J.
    1971   How dextrous was Neandertal man? *Nature (London)* **233**, 538–541.
Napier, J. R.
    1961   Prehensibility and opposability in the hands of primates. *Symp. Zool. Soc. London* **5**, 115–32.
    1962   The evolution of the hand. *Sci. Amer.* **207**, 56–62.
    1964   The evolution of bipedal walking in the hominids. *Arch. Biol. (Liège)* **75**, 673–708.
Napier, J. R. and Davis, P. R.
    1959   *The forelimb skeleton and associated remains of Proconsul Africanus.* London: British Museum (Natural History).
Napier, J. R. and Napier, P. H.
    1967   *A handbook of living primates.* London and New York: Academic Press.
Napier, J. R. and Walker, A.
    1967   Vertical clinging and leaping: a newly recognized category of locomotor behavior of primates. *Folia Primatol.* **7**, 204–219.
Noback, C. R. and Moskowitz, N.
    1963   The primate nervous system: functional and structural aspects in phylogeny, *in* J. Buettner-Janusch (Ed.), *Evolutionary and genetic biology of primates*, Vol. 1, pp. 131–177. New York: Academic Press.
Oakley, K. P.
    1951   A definition of man. *Penguin Sci. News* **20**, 69–81. Also *in* M. F. Ashley Montagu (Ed.), *Culture and evolution of man*, pp. 3–12. New York: Oxford Univ. Press, 1962.
    1961   On man's use of fire, with comments on toolmaking and hunting, *in* S. L. Washburn (Ed.), *The social life of early man*, pp.176–93. (Viking Fund Publs. Anthrop., No. 31.) Chicago: Aldine.
    1963   *Man the toolmaker*, 5th ed. London: British Museum (Natural History).
    1969   *Frameworks for dating fossil man*. 3rd ed. London: Weidenfeld and Nicholson; Chicago: Aldine.
Oakley, K. P., Campbell, B. G., and Mollison, T. I.
    1967–74 Catalogue of fossil hominids. 3 Vols. London: British Museum (Natural History).

Ornstein, R. E.
    1972   *The psychology of consciousness.* San Francisco, Calif.: Freeman.
Oxnard, C. E.
    1963   Locomotor adaptations in the primate forelimb. *Symp. Zool. Soc. London* **10**, 165–82.
    1968   A note on the Olduvai clavicular fragment. *Amer. J. Phys. Anthropol.* **29**, 429–432.
    1969   Evolution of the human shoulder; some possible pathways. *Amer. J. Phys. Anthropol.* **30**, 319–322.
Passingham, R. E.
    1978   Brain size and intelligence in primates, *in* D. Chivers and K. A. Joysey (Eds.), *Recent advances in primatology,* Vol. 3. London: Academic Press.
    1979   Brain size and intelligence in man. *Brain Behav. Evol.* **16**, 253–270.
    1982   *The human primate.* Oxford and San Francisco, Calif.: Freeman.
Patterson, B. and Howells, W. W.
    1967   Hominid humeral fragment from early Pleistocene of northwest Kenya. *Science* **156**, 64–66.
Penfield, W. and Rasmussen, T. B.
    1957   *The cerebral cortex of man.* New York: Macmillan.
Penfield, W. and Roberts, L.
    1959   *Speech and brain mechanisms.* Princeton, N.J.: Princeton Univ. Press.
Perles, C.
    1977   *Préhistoire du Feu.* Paris: Masson.
Pilbeam, D. R.
    1972   *The ascent of man.* New York: Macmillan.
    1982   New hominoid skull material from the Miocene of Pakistan. *Nature (London)* **295**, 232–234.
Pocock, R. I.
    1925   The external characters of the catarrhine monkeys and apes. *Proc. Zool. Soc. London,* pp. 1479–1579.
Polyak, S.
    1957   *The vertebrate visual system.* Chicago: Univ. of Chicago Press.
Portmann, A.
    1962   Cerebralisation und Ontogenese. *Medizin Grundlagenfor* **4**, 1–62.
Premack, D.
    1976   *Intelligence in ape and man.* Hillsdale, NJ: Erlbaum.
Prost, J.
    1965   A definitional system for the classification of primate locomotion. *Amer. Anthropol.* **67**, 1198–1214.
Rasmussen, T. B. and Penfield, W.
    1947   Further studies of the sensory and motor cerebral cortex in man. *Fed. Proc. Amer. Soc. Expt. Biol.* **6**, 452–560.
Rensch, B.
    1956   Increase of learning capability with increase in brain size. *Amer. Natur.* **90**, 81–95.
    1959   Trends toward progress of brains and sense organs. *Cold Spring Harbor Symp. Quant. Biol.* **24**, 291–303.
    1960   *Evolution above the species level.* New York: Columbia Univ. Press.
Ripley, S.
    1967   The leaping of langurs: a problem in the study of locomotor adaptation. *Amer. J. Phys. Anthropol.* **26**, 149–170.

Robinson, J. T.
1956 *The dentition of the Australopithecinae.* Transvaal Mus. Mem., No. 9. Pretoria.
1963 Australopithecines: culture and phylogeny. *Amer. J. Phys. Anthropol.* **21**, 595–605.
1972 *Early hominid posture and locomotion.* Chicago: Univ. of Chicago Press.
Roe, A.
1963 Psychological definitions of man, *in* S. L. Washburn (Ed.), *Classification and human evolution,* pp. 320–31. (Viking Fund Publs. Anthrop., No. 37.) Chicago: Aldine.
Romer, A.
1945 *Vertebrate paleontology.* Chicago: Univ. of Chicago Press.
1956 *The osteology of reptiles.* Chicago: Univ. of Chicago Press.
1962 *The vertebrate body.* Philadelphia and London: Saunders.
Rose, M. D.
1973 Quadrupedalism in primates. *Primates* **14**, 337–358.
1983 Miocene hominoid postcranial morphology, *in* R. L. Ciochon and R. S. Corruccini (Eds.), *New interpretations of ape and human ancestry.* New York, Plenum, pp. 405–417.
Rosen, S. I. and McKern, T. W.
1971 Several cranial indices and their relevance to fossil man. *Amer. J. Phys. Anthropol.* **35**, 69–74.
Rumbaugh, D. M.
1977 *Language learning by a chimpanzee.* New York: Academic Press.
Sacher, G. A.
1975 Maturation and longevity in relation to cranial capacity in hominid evolution, *in* R. Tuttle (Ed.), *Primate functional morphology and evolution.* The Hague: Mouton.
1976 Evaluation of the entropy and information terms governing mammalian longevity. *Interdisc. Topics Gerontol.* **9**, 69–82.
Sahlins, M. D.
1959 The social life of monkeys, apes and primitive man. *Human Biol.* **31**, 54–73.
Sarich, V. M. and Cronin, J. E.
1977 Generation lengths and rates of hominoid molecular evolution. *Nature (London)* **269**, 354–355.
Schaller, G. B.
1963 *The mountain gorilla: Ecology and behavior.* Chicago: Univ. of Chicago Press.
Schaller, G. B. and Lowther, G. R.
1969 The relevance of carnivore behavior to the study of early hominids. *Southwest. J. Anthropol.* **25**, 307–341.
Schenkel, R.
1947 Ausdrucks-studien an wölfen. *Behavior* **1**, 81–129.
Schultz, A. H.
1924 Growth studies on primates bearing on man's evolution. *Amer. J. Phys. Anthropol.* **7**, 149–64.
1930 The skeleton of the trunk and limbs of higher primates. *Human Biol.* **2**, 303–438.
1935 Eruption and decay of teeth in primates. *Amer. J. Phys. Anthropol.* **19**, 489–581.

1936   Characters common to higher primates and characters specific for man. *Quart. Rev. Biol.* **11**, 259–83, 425–55.

1937   Proportions of long bones in man and apes. *Human Biol.* **9**, 281–328.

1941   Length of spinal regions in primates. *Amer. J. Phys. Anthropol.* **24**, 1–22.

1942   Conditions for balancing the head in primates. *Amer. J. Phys. Anthropol.* **29**, 483–97.

1948   The number of young at birth and the number of nipples in primates. *Amer. J. Phys. Anthropol.* **6**, 1–23.

1950   Man and the catarrhine primates. *Cold Spring Harbor Symp. Quant. Biol.* **15**, 35–53.

1953   Relative thickness of bones in primates. *Amer. J. Phys. Anthropol.* **11**, 277–311.

1955   The position of the occipital condyles and of the face relative to the skull base in primates. *Amer. J. Phys. Anthropol.* **13**, 97–120.

1956   Post-embryonic age changes. *Primatologia* **1**, 887–964.

1960   Einige Beobachtungen und Masse am Skelett von *Oreopithecus*. *Z. Morphol. Anthropol.* **50**, 136–49.

1961   Some factors influencing the social life of primates in general and early man in particular, *in* S. L. Washburn (Ed.), *The social life of early man*, pp. 58–90. (Viking Fund Publs. Anthrop., No. 31.) Chicago: Aldine.

1963a The foot skeleton in primates. *Symp. Zool. Soc. London* **10**, 199–206.

1963b Age changes, sex differences, and variability as factors in the classification of primates, *in* S. L. Washburn (Ed.), *Classification and human evolution*, pp. 85–115. (Viking Fund Publs. Anthrop., No. 37.) Chicago: Aldine.

1969   Observations on the acetabulum of Primates. *Folia Primatol.* **11**, 181–199.

Schultz, A. H. and Straus, W. L.

1945   The numbers of vertebrae in primates. *Proc. Amer. Phil. Soc.* **89**, 601–626.

Seyfarth, R. M., Cheney, D. L., and Marler, P.

1980   Monkey responses to three different alarm calls: Evidence of predator classification and semantic communication. *Science* **210**, 801–803.

Sherfey, M. J.

1972   *The nature and evolution of female sexuality.* New York: Random House.

Shock, N. W.

1951   Growth curves, *in* S. S. Stevens (Ed.), *Handbook of experimental psychology*, pp. 330–46. New York: Wiley.

Short, R. V.

1976a Lactation—the central control of reproduction, *in: Breast-feeding and the mother.* Ciba Foundation Symposium No. 45.

1976b The evolution of human reproduction. *Proc. Roy. Soc. London* **B195**, 3–24.

1977   Sexual selection and the descent of man, *in* J. H. Calaby and C. H. Tyndale-Biscoe (Eds.), *Reproduction and evolution*, Proc. 4th Symp. Comp. Biol. Reproduction. Australian Academy of Science.

Sigmon, B. A.

1982   Comparative morphology of the locomotor skeleton of *Homo erectus* and other fossil hominids, with special reference to the Tautavel innominate and femora. *Proc. 1st Congr. Intern. Paleontol. Human* **1**, 422–446.

Simpson, G. G.
1935 The first mammals. *Quart. Rev. Biol.* **10**, 154–180.
1944 *Tempo and mode in evolution.* New York: Columbia Univ. Press.
1945 The principles of classification and a classification of mammals. *Bull. Amer. Mus. Natur. History* **85**, 1–350.
1949 *The meaning of evolution.* New Haven, Conn.: Yale Univ. Press.
1953 *The major features of evolution.* New York: Columbia Univ. Press.
1961 *The principles of animal taxonomy.* New York: Columbia Univ. Press.
1963 The meaning of taxonomic statements, *in* S. L. Washburn (Ed.), *Classification and human evolution.* Chicago: Aldine.

Simpson, M. J. A. and Simpson, A. E.
1982 Birth sex ratios and social rank in rhesus monkey mothers. *Nature (London)* **300**, 440–441.

Singer, P.
1981 *The expanding circle: Ethics and sociobiology.* Oxford: Clarendon.

Sivertsen, E.
1941 On the biology of the harp seal. *Hvalradets Skr.* **26**, 1–166.

Smith, G. E.
1924 *The evolution of man.* London: Oxford Univ. Press.

Sneath, P. H. A. and Sokal, R. R.
1963 Numerical taxonomy. *Nature (London)* **193**, 855–60.

Soules, M. R. and Bremer, W. J.
1982 The menopause and climacteric. *J. Amer. Geriat. Soc.* **30**, 547–552.

Sperry, R. W.
1955 On the neural basis of the conditioned response. *Brit. J. Animal Behav.* **3**, 41–44.
1966 Brain bisection and consciousness, *in* J. C. Eccles (Ed.), *Brain and conscious experience.* Heidelberg: Springer-Verlag.

Spuhler, J. N.
1959 Somatic paths to culture. *Human Biol.* **31**, 1–13.

Stanley, S. M.
1979 *Macroevolution: pattern and process.* San Francisco, Calif.: Freeman.

Stebbins, W. C.
1976 Comparative hearing function in vertebrates, *in* B. Masterton, M. E. Bitterman, C. B. G. Campbell, and N. Hotton (Eds.), *Evolution of brain and behavior in vertebrates.* Hillsdale, NJ: Erlbaum.
1978 Hearing of the primates, *in* D. J. Chivers and J. Herbert, *Recent advances in primatology.* London and New York: Academic Press.

Stern, J. T.
1971 Functional myology of the hip and thigh of cebid monkeys and its implications for the evolution of erect posture. *Bibl. Primatol. No.* **14**, Basel, Karger.
1975 Before bipedality, *Yearb. Phys. Anthropol.* **19**, 59–68.

Stern, J. T. and Susman, R. L.
1983 The locomotor anatomy of *Australopithecus afarensis. Amer. J. Phys. Anthropol.* **60**, 279–317.

Stewart, T. D.
1962 Neanderthal cervical vertebrae. *Bibl. Primatol. Fasc.* **1**, 130–54.

Stott, D. H.
    1969   Cultural and natural checks on population growth, *in* A. P. Vayda (Ed.), *Environment and cultural behavior*. New York: Natural History Press.
Straus, W. L.
    1936   The thoracic and abdominal viscera of the primates. *Proc. Amer. Phil. Soc.* **76**, 1.
    1949   The riddle of man's ancestry. *Quart. Rev. Biol.* **24**, 200–23.
Straus, W. L. and Cave, A. J. E.
    1957   Pathology and posture of Neanderthal man. *Quart. Rev. Biol.* **32**, 348–63.
Symons, D.
    1979   *The evolution of human sexuality*. New York and Oxford: Oxford Univ. Press.
Szalay, F. S. and Delson, E.
    1979   *Evolutionary history of the primates*. New York and London: Academic Press.
Terrace, H. S.
    1979   *Nim*. London: Eyre Methuen.
Testut, L.
    1928   *Traité d'Anatomie humaine*. 8th ed. Paris: Doin and Co.
Thompson, D. W.
    1942   *On growth and form*. New ed. Cambridge: Cambridge Univ. Press.
Thorpe, W. H.
    1965   *Science, man and morals*. London: Methuen.
Tobias, P. V.
    1967   *Olduvai Gorge*, Vol. 2; *The cranium of Australopithecus (Zinjanthropus) boisei*. Cambridge: Cambridge Univ. Press.
    1971   *The brain in hominid evolution*. New York: Columbia Univ. Press.
    1980   *Australopithecus* and early *Homo, in* R. E. Leakey and B. A. Ogot (Eds.), *Proc. 8th Panafr. Congr. Prehist. Nairobi*, pp. 161–165.
    1982   The evolution of man's upright posture. *Trans. Colloq. Med. S. Africa*.
Tobias, P. V. and Von Koenigswald, G. H. R.
    1964   A comparison between the Olduvai hominines and those of Java and some implications for hominid phylogeny. *Nature (London)* **204**, 515–518.
Trinkhaus, E.
    1981   Neandertal limb proportions and cold adaptation. *in* C. B. Stringer (Ed.), *Aspects of human evolution*. London: Taylor and Francis.
    1983   *The Shanidar neandertals*. New York and London: Academic Press.
Trivers, R.
    1971   The evolution of reciprocal altruism. *Quart. Rev. Biol.* **46**, 35–57.
    1972   Parental investment and sexual selection, *in* B. G. Campbell (Ed.), *Sexual selection and the descent of man, 1871–1971*. Chicago: Aldine.
Tutin, C.
    1975   Sexual behaviour and mating patterns in a community of wild chimpanzees. Ph.D. Thesis, Univ. Edinburgh.
Tuttle, R. H.
    1967   Knuckle-walking and the evolution of hominoid hands *Amer. J. Phys. Anthropol.* **26**, 171–206.

1969 Knuckle-walking and the problem of human origins. *Science* **166**, 953–961.

1970 Postural, propulsive, and prehensile capabilities in the cheiridia of chimpanzees and other great apes, *in* G. Bourne (Ed.), *The chimpanzee.* New York: Karger.

1981 Evolution of hominid bipedalism and prehensile capabilities. *Phil. Trans. Roy. Soc. London,* **B292**, 89–94.

Tuttle, R. H. and Basmajian, J. V.

1974 Electromyography of brachial muscles in *Pan gorilla* and hominoid evolution. *Amer. J. Phys. Anthropol.* **41**, 71–90.

1976 Electromyography of pongid shoulder muscles and hominoid evolution, I. Retractors of the humerus and scapula. *Yearb. Phys. Anthropol.* **20**, 491–497.

Tuttle, R. H., Basmajian, J. V., and Ishida, H.

1975 Electromyography of the Gluteus Maximus muscle in gorilla and the evolution of hominid bipedalism, *in* R. H. Tuttle, (Ed.) *Primate functional morphology and evolution,* pp. 253–269. The Hague: Mouton.

1979 Activities of pongid thigh muscles during bipedal behavior. *Amer. J. Phys. Anthropol.* **50**, 123–136.

Tylor, E. B.

1871 *Primitive culture.* London: Murray.

Udry, J. R. and Morris, N. M.

1968 Distribution of coitus in the menstrual cycle. *Nature (London)* **220**, 593–596.

Van Valen L.

1974 Brain size and intelligence in man. *Amer. J. Phys. Anthropol.* **40**, 417–424.

Van Valen, L. and Sloan, R. E.

1965 The earliest primates. *Science* **150**, 743–745.

Verner, J. and Willson, M. F.

1969 Mating systems, sexual dimorphism and the role of the male North American passerine birds in the nesting cycle. *Ornithol. Monogr.* **9**, 1–76.

Vrba, E.

1979 A new study of the scapula of *Australopithecus africanus* from Sterkfontein. *Amer. J. Phys. Anthropol.* **51**, 117–129.

Waal, F., de

1982 *Chimpanzee politics; power and sex among apes.* New York: Harper & Row.

Wagner, P.

1960 *The human use of the earth.* Glencoe, Ill.: Free Press.

Walker, A. C.

1969 The locomotion of the lorises with special reference to the potto. *E. Afr. Wildl. J.* **7**, 1–5.

Walker, A. C. and Pickford, M.

1983 New postcranial fossils of *Proconsul africanus* and *Proconsul nyanzae, in* R. L. Ciochon and R. S. Corruccini (Eds.), *New interpretations of ape and human ancestry,* pp. 325–351. New York: Plenum.

1973 *New Australopithecus* femora from East Rudolf, Kenya. *J. Human Evol.* **2**, 545–555.

Walls, G. L.
    1963 *The vertebrate eye and its adaptive radiation.* New York and London:
          Haffner.
Washburn, S. L.
    1950 The analysis of primate evolution with particular reference to the
          origin of man. *Cold Spring Harbor Symp. Quant. Biol.* **15**, 57–78.
    1963a The study of race. *Amer. Anthropol.* **65**, 521–531.
    1963b Behavior and human evolution, *in* S. L. Washburn (Ed.), *Classification
          and human evolution*, pp. 190–203. (Viking Fund Publs. Anthrop., No.
          37.) Chicago: Aldine.
Washburn, S. L. and Devore, I.
    1961 Social behavior of baboons and early man, *in* S. L. Washburn (Ed.),
          Social life of early man, pp. 91–105. (Viking Fund Publs. Anthrop., No.
          31.) Chicago: Aldine.
Washburn, S. L. and Hamburg, D. A.
    1965 The implications of primate research, *in* I. Devore (Ed.), *Primate be-
          havior*, pp. 607–622. New York: Holt, Rinehart and Winston.
Weidenreich, F.
    1938 Discovery of the femur and humerus of *Sinanthropus. Nature (London)*
          141, 614–617.
    1939–1941 The brain and its role in the phylogenetic transformation of the
          human skull. *Trans. Amer. Phil. Soc.* **31**, 321–442.
Weiner, J. S.
    1954 Nose shape and climate. *Amer. J. Phys. Anthropol.* **12**, 615–618.
Westoff, C. F.
    1974 *Family Planning Perspect.* **6**, 136–141.
White, L. A.
    1959 The concept of culture. *Amer. Anthropol.* **61**, 227–251.
White, T. D., Johanson, D. C., and Kimbel, W. H.
    1981 *Australopithecus africanus:* its phyletic position reconsidered. *S. Afr. J.
          Sci.* **77**, 445–470.
Whitehead, A. N.
    1920 *The concept of nature.* Cambridge: Cambridge Univ. Press.
Whyte, L. L.
    1965 *Internal factors in evolution.* London: Tavistock.
Wilkinson, R. G.
    1973 *Poverty and progress.* New York: Praeger.
Williams, B. J.
    1973 *Evolution and human origins.* New York: Harper and Row.
Wilson, A. C. and Sarich, V. M.
    1969 A molecular time scale for human evolution. *Proc. Nat. Acad. Sci. U.S.*
          63, 1088–1093.
Wilson, E. O.
    1975 *Sociobiology.* Cambridge: Harvard Univ. Press.
    1978 *On human nature.* Cambridge: Harvard Univ. Press.
Wolberg, D. L.
    1970 The hypothesized osteodontokeratic culture of the Australo-
          pithecinae: a look at the evidence and opinions. *Curr. Anthropol.* **11**,
          23–37.
Wolin, L. R. and Massopust, L. C.
    1970 Morphology of the primate retina, *in* C. R. Noback and W. Montagna
          (Eds.), *The primate brain.* New York: Appleton-Century-Crofts.

Wolpoff, M. H.
    1968   Climatic influence on the skeletal nasal aperture. *Amer. J. Phys. An-thropol.* **29**, 405–424.
    1983   Lucy's little legs. *J. Human Evol.* **12**, 443–453.
Woollard, H. H.
    1927   The retina of primates. *Proc. Zool. Soc. London*, pp. 1–17.
Wynne-Edwards, V. C.
    1965   Self-regulating systems in populations of animals. *Science* **147**, 1543–48.
Ziegler, A. C.
    1964   Brachiating adaptations of chimpanzee upper limb musculature. *Amer. J. Phys. Anthropol.* **22**, 15–32.
Zuckerman, S.
    1930   The menstrual cycle of the primates. *Proc. Zool. Soc. London*, pp. 691–754.
Zuckerman, S., Ashton, E. H., Flinn, R. M., Oxnard, C. E., and Spence, T. F.
    1973   Some locomotor features of the pelvic girdle in primates, *in* S. Zuckerman (Ed.), *The concepts of human evolution.* London and New York: Academic Press.

APPENDIX Fossil Evidence for Human Evolution: Some Important Fossil Remains of Early Man

Country	Locality	Remains	Approximate age in years (BP)
Homo Sapiens			
Europe			
Belgium	Spy, Province de Namur	Parts of 2 male skeletons and tibial fragment*	35,000–70,000
Britain	Swanscombe, Kent	Occipital and parietal bones	150,000–250,000
Czechoslovakia	Predmosti, Moravia	29 Skeletons	ca. 26,000
France	Biache	Occipital and parietal fragments	ca. 150,000
	Cro-Magnon, Dordogne	Five adult skeletons	20,000–30,000
	Fontéchevade, Charente	Frontal bone of 1 individual; calotte of second	70,000–150,000
	La Chaise, Charente	Fragments of 16 individuals	ca. 150,000
	La Chapelle-aux-Saints, Correze	Male skull and nearly complete skeleton*	40,000–55,000
	La Ferrassie, Dordogne	Parts of 6 individuals, 1 nearly complete (male)*	40,000–55,000
	La Quina, Charente	Parts of female skull and skeletal fragments*	40,000–55,000
	Regourdou, Dordogne	Mandible and skeletal fragments*	55,000–75,000
	St. Césaire	Half cranium and mandible	ca. 33,000
GDR	Ehringsdorf, nr. Weimar	Parts of female cranium and 2 mandibles plus remains of child's skeleton*	60,000–120,000
Germany	Neandertal, nr. Dusseldorf	Parts of male skull and skeleton*	35,000–70,000
	Steinheim, nr. Stuttgart	Female cranium*	150,000–250,000
Gibraltar	Forbes Quarry	Female cranium*	35,000–70,000

continued

APPENDIX (*Continued*)

Country	Locality	Remains	Approximate age in years (BP)
Italy	Circeo, Latina	Cranium of male individual; and mandible of another*	35,000–55,000
	Saccopastore, nr. Rome	Female cranium and parts of male skull*	60,000–100,000
Yugoslavia	Kaprina, nr. Zagreb	Fragmentary remains of c. 13 individuals	55,000–75,000
Asia			
Iraq	Shanidar, N. Iraq	5 Crania with skeletal fragments and remains of two others*	45,000–70,000
Israel	Amud	Almost complete skull and skeleton*	35,000–50,000
	Djebel Kafzeh	7 Crania, 4 with skeletal remains and fragments of 2 other individuals	ca. 40,000
	Galilee	Parts of cranium and fragments of other indivuals*	ca. 70,000
	Mount Carmel		
	Cave of Mugharet es-Skhul	7 Crania with skeletal remains and fragments of 3 others	35,000–40,000
	Cave of Mugharet et-Tabun	Skull and skeleton of female, mandible; fragments*	40,000–45,000
Sarawak	Niah Cave	Cranium	ca. 40,000
Africa			
Ethiopia	Omo (Kibish)	2 Calvariae and skeletal fragments	50,000–100,000

Kenya	Kanjera	Fragments of 4 crania	ca. 60,000
Morocco	Jebel Irhoud	Two crania*	ca. 40,000
	Border Cave	2 Calvariae, mandible and skeletal fragments	50,000–90,000
South Africa	Florisbad	Calotte with facial bones	ca. 39,000
Sudan	Singa	Cranium	ca. 20,000

Homo sapiens/erectus boundary

Europe

France	Arago	Cranium, 2 mandibles, and innominate	ca. 400,000
	Montmaurin	Mandible	ca. 200,000
GDR	Bilzingsleben	Cranial fragments	ca. 400,000
Germany	Mauer	Mandible	400,000–600,000
Greece	Petralona	Cranium	400,000–800,000
Hungary	Vértesszöllös	Cranial and mandibular fragments	300,000–400,000
USSR	Azykh	Mandible	ca. 250,000

Asia

China	Hsuchiayao	Cranial fragments of several individuals	150,000–200,000
	Mapa	Calotte	150,000–200,000
	Tali	Cranium	100,000–200,000

Africa

Ethiopia	Bodo	Cranium	ca. 350,000
Morocco	Rabat	Mandible fragments	ca. 160,000
	Salé	Cranial fragment	ca. 400,000
South Africa	Hopefield	Calotte	~50,000–250,000
Zambia	Broken Hill	Cranium and skeletal fragments	150,000–250,000

continued

433

APPENDIX (*Continued*)

Country	Locality	Remains	Approximate age in years (BP)
Homo erectus			
Algeria	Ternifine	Parietal and three mandibles	ca. 400,000
China	Chenchiawo	Mandible	ca. 650,000
	Choukoutien	Crania, teeth and skeletal fragments of 48 individuals	450,000–500,000
	Gongwangling	Calvaria	500,000–750,000
	Yuanmou	Incisor teeth	ca. 0.7 mya
Ethiopia	Gomboré	Parietal and humerus fragments	1.2–1.5 mya
Java	Trinil and Sangiran	10 Calvariae, calottes, 9 fragmented crania, 6 mandibular and maxillary fragments	ca. 0.8 mya
Kenya	Koobi Fora	Skull and skeleton, 2 crania and skeletal fragments	ca. 1.5 mya
Tanzania	Olduvai Bed IV	Calotte, pelvis, femur	400,000–600,000
	Bed II	Calotte	ca. 1.2 mya
	Ndutu	Cranium	400,000–600,000
Homo habilis			
Ethiopia	Omo (Shungura)	Calotte and teeth	1.84 mya
Kenya	Koobi Fora	Cranium, calotte, mandibular and skeletal fragments	1.8–2.1 mya
South Africa	Sterkfontein	Fragmentary cranium	1.5–2.0 mya
	Swartkrans	Cranium and mandible	1.5–2.0 mya
Tanzania	Olduvai	3 Calvariae, 3 calottes, and skeletal fragments	1.7–1.8 mya

Australopithecus afarensis			
Ethiopia	Hadar	1 Incomplete skeleton and cranial and skeletal remains of at least 25 individuals	2.8–3.3 mya
Tanzania	Laetoli	4 Mandibles and isolated teeth	3.5–3.75 mya
Australopithecus africanus			
Ethiopia	Omo (Shungura)	Isolated teeth	2.3–3.0 mya
South Africa	Taung	Child's cranium and mandible	ca. 2.0 mya
	Sterkfontein	2 Crania and many cranial, mandibular, and skeletal fragments	2.0–3.0 mya
	Makapansgat	2 Calvariae, mandibular and skeletal fragments	2.0–3.0 mya
Australopithecus robustus			
South Africa	Kromdraai	Part cranium, mandibular, and skeletal fragments	1.5–2.0 mya
	Swartkrans	Cranium, cranial, mandibular, and skeletal fragments	1.5–2.0 mya
Australopithecus boisei			
Ethiopia	Omo (Shungura)	Isolated teeth and mandibular fragments	1.9–3.0 mya
Kenya	Koobi Fora	3 Calvariae, many mandibular and skeletal fragments	1.3–2.0 mya
Tanzania	Olduvai (Bed I)	Cranium	ca. 1.7 mya
	Peninj	Mandible	ca. 1.2 mya

[a]The science of chronology is in a state of rapid development. Many of the dates given require revision. Those about which there is most doubt belong to the species *Homo erectus*. For a full catalog of fossil Hominidae see Oakley, Campbell, and Mollison (1967–1974). The most important sites are shown on the map of the Old World, Fig. 4.13.

*Large-jawed forms of *Homo sapiens* (by implication when jaws not present).

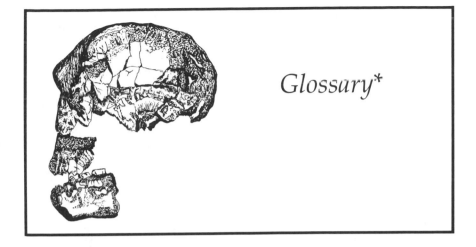

Glossary*

Abductor. Term applied to muscles that move a part of the body away from the midline (median plane) of the body or of a specific organ, such as the hand. (Compare ADDUCTOR.)

Absolute Dating. See CHRONOMETRIC DATING.

Acetabulum. A cup-shaped depression on the external surface of the *hip bone* into which the head of the *femur* fits.

Achilles Tendon. Tendon by which the calf (the *gastrocnemius* and *soleum* muscles) is attached to the heelbone (*calcaneus*).

Acromioclavicular Joint. *Articulation* between the *acromion* process of the *scapula* and the *clavicle* (collar bone).

Acromion. The outer end of the spine of the *scapula* that projects over the *glenoid fossa*. It *articulates* with the *clavicle* at the *acromioclavicular joint*.

Adaptability. The ability of an individual organism to alter its phenotype during its life-span in adaptation to its environment, for example, sun-tanning in Europeans.

Adaptation (adj., Adaptive). A *character* or set of *characters* of a *population* selected by the *environment* that by its existence improves the chance of survival of the population. Adaptations also allow exploitation of the *environment*. Adaptation is also used to describe the process of adapting to an environment.

Adductor. Term applied to muscles that move a part of the body toward the midline (median plane) of the body, or of a specific organ. (Compare ABDUCTOR.)

Adductor Magnus. A muscle, part of which extends the thigh (in man) and part of which adducts it. This muscle arises on the lower borders of the *pubis* and *ischium*, including the *ischial tuberosity*, and is inserted in the shaft and *distal* end of the *femur*.

Adrenaline. A *hormone* produced by the adrenal *glands* during conditions of stress or fear.

Aegyptopithecus. A *genus* of fossil apes found in the Oligocene of northern Egypt.

Italicized words appearing in definitions are cross-references to other entries in the Glossary.

Allantois. A *fetal* membrane of egg-laying *vertebrates* that lies close to the shell and forms an *organ* of *respiration*; in *mammals* it contributes to the formation of the *placenta*.

Altricial. Literally, nursing; describes species wherein the young are born helpless and require considerable parental care. (Compare PRECOCIAL.)

Altruism. Truly unselfish, even self-destructive, behavior performed for the benefit of others. (Compare BIOALTRUISM.)

Amino acid. An organic compound based on a carbon chain or ring composed of (NH_2) and carboxyl (COOH) groups of atoms. All amino acids are synthesized by green plants and some are synthesized by animals. In the case of humans eight out of about twenty must be obtained "ready made" and cannot be synthesized. *Proteins* are synthesized from amino acids.

Amnion. The membrane that forms the fluid-filled sac in which the embryo grows.

Amniote Egg. An egg containing an *embryo* provided with an *amnion*.

Amygdala. A component of the *limbic system* believed to be responsible for the expression of rage, and the level of activity of the *hypothalamus*.

Anatomy. The science of the *morphology* and structure of *organisms*.

Anestrus (adj., Anestrous). A period of sexual quiescence between the mating seasons of *mammals*; a condition not regularly found in humans except following *parturition*.

Anterior. The front of, or front part of, any object or *organism*.

Anterior Inferior Iliac Spine. A protuberance of bone on the *anterior* edge of the *ilium* to which is attached the *iliofemoral ligament* and *rectus femoris* muscle, which flexes the thigh and extends the knee. (See Figure 6.8.)

Anterior Superior Iliac Spine. A small protuberance of bone on the *anterior* edge of the *ilium* to which is attached the *sartorius muscle*.

Anthropocentric. A view of nature that places humans as central and examines the world from a human point of view.

Anthropoidea (adj., Anthropoid). A *suborder* of the *order Primates* containing all monkeys, apes, and *hominids*.

Antibodies. *Proteins* in the blood of vertebrates that are produced in response to *antigens* and protect the body from their toxic action.

Antigens. Substances which are foreign to the body of an animal and when introduced are capable of stimulating the production of *antibodies*.

Apocrine. The term applied to *glands* found in the skin that *secrete* the odorous components of sweat.

Apomorphic. Derived characters; a term applied to those characters of a species that have evolved only in the taxonomic group under consideration. (Compare PLESIOMORPHIC.)

Arboreal. Tree-living.

Archipallium. The older part of the *cortex* of the *cerebral hemispheres*, distinguished from the *neopallium*, which evolved more recently and lies above it.

Archosauria. The dinosaurs (e.g., *Tyrranosaurus rex*) which dominated the earth during *Cretaceous* and *Jurassic* times before the great *radiation* of the *mammals*.

Areola (pl., Areolae). A ring of pigmented skin surrounding the nipple.

Articulation (adj., Articular). A connection between two bones allowing movement; colloquially, a joint.

Artifacts. Used here to refer to objects formed by the purposeful activity of animals, especially humans.

Artiodactyls. The even-toed *ungulates*—an order of *mammals*—including pigs, sheep, antelopes, giraffes, and cattle.

Association Cortex. That part of the outer layer of the *cerebral hemispheres* believed to function as a result of the connection there of two or more other areas of the *cortex* concerned with sensory or motor activity.

Auditory Cortex. That part of the outer layer of the *cerebral hemispheres* believed to be responsible for the reception and transmission of auditory input to the brain.

Australopithecus. A genus of fossil *Hominidae* first discovered in South Africa in 1924 and named by Raymond Dart in 1925. It contains four species: *A. afarensis, A. africanus, A. robustus,* and *A. boisei.*

Axillary. Pertaining to the armpit.

Axons. Nerve fibers.

Baculum. A small rod-shaped bone which serves to stiffen the *penis* of many nonhuman *primates.*

Bar-Glenoid Angle. The angle between the plane of the *glenoid cavity* and the *ventral* bar of the *scapula.* (See Figure 7.8.)

Basal Ganglia. Part of the brain which forms a component of the *limbic system.*

Basal Metabolic Rate. The rate of heat production in an individual at the lowest level of body activity in the waking state.

Behavior. The totality of motor activity of an animal.

Biceps Brachii. One of the two muscles that flex the forearm in relation to the arm at the elbow joint. Originating on the *scapula,* it is *inserted* upon the *bicipital tuberosity* of the *radius.*

Bicipital Tuberosity. A protuberance upon the shaft of the *radius* where the *biceps brachii* muscle is *inserted.*

Bicuspid. Having two *cusps;* in humans, a premolar tooth.

Bilophodont. Having the *cusps* of the molar tooth arranged to form two ridges with a valley between, characteristic of *Cercopithecoidea* among the *primates.* (See Figure 9.13.)

Bioaltruism. Behavior which appears *altruistic,* but which can be explained by current theories of *sociobiology* to enhance the chances of survival of the individual or its kin. (Compare ALTRUISM.)

Biological Efficiency. The efficiency of a species in converting food into *biomass.*

Biology. The branch of science that deals with living *organisms.*

Biomass. The total mass of living organisms that constitutes a particular species.

Biospecies. A group of naturally or potentially interbreeding *populations* of animals or plants between which *gene flow* can occur, but which in nature is reproductively isolated from other such groups or populations.

Bipedalism. The act of standing or moving on the hindlimbs.

BP: Literally, years Before Present; in practice the ages of archaeological sites termed BP are calculated as years before 1950 AD.

Brachiation. The act of locomotion by means of the forelimbs, characteristic of gibbons.

Brachycephalic. Having a head that is relatively short and broad. (Compare DOLICHOCEPHALIC.)

Brain Stem. That part of the base of the brain which leads to the *spinal cord.*

Broca's Area. An area of the *cerebral cortex* discovered by Broca which is believed to be responsible for the production and articulation of speech.

Calcaneus. The heel bone.

Calotte. The skull cap; the bones of the roof of the skull.

Calvaria (pl., Calvariae). The *neurocranium* without the facial bones.

Cantilever. A beam supported from one end.

Carnivora. An order of flesh-eating *mammals*, including lions, cats, hyenas, and dogs.

Carnivore. An animal living exclusively or almost exclusively on the flesh of other animals. (Compare HERBIVORE.)

Carpal Bones. The bones of the wrist.

Cartilage. A skeletal tissue of *vertebrates* distinct from bone, consisting of a resilient translucent matrix containing fibers.

Catarrhini. Higher primates of the Old World, including monkeys, apes, and humans.

Ceboidea (adj., Ceboid). The *superfamily* of *Primates* found in Central and South America, containing 16 *genera*; also known as New World monkeys.

Cecum. A short and blind-ended branch of the intestine, found in certain *herbivorous* mammals, that has evolved to facilitate the digestion of cellulose.

Center Of Gravity. A point, the support of which allows a body to remain balanced in any position.

Central Nervous System (CNS). The brain and *spinal cord*.

Cercopithecoidea (adj., Cercopithecoid). A *superfamily* of *primates* found in the Old World, containing 14 *genera*; also known as Old World monkeys.

Cerebellum. The *posterior* part of the brain lying, in humans, beneath the *occipital* lobe of the *cerebral hemispheres* and attached to the *brain stem*.

Cerebral Cortex. The outer layer of the *cerebral hemispheres* or *cerebrum*.

Cerebral Hemispheres. The two parts of the *cerebrum* connected by the *corpus callosum*.

Cerebrum. The principal and uppermost portion of the human brain, which is divided into two hemispheres.

Cervical Region. The neck bones of the vertebral column.

Cetaceans. *Mammals* that evolved *adaptations* to a marine *environment*, including whales, porpoises, and dolphins.

Channel. Used here to denote the precise mode of communication between individuals, e.g., vocal-auditory, or chemical-olfactory.

Characteristic. Refers here to a particular attribute of an animal.

Chemoreceptor. A sensory *organ* of the nervous system that is stimulated by certain chemical substances (e.g., *organ* of taste or smell.)

Cheiridia. *Organs* that terminate the limbs; a collective term for paws, hands, and feet.

Chromosome (adj., Chromosomal). Thread-shaped structures occurring in the nucleus of every plant and animal cell, consisting largely of deoxyribonucleic acid (DNA) and carrying the genetic code of the individual or *gamete*.

Chronology. The science of the study of time, age, and sequence.

Chronometric Dating. Methods for dating minerals which give an age in years since their formation. Also called ABSOLUTE DATING.

Chronospecies. A *biospecies* with a temporal dimension, that is, a lineage of interbreeding *populations* during a certain defined period of time; a fossil species.

Cingulum (Cingulate Gyrus). A part of the *cerebral cortex* which lies low on the *medial* surface of the hemispheres; a component of the *limbic system*.

Cladistics. Method of taxonomic analysis in which characteristics of a species are divided: derived or *apomorphic* and ancestral or *plesiomorphic*. This method reduces subjectivity and is believed to generate a more reliable taxonomy than *phenetics*.

Clan. Large unilateral (one-line) descent group.

Class. In biology, a taxonomic rank. (See TAXONOMY.)

Clavicle. The collar bone, which forms part of the *pectoral girdle* and connects the *scapula* and forelimb to the *sternum*.

Clitoris. A penis-like *organ* situated at the anterior convergence of the *labia* in female *Primates*. Its stimulation effects erection and erotic response.

CNS. See CENTRAL NERVOUS SYSTEM.

Coccyx. Much reduced *vertebrae* at the end of the vertebral column present in the *Hominoidea*.

Cochlea. A cone-shaped cavity in the ear region of the skull consisting of a spinal canal containing the *organ of Corti*.

Coitus. Sexual intercourse.

Comparative Anatomy. The study and comparison of the form and structure of different animals.

Compound Girder. A beam or girder constructed of smaller units with spaces between them.

Concept. A mental abstraction generalized from particular experience.

Conceptual Memory. That part of the memory which records *concepts* as an analysis and classification of experience.

Conditioning. A kind of learning from which the resulting behavior pattern becomes fully automatic and involves no conscious thought processes.

Condyle. A kind of joint in the skeleton in which paired but separate bearing surfaces allow a limited hingelike movement, e.g., *occipital condyles, mandibular condyles*, knee joint.

Cone. A sensory *organ* of the nervous system stimulated by light and able to distinguish differences in wave frequency; it is found in the *retina* of animals able to see color.

Contact Receptor. A sensory *organ* of the nervous system stimulated by contact with objects, including *organs* sensitive to touch and taste.

Copulation. Sexual union between male and female individuals; see COITUS.

Coracoid Process. A prominence on the *mandibular ramus* in which the *temporalis* muscle is inserted.

Corpus Callosum. A mass of nerve fibers running across the *cerebrum* between the *cerebral hemispheres*.

Corpus Luteum (pl., Corpora Lutea). A small yellowish ductless *gland* that develops in the *ovary* immediately after *ovulation*. Its most important product is the *hormone progesterone*. If pregnancy does not occur, the corpus luteum rapidly degenerates.

Cortex (adj. Cortical). The outer layer or mantle of an *organ*—used here to describe the important outer layer of the *cerebrum*.

Corticospinal Fibers. Nerve fibers connecting the *cerebral cortex* and the *spinal cord*.

Cranium (adj., Cranial). The skull, excluding the mandible.

Creationism. A system of belief, not a scientific theory, that posits the separate creation of every living animal and plant species; sometimes misnamed "creation-science."

Cretaceous. A *period* of the earth's history (the last period of the Mesozoic Era) believed to have occurred about 135–63 million years ago. (See Table 2.1.)

Culture. The totality of *behavior* patterns of a social group of animals that are passed between generations by learning and usually associated with a symbolic mode of communication or expression; socially determined *behavior* learned by observation, imitation, or instruction. (Compare PROTOCULTURE, MATERIAL CULTURE.)

Cusp. A protruberance on the occlusal surface of a tooth.

Deciduate. Applied here to a *placenta*; the term indicates that when the *placenta* is shed, it brings with it the outer layers of the *endometrium* with which it was intimately associated.

Deltoid. A muscle of the shoulder, with its *origin* on the *clavicle* and *scapula* and its *insertion* in the shaft of the *humerus*.

Dental Arcade. The curved line formed by teeth in the *maxilla* and *mandible*; the dental arch.

Dentition. The teeth.

Deoxyribonucleic Acid (DNA). A chemical substance from which *chromosomes* are constituted. It carries the genetic code, the *genotype*, of each individual within every living cell.

Developmental Homeostasis. The capacity of the developmental pathways to produce a normal *phenotype* in spite of developmental or environmental disturbances.

Diastema (pl., Diastemata). The gap formed in each jaw to receive the canine tooth of the opposite jaw, usually found in animals with canines much larger than their other teeth.

Digastric Muscles: Muscles that lower the jaw and stabilize the *hyoid bone*. Their *origin* lies near the *mastoid process* at the base of the skull; they are *inserted* into the lower border of the *symphyseal* (anterior) region of the *mandible*.

Digital Formula. A simple formula indicating the order of length of the digits. The human formula for the hand is 3,2,4,5,1.

Digitigrade. A type of locomotion in which the ventral surfaces of the fingers or toes only carry the weight of the animal, and the proximal bones (i.e., metatarsals, metacarpals, tarsal, and carpal bones) are held clear of the ground, as in the dog or cat. (See PALMIGRADE, PLANTIGRADE.)

Digits. Fingers or toes.

Diphyodontism. The property of producing two sets of teeth during an individual's life—the milk teeth or deciduous teeth, and the permanent teeth.

Diplöe. A spongelike bony structure separating and joining the two layers of compact bone (the *inner* and *outer tables*) of the *cranium*.

Discrimination. The act of perceiving differences.

Displacement. Used here in the linguistic sense of implying the ability to refer to objects or events out of sight.

Distal. In anatomy, that part of a structure (such as a limb) furthest from the root of that structure or from the trunk. (Compare PROXIMAL.)

Distance Receptor. A sensory organ of the nervous system stimulated by physical disturbances emitted by objects at a distance (e.g., light, sound, scent).

DNA. See DEOXYRIBONUCLEIC ACID.

Dolichocephalic. Having a head that is relatively long and narrow. (Compare **Brachycephalic**.)

Dorsal. Relating to the back; the opposite of *ventral*.

Dryopithecidae. A *family* of fossil apes, including *Dryopithecus, Proconsul, Sivapithecus, Ramapithecus,* and *Gigantopithecus.*

Dryopithecus. A fossil *genus* of the *family Dryopithecidae.*

Eccrine. *Glands* of the skin that secrete sweat.

Ecological Niche. See NICHE.

Ecology. The science of the mutual relations of different *organisms* and their *environments*.

Effector. An organ that receives nerve impulses and reacts by movement or *secretion* (e.g., a muscle or a *gland*).

Embryo (adj., Embryonic). Term applied to organisms during the early stages of *ontogeny* while dependent on maternal food supplies; within the egg or during the early stages of gestation. (See also FETUS.)

Endocast. A cast of the interior of the *cranial* cavity.

Endocrine. Describes a system and its components which control bodily activity by chemical, as distinct from nervous, signals. The chemicals produced are termed *hormones*, and the *glands* that secrete them straight into the bloodstream are termed "ductless" or endocrine glands.

Endogamy. Inbreeding; the selection of a mate from within a small group by another member of that group. (Compare EXOGAMY.)

Endometrium. The *vascular* lining of the *mammalian uterus*, into which the *placenta* penetrates.

Entropy. The phenomenon of randomness, which, according to the second law of thermodynamics, can, in any closed system, only increase, and never decrease.

Environment. The total surroundings of an individual.

Epidermis. The outer layers of the skin.

Epigamic. Relating to sexual reproduction or to *copulation*, and serving to attract or stimulate individuals of the opposite sex during courtship.

Epiglottis. A small plate of *cartilage* at the root of the tongue that folds back over the aperture of the *larynx*, covering it during the act of swallowing.

Epitheliochorial Placenta. A *placenta* in which the maternal and *fetal* bloodstreams are separated by both maternal and *fetal tissues*. (Compare HEMOCHORIAL PLACENTA.)

Equability. A relative absence of fluctuation or *variation*; refers here to climates with relatively small seasonal variation in temperature.

Era. A span of time; a geological term referring to certain major subdivisions in the earth's *chronology*.

Erector Spinae. Long muscles of the back that help maintain the erect trunk in man. They originate on the *sacrum* and *ilium* and *insert* into the ribs and *vertebrae*.

Esophagus. Part of the digestive tract between the *pharynx* and the stomach.

Estrogen. *Hormones*, mainly produced in the *ovary*, which induce *estrus* and *ovulation* in a mature female individual. During growth, the increasing production of estrogen brings about the development of *secondary sexual characters* and *menarche*.

Estrous Cycle. The series of *uterine*, *ovarian*, and other changes that occur in mammals and are responsible for *copulation*, pregnancy, etc.

Estrus. The stage of the *estrous cycle* occurring around the time of *ovulation*, when the female is sexually receptive and in some species encourages males to *copulate*.

Evolution. Cumulative change in the gene frequencies of populations of *organisms*, occurring in the course of successive generations related by descent, i.e., descent with change.

Exogamy. Outbreeding; the selection of a mate from outside a small group of individuals considered to be close kin. (Compare ENDOGAMY.)

Experiential Memory. That part of the memory which records experience in temporal sequence.

Extensor. A muscle, the contraction of which tends to move a limb posteriorly and to straighten it. (Compare FLEXOR.)

External Pterygoids. See LATERAL PTERYGOIDS.

Extrapyramidal System. Descending *nerve tracts* that are interrupted by *synapses* and form relatively indirect links between the *brain stem* and the muscles.

Facilty. An *artifact* designed and used to restrict or prevent motion or energy exchanges (such as dams or insulation); the simplest include all heat-retaining structures and containers of liquids or solids (Wagner, 1960). (Compare TOOLS.)

Fallopian Tubes. Tubular structures leading from each of the two *ovaries* to the *uterus*. Eggs are shed into the open ovarian end of the tubes and pass into the *uterus*. In humans, fertilization occurs either in the fallopian tubes or (less commonly) in the uterus.

Family. Although this term cannot be strictly defined, it usually refers to a group of individuals related by blood and/or marriage including at least one adult male, one adult female, and one or more young. It is also a taxonomic rank. (See TAXONOMY.)

Femur (pl., Femora). The *proximal* bone of the leg, the thighbone. The head of the *femur* fits into the *acetabulum*.

Fenestra Ovalis. A small opening, sealed by a *membrane*, through which vibration is transmitted from the *middle ear* to the *cochlea*.

Fetus (adj., Fetal). The unborn young of a *viviparous* animal after it has taken form in the *uterus*.

Fibula (Pl., Fibulae). One of the two *distal* bones of the leg. (See also TIBIA.)

Fitness. In an evolutionary context, this term refers to the possibility of survival of a *population* over a long period of time in a changing environment.

Flexor. A muscle, the action of which is to carry the limb anteriorly and fold it. (Compare EXTENSOR.)

Follicle, Ovarian (or Graafian Follicle). A *vesicular* body in the *ovary* containing the egg.

Follicle, Hair. Pit within mammalian skin surrounding and supporting the hair root.

Follicle-Stimulating Hormone (FSH). A hormone produced by the *anterior pituitary gland* that stimulates the *ovarian follicles* inducing their maturation and the liberation of *estrogen*.

Fossa (pl. Fossae). Refers here to certain recognized hollows in bones, such as the fossa iliaca.

Fossil Man. Descriptive of fossil bones, skulls, etc., of great age, belonging to early humans and other early *hominids*.

Fovea Centralis. A small pit in the surface of the *retina* where the *photoreceptor* cells are not overlaid by nerve fibers and blood capillaries—an area of the *retinal* surface permitting optimum optical *discrimination* (also called "yellow spot").

Frontal Bone. The bone of the skull that rises above and behind the *orbits* and forms the *anterior* part of the *neurocranium*, as well as the roof of the orbit.

Frontal Lobes. The *anterior* lobes of the *cerebral hemispheres*.

FSH. See FOLLICLE-STIMULATING HORMONE.

Function. In *biology*, the activity of a biological mechanism.

Functional Complex. A group of *anatomical* and *physiological* characteristics that jointly bring about a particular necessary and integrated activity in an animal.

Gamete. The germ or sex cells that are produced by male and female sexually reproducing *organisms*; gametes of each sex must fuse together at the time of fertilization to initiate the growth of a new individual.

Ganglion. An aggregation of nerve cells.

Gastrocnemius. A muscle of the calf of the leg that has its *origin* on the *distal* end of the *femur* and its insertion by the *achilles tendon* onto the *calcaneus*.

Gene (adj., Genetic). The unit of inheritance.

Gene Flow. The passage of genes through a population or between *populations* over a period of time which is the result of sexual reproduction. *Endogamy* will restrict gene flow between populations, *exogamy* will increase gene flow.

Gene Pool. The totality of *genes* of a given population existing at a given time (Mayr, 1963).

Gene Selection. Selection favoring individuals having similar genotypes, e.g., relatives; terms proposed by Dawkins (1976). See KIN SELECTION; INCLUSIVE FITNESS.

Genetic Drift. Random changes in the *gene pool* not due to *selection* or immigration and characteristic of small *populations*.

Genetic Homeostasis. The property of a population of balancing its *genetic* composition so as to resist sudden changes (Mayr, 1963).

Genial Tubercles. Small protuberances of bone on the inner surface of the *symphyseal* region of the *mandible* onto which are *inserted* the *genioglossus muscles* which anchor the tongue to the *mandible*, and the *geniohyoid* muscles.

Geniohyoid. A muscle that lies above the *mylohyoid* in the floor of the mouth, and like it raises the *hyoid bone* or lowers the jaw. It has its *origin* on the *genial tubercles* of the *mandible* and is *inserted* on the *hyoid bone*.

Genotype. The totality of *genetic* factors that make up the *genetic* constitution of an individual.

Genus (pl., Genera). An important taxonomic rank including a group of *species* which have more in common with each other than they have with other similar groups. (See TAXONOMY.)

Gestation. The period of pregnancy during which the *embryo* and *fetus* develop and grow. It begins with *implantation* and ends with *parturition*.

Gland. An *organ* producing one or more *secretions*, either onto the surface (as sweat glands) or into a cavity (as digestive glands) or into the bloodstream (the *endocrine* glands).

Glans Penis. An expansion that forms the end of the *penis*; it is sensitive to mechanical stimulation.

Glenoid Cavity. The socket on the *scapula* with which the head of the *humerus* articulates.

Gluteus Maximus. A muscle that *abducts* the thigh in nonhuman primates but *extends* it in humans; it *originates* on the posterior external surface of the ilium, the back of the sacrum and coccyx, and the sacrotuberous ligament, and is *inserted* into the upper part of the shaft of the *femur* (the gluteal tuberosity) and the iliotibial tract.

Gluteus Medius. A muscle that *abducts* the thigh in humans; it *originates* on the external surface of the *ilium* and is *inserted* on the posterosuperior angle of the *greater trochanter* of the *femur*.

Gluteus Minimus. A muscle that *abducts* the thigh in humans; it *originates* on the external surface of the *ilium* and *inserts* on the front of the *greater trochanter* of the *femur*.

Gonads. *Organs* that act both as *endocrine* glands and in the generation of *gametes*—the *testes* in the male and the *ovaries* in the female.

Gorillidae. The family that comprises the African apes.

Greater Trochanter. A large bony prominence on the upper end of the *femur* onto which are *inserted* the *gluteus medius, gluteus minimus* muscles, and some lateral rotators.

Group Selection. Selection that operates on two or more members of a social group as a unit; includes *kin selection* but usually implies natural selection operating on entire breeding populations or social groups.

Hallux. The great toe; the first *digit* of the *pes*.

Hand-Axes. Stone artifacts of a particular shape and with certain functions first produced in the Middle *Pleistocene*.

Hemochorial Placenta. A *placenta* in which the maternal and *fetal* bloodstreams are separated only by *fetal* tissues; the maternal tissues have disintegrated. (Compare EPITHELIOCHORIAL PLACENTA.)

Herbivore. An animal feeding more or less exclusively on vegetable matter. (Compare CARNIVORE.)

Heterodont. Having teeth of varying shapes and functions—a characteristic, typical of *mammals*, that facilitates *mastication*. (Compare HOMODONT.)

Heterosis. The *selective* superiority of individuals with dissimilar paired *genes*.

Hindlimb Dominance. The term refers to the balance of an animal when the center of gravity lies closer to the hind than to the forelimbs. Under these circumstances the hindlimbs develop greater power.

Hip Bone. A pelvic bone formed by the fusion of the *ilium, ischium,* and *pubis,* and constituting the lateral and inferior (ventral) parts of the pelvis. The two hip bones articulate with each other inferiorly (ventrally) and with the sacrum superiorly (dorsally).

Hippocampus. A small structure which lies on the floor of the lateral ventricle of the brain and is a component of the *limbic system*.

Homeostasis (adj., Homeostatic). The maintenance of a dynamic equilibrium in living processes; the self-regulating property of *organic* systems.

Hominidae (adj., Hominid). The *family* of *Primates* including humans and related fossil *species*. Two *genera* are usually recognized: *Australopithecus* and *Homo*.

Hominoidea (adj., Hominoid). A *superfamily* of the *primates* containing the *families* of apes and humans.

Homo. A *genus* of the family *Hominidae* containing three *species: Homo sapiens,* which includes modern humans and Neandertals, *Homo erectus,* and the earliest species, *Homo habilis* (both known only from fossils). Its earliest members probably date from about 2.3 million years ago.

Homo erectus. A *species* of the *genus Homo* known only from fossils and probably extant from about 1.5 million to about 300,000 years ago.

Homo habilis. A *species* of the *genus Homo* including fossils discovered in Tanzania and Kenya. It was extant from about 2.3 to 1.5 million years ago.

Homo sapiens. The most recent *species* of the genus *Homo,* which includes modern humans and Neandertals. It is believed to date from about 300,000 years ago.

Homodont. Having teeth of similar shape and function characteristic of lower vertebrates. (Compare HETERODONT.)

Homoiothermy. The property, characteristic of birds and *mammals*, of maintaining the body at a constant temperature by means of a complex *homeostatic* mechanism.

Hormone. A chemical substance formed in one part of the body by a ductless *gland* and carried by the blood to another part, which it stimulates to functional activity—the chemical messenger of the *endocrine* system.

Humerus (pl., Humeri). The bone of the arm, *articulating* with the *scapula* proximally and with the *radius* and *ulna* at its *distal* end.

Hylobatinae. A *subfamily* of the *Pongidae* including the smaller *genus* of apes that are currently found in Asia, the gibbons (*Hylobates*), which includes the siamang.

Hyoid. The bone to which the tongue is anchored at its base.

Hypertrophy. Increase in size of an *organ* or part of an organ.

Hypoconulid. The unpaired fifth *cusp* found on the back of some lower molar teeth in most *Hominoidea*.

Hypothalamus. A small mass of nervous tissue at the base of the brain acting as an important control center in nervous and *endocrine* function.

Hypothesis. A supposition advanced as a basis for reasoning or argument or as a guide to experimental investigation; a tentative theory.

Iliac Pillar. A thickening of the *ilium* that helps to resist the compression of the bone produced between the *acetabulum* and the *origin* of the *gluteus* muscles that extend the thigh; also involved in weight transference from the sacrum to the acetabulum.

Ilio-psoas. A composite muscle that flexes the *femur* at the hip joint; it *originates* from the *lumbar* vertebrae and internal surface of the *ilium* (iliac fossa) and *inserts* on the *lesser trochanter* of the *femur*.

Ilium (adj., Iliac). The blade of the hip bone that forms part of the *pelvis*.

Implantation. The entry of the fertile egg of *mammals* into the *endometrium* of the *uterus*, thus inaugurating pregnancy.

Incest. Forbidden *copulation* with an individual within a group of close kin.

Inclusive Fitness. The sum of an individual's own fitness plus its influence on the fitness of its relatives (which share its genes) through its behavior.

Inferior. Lower, nearer the feet; caudal. (Compare SUPERIOR.)

Innate Behavior. Behavior which arises spontaneously in individuals without learning. Sometimes referred to as unlearned, such behavior is difficult to detect in higher *Primates* and can be identified by rearing an animal away from its kind but in an otherwise natural environment.

Inner Ear. That part of the ear which contains the *mechanoreceptors* for sound, balance, and movement.

Inner Table. The layer of compact bone that forms the inner surface of the *neurocranium*.

Insectivora. An order of *mammals* that contains small insect-eating creatures in both living (see Figure 3.1) as well as certain fossil forms.

Insertion. Refers to the more movable of the two points of attachment of a muscle. (Compare ORIGIN.)

Instinct. Sometimes defined as an unlearned behavior pattern, sometimes as a conditioned behavior pattern, and sometimes as a psychological drive.

Institution. An established form of *cultural* group *behavior*.

Intermembral Index. An index relating the different lengths of the hind- and forelimbs. Index = length of *humerus* + *radius* × 100 over length of *femur* + *tibia*.

Internal Environment. The *environment* of the body cells, consisting primarily of a complex fluid whose composition and temperature is maintained in a constant state by a wide range of *homeostatic* mechanisms.

Internal Fertilization. The fertilization of the egg within the body of the female rather than in water. This is an important adaptation of *terrestrial vertebrates* to dry land.

Internal Pterygoids. See MEDIAL PTERYGOIDS.

Internuncial Neuron. A nerve cell that relays impulses from the descending nerve fibers to the *motor neuron*, which effects muscular contraction.

Interosseus Membrane. Refers here to the *ligamentous membrane* connecting the *ulna* to the *radius* and transmitting forces of tension between them.

Interspinous Ligaments. *Ligaments* connecting the *spinous processes* of the *vertebrae* and transmitting tension between them.

Intervertebral Disk. A disk interposed between the bodies of adjacent *vertebrae*, consisting of an outer fibrous part and an inner gelatinous mass.

Intromission. The insertion of the *penis* into the *vagina*.

Involution. The opposite of *evolution*; the loss of organic *variety*; in contrast to evolutionary radiation, organic life becomes limited and *specialized*.

Ischial Callosities. Thickenings of the outer layer of the skin overlying the *ischial tuberosities*, found in Old World monkeys and *Hylobates*.

Ischial Spine. A small bony protuberance on the *posterior* margin of the *ischium* on which *originate* the sacrospinous ligament and the levator ani and coccygeal muscles that form the floor of the abdomen in humans.

Ischium (adj. Ischial). One of the three components of the *hip bone*. In adult *mammals* it is fused to the *ilium* and *pubis*.

Isoimmunization. The development of an *antibody* as a result of *antigens* introduced from another individual; for example, *antigens* may pass from the *fetal* to the maternal bloodstreams during pregnancy and cause isoimmunization of the mother.

K-**Selection.** Selection favoring superiority in stable, predictable environments in which rapid population growth is unimportant. (Compare *r*-SELECTION.)

Kin Selection. The selection of genes resulting from an individual contributing to the survival of relatives who possess the same genes.

Labia. Refers here to the two pairs of lips, the *labia majora* and *labia minora*, which bound and protect the *clitoris, urethra,* and *vagina*.

Lactation. The production of milk by *mammals* from the *mammae* or breasts following *parturition*.

Larynx. The organ of voice production; the upper part of the respiratory tract between the *pharynx* and the *trachea*.

Lateral. On the side; to the side of the midline or median plane. (Compare MEDIAL.)

Lateral Abdominal Muscles. Muscles forming the sidewalls of the abdominal cavity attached to the lower ribs and the crest of the ilium.

Lateral Geniculate Body. A swelling at the angle of each optic nerve tract, at the base of the brain, to the side of the midline.

Lateral Pterygoids. Also known as external pterygoids, these paired muscles *originate* on the *sphenoid bone* at the base of the skull and *insert* into the *mandible*. Contracted together, they move the *mandible* forward.

Latissimus Dorsi. These broad muscles of the back *originate* in the *thoracic* and *lumbar vertebrae, sacrum,* and medial *iliac crest*, and insert into the *humerus*. They *adduct*, medially rotate, and extend the upper limb in humans.

Law. A politically formulated rule supported by sanctions.

Lemniscal System. A system of *nerve tracts* making a relatively direct link between the *peripheral* sensory nerve cells and the *cerebral cortex.*

Lesser Trochanter. A bony prominence near the upper end of the *femur,* to which are *inserted* the *iliopsoas* muscles that flex the thigh.

LH. See LUTEINIZING HORMONE.

Ligament (adj., Ligamentous). A strong band of collagen fibers connecting two bones at a joint, to guide movements and prevent dislocation.

Limb Girdle. A complex structure of bone that connects the limbs and trunk and transmits forces of compression and tension between them. (See PECTORAL GIRDLE and PELVIC GIRDLE.)

Limbic System. A group of interconnected structures in the brain which lie below the *cerebral cortex* and which jointly generate motivation and mediate emotional responses, such as rage and fear.

Lingual. Adjective referring to the tongue.

Load Arm. That part of a beam which lies between the pivot and the point at which the load is applied.

Load Line. Line of action of forces through a structure.

Lumbar Region. The third region of the spine; the lower back between *thorax* and *pelvic girdle.*

Lunate Sulcus. A furrow on the surface of the *cerebral hemispheres,* readily identifiable in *primates,* which forms a boundary to the *visual cortex;* sometimes termed *simian sulcus.*

Luteinizing Hormone (LH). A hormone *secreted* by the *anterior pituitary gland* whose prime function is to stimulate development of the *corpus luteum.*

Mammae. Milk-secreting *glands* of *mammals;* breasts.

Mammalia. A class of *vertebrate* animals the majority of which are characterized by *homoiothermy, mastication, viviparity,* and the secretion of milk for the nourishment of their young. They are divided into three *subclasses: Placentalia, Marsupialia,* and *Monotremata.*

Mandible (adj., Mandibular). The fused bones of the lower jaw that bear the lower *dentition.*

Mandibular Condyles. Bony eminences on each side of the *mandible* that fit into *fossae* in the base of the skull and about which the jaw pivots.

Mandibular Corpus. The horizontal part of the *mandible* that carries the teeth.

Mandibular Ramus. That part of the *mandible* which carries no teeth and to which the masticatory muscles are attached; in higher *primates* it is formed at right angles to the *mandibular corpus.*

Mandibular Torus. A thickening in the *symphyseal* region of the *mandibular corpus* on the inner side, which helps to strengthen it.

Manubrium. The upper segment of the *sternum.*

Manus. The hand; in other vertebrates it is the forepaw or front foot.

Marsupialia. A *subclass* of *mammals* (including the opossum and kangaroo) living in North and South America and Australia in which a *placenta* of a peculiar kind is developed, with the young born in a very undeveloped state. After birth they continue growth in a pouch containing the milk glands.

Masseter. Paired muscles of *mastication;* each *originates* upon the *zygomatic arch* and is *inserted* on the outer surface of the *ramus.*

Mastication. The breakdown of foodstuffs by dentition, involving cutting, chewing, grinding, tearing, etc.

Mastoid Process. A bony prominence at each side of the base of the skull behind the ear from which is inserted the *sternomastoid muscle.*

Material Culture. The totality of *artifacts* produced by a population of humans.

Maxilla. The bone of the face constituting the upper jaw.

Mechanoreceptor. Sensory *organ* of the nervous system that is stimulated by pressure or movement, e.g., organs of touch, hearing, and balance.

Medial. Nearer the midline or *median plane*.

Median Plane. An imaginary plane that divides a bilaterally symmetrical organism into two halves.

Medial Pterygoids. Also known as internal pterygoids, these paired muscles *originate* in the base of the skull (at the *maxillae* and *sphenoid* bones) and *insert* in the inner surface of the *mandibular ramus*; they help to close the jaw.

Medulla (adj., Medullar). The center of a structure or *organ*.

Meganthropus. A *genus* of the *Hominidae*, now believed to be indistinct from the *genus Homo*. The name was given to a *mandibular* fragment discovered in Java in 1941.

Melanin. A dark brown or black pigment.

Melanocytes. Special cells, the function of which is to produce *melanin* and which are normally present in the skin of *primates*.

Membrane (adj., Membranous). Fine sheet-like tissue lining parts of the body.

Menarche. The establishment, during growth, of the *menstrual cycle* in higher *primates*, as shown by menstruation.

Menstrual Cycle. A form of the *estrous cycle* found in humans and some other *primates*. It involves menstruation, which occurs more or less monthly, and is part of the estrous cycle.

Mentifacts. Assumptions, ideas, values, intentions (Huxley, 1958).

Mesozoic. A geological *era*, including the *Cretaceous, Jurassic*, and *Triassic epochs*.

Metacarpus (adj., Metacarpal). Palm bones of the *manus* (between the *carpal* bones and fingers), common to all *primates*.

Matatarsus (adj., Metatarsal). Sole bones of the *pes* (between the *tarsal* and toe bones), common to all *primates*.

Midbrain. A component of the *limbic system* lying below the *cerebral hemispheres*.

Middle Ear. Small cavity in the bones of the skull through which sound vibrations are transmitted by small bones, between the eardrum and the *fenestra ovalis*.

Miocene. A geological *period* of the *Tertiary era*, believed to have occurred from about 25 million to 12 million years ago.

Mobility. The ability to be moved by an outside force. (Compare MOTILITY.)

Moment of Bending. A moment (rotational force) tending to cause a beam to bend; a product of the magnitude of the force upon the beam and the perpendicular distance between the pivot and the line of action of the force.

Monogamy. Normal usage refers to marriage to a single spouse only; in zoology the term refers to a permanent or temporary pair bond.

Monotremata. A *subclass* of *mammals* confined to Australia and New Guinea, clearly distinguished from other *mammals*, since they lay eggs and possess other reptilian features.

Morphological Status. An estimate, based on *morphology*, of the *phylogenetic* position of an *organism* or *species* in its relation to other *organisms*.

Morphology. The science of the form and structure of animals.

Mortality Rate. The death rate, usually calculated as the number of deaths per annum per 1000 of the population.

Motility. The ability to move actively that is peculiar to animals. (Compare MOBILITY.)

Motor Cortex. A part of the *cerebral cortex* that has been discovered to function as a transmitter and receiver of nerve impulses associated with muscular contraction.

Motor Gyrus. A prominent rounded elevation on the surface of the *cerebral hemispheres* that includes the *motor cortex*.

Motor Neuron. A *neuron* whose *axon* connects to a muscle fiber and that transmits impulses from the *central nervous system* to effect muscular contraction

Mucoperiosteum. The fine *membranous* lining of bone characteristic of the roof of the mouth and the interior of the nose, which, in the latter, carries the *olfactory* receptors.

Mutagenic. Having the property of inducing *mutation*.

Mutation. Refers in *biology* to a sudden and relatively permanent change in a particular *gene* or *chromosomal* structure.

Muzzle. The lower part of the face containing the nasal cavity, the *turbinal bones*, and *olfactory* organ; the protruding snout, typical of dogs.

Mylohyoid. Paired muscles that form the floor of the mouth, and either raise the *hyoid bone* or lower the jaw. They arise on the inner margin of the *mandible* and are *inserted* on the *hyoid bone*, and are fused medially by a raphe (fibrous suture).

Myotomes. Segmentally arranged blocks of muscle lying to the side of the spine that, on contraction, cause *lateral* curvature of the spine; characteristic of lower *vertebrates*, fish, etc. In the primate embryo they give rise to the voluntary muscles of the adult.

MYA. Millions of years ago. Equivalent to BP.

Natural Selection. The principal mechanism of *evolutionary* change described by Darwin in 1859; the mechanism whereby those individuals best *adapted* to the environment contribute more offspring to succeeding generations than do the remainder, so that as their characteristics are inherited, the composition of the population is changed. See also GENE SELECTION; GROUP SELECTION; KIN SELECTION.

Neandertal. A valley in western Germany famous for the discovery in a limestone cave in 1856 of the first fossil man, clearly identifiably distinct from modern man.

Negative Feedback. A concept that refers to the mechanism of *homeostasis*, whereby changes in one direction effect adjustment in an opposite direction.

Neoteny. An evolutionary change in which formerly juvenile characteristics are retained by adult descendents as a result of the retardation of somatic development.

Nerve Pathways. See NERVE TRACTS.

Nerve Tracts. Bundles of nerve fibers connecting different points of the nervous system.

Neurocranium. That part of the skull which encloses the brain; the brain box, excluding the jaws and facial bones.

Neurons. Basic cells of the nervous system composed of a cell body and two or more long fibers. Impulses are carried along one or many of the fibers (or dendrites) to the cell body; only one fiber, the *axon*, carries impulses away from the cell.

Niche. That part of the *environment* occupied by a *species* or *subspecies* with particular reference to food and other natural resources upon which the species depends for survival.

Nomen (pl., Nomina). Refers here to the Latin names given to organic *species* according to the International Code of Zoological Nomenclature.

Nubility. The stage in the growth of a girl at which eggs are regularly produced; the beginning of the reproductive span of life.

Nuchal. Adjective referring to the neck; the nuchal area on the *occipital* bone is where the nuchal muscles are inserted.

Nuchal Crest. A transverse bony ridge that develops on the skull of some *primates* at the upper boundary of the *nuchal* area and serves to increase the area of attachment of the *nuchal* muscles upon the *occipital* bone.

Occipital. Adjective referring to the back of the head and, in particular, to the bone that forms the back and posterior base of the skull.

Occipital Condyles. A pair of rounded *articular* surfaces on the *occipital* bone at the base of the skull that form the joint between the skull and the first *cervical vertebra* or atlas.

Occipital Lobe. Part of the *cerebral hemispheres* which lies at the back of the head and is related to vision.

Occlusal Plane. The plane of *occlusion* of the teeth; the plane on which the teeth meet when the jaw is closed.

Occlusion. The way in which the dentitions of the upper and lower jaws articulate when the jaws are closed.

Olecranon Process. Proximal end of the ulna which embraces the trochlea of the humerus.

Olfactory Apparatus. The totality of structures that contribute to the sense of smell.

Olfactory Bulbs. The two small bulblike extremities of the olfactory region of the brain that receive the olfactory nerves from the nose.

Oligocene. A geological *period* of the *Tertiary era*, believed to have lasted from about 36 to 25 million years ago.

Ontogeny. The course of development and growth during the life of an individual. (Compare PHYLOGENY.)

Optic Chiasma. The structure formed beneath the forebrain by nerve fibers from the right eye crossing to the left side of the brain and vice versa.

Orbit. A cavity in the skull surrounded by a ring of bone that contains and protects the eyeball.

Order. A taxonomic rank. See TAXONOMY.

Oreopithecus. A *genus* of the *Hominoidea* from the *Pliocene period*, not clearly related to either *Hominidae* or *Pongidae*.

Organ. Any part of an animal that forms a structural or functional unit.

Organ of Corti. The auditory *receptor*.

Organism. An individual living thing.

Orgasm. The culmination of *copulation*, characterized by the pleasurable release of nervous tension, by muscular contraction, and ejaculation of semen by the male.

Oriented. Describes the direction in which an object lies (past participle of verb "to orient").

Origin. Refers here to the less movable of the two points of attachment of a muscle. (Compare INSERTION.)

Orthogenesis. *Evolution* of lineages supposedly following a predetermined pathway not subject to natural selection.

Os Calcis. See CALCANEUS.

Os Coxae. See HIP BONE.

Os Penis. See BACULUM.

Osteology. The science of the skeleton and its structure; the study of bones.

Outer Ear. The pinna; that part of the ear which is visible externally and consists of a flap of skin upon *cartilage*.

Outer Table. The layer of compact bone that forms the outer surface of the *neurocranium*.

Ovarian Follicle. See FOLLICLE, OVARIAN.

Ovary. The *gonad* of female animals that produces eggs and, in *mammals*, hormones.

Oviduct. The duct, generally found in vertebrates, that carries the eggs from the *ovary* to the exterior; in *mammals* it consists of three parts: the *fallopian tubes*, the *uterus*, and the *vagina*.

Ovulation. The discharge of a ripe egg from the *ovarian follicle* into the opening of the *fallopian tubes*.

Paedomorphosis. The retention of ancestral juvenile characteristics into later ontogenetic stages of descendent populations. (See NEOTENY, PROGENESIS.)

Palmar. Relating to the palms of the hands, the *volar* area.

Pan. The only *genus* of the family, found in equatorial Africa, comprising the chimpanzee and gorilla.

Papilla. A small projection on the skin.

Papio. A *genus* of the Old World monkeys, the *Cercopithecoidea* which includes the baboon, a common *quadrupedal*, ground-living form found throughout much of Africa.

Parabolic Girder. A girder or beam of parabolic form used in the construction of *cantilevers* of certain types.

Parietal Bone. The bone that forms the side of the *neurocranium*; the parietal bones meet at the midline on top of the skull and are elsewhere fused, in adults, with the *occipital, temporal, sphenoid*, and *frontal bones*.

Parietal Lobe. Those parts of the *cerebral hemispheres* which lie approximately under the *parietal bones* of the skull, between the *frontal* and *occipital lobes*.

Parturition. Giving birth, childbirth.

Patella. The kneecap; a small sesamoid bone lying over the knee joint that transmits the tension developed by the *quadriceps femoris* muscle to the *tibia*.

Pectoral Girdle. The bones that suspend the body between the forelimbs in quadrupeds and that suspend the arms from the body in humans; in *primates*, it consists of two bones on each side, a *clavicle* and a *scapula*.

Pectoralis Major. A muscle that *originates* on the *clavicle, sternum*, and ribs and *inserts* into the crest of the greater tubercle of the *humerus*. It *adducts* and medially rotates the arm, and can also return the extended humerus to the verticle position.

Pelvic Canal. The cavity formed by the bony ring of the *pelvis* through which the young must pass at birth.

Pelvic Girdle. The right and left hip bones that transmit the weight of the body to the hindlimbs. Unlike the *pectoral girdle*, it is fused to the vertebral column, where the first three *sacral vertebrae* themselves are fused together. (See PELVIS.)

Pelvis (pl., Pelves; adj., Pelvic). A bony structure consisting of the two hip bones fused together *anteriorly* and to the *sacrum posteriorly*, forming a basinlike ring of bone in humans. (See also PELVIC GIRDLE.)

Penis. The male sex organ evolved as an organ of internal fertilization containing erectile tissues and, in most primates, a small bone, the *baculum*.

Pentadactyly. The possession of five *digits* on the *manus* and *pes*.

Percept. The mental product of perception; a mental construct quite distinct from the thing perceived.

Period. Refers here to subdivisions in the earth's *chronology*.

Peripheral. The opposite of central; applied to the surface of an *organ* or the body of an animal.

Perissodactyls. The odd-toed *ungulates*, and *order* of *mammals* including the rhinoceros, tapir, and horse.

Pes. The foot; in animals the paw or hindlimb.

Phalanges. See PHALANX.

Phalanx (pl. Phalanges). The small bones of the *digits distal* to the *metacarpals* and *metatarsals*; the finger and toe bones.

Pharynx. The throat; the area connecting the nasal and mouth cavities with the voice box or *larynx*.

Phenetics. Taxonomic analysis in which relationships are established by giving all characters equal weight. (Compare CLADISTICS.)

Phenotype. The sum of the characteristics manifest in an *organism*, to be contrasted with the *genotype*; the phenotype is formed by the interaction of the fertilized egg and its *environment* in the process called growth.

Pheromone. A chemical substance produced by an animal, either by a scent *gland*, or as a waste product, which acts as a signal in communication.

Phonation. The act of making speech sounds.

Photoreceptor. The sensory organ of the nervous system that is stimulated by electromagnetic waves of certain frequencies (380–760 μm), the nerve impulses from which are interpreted in the brain as light.

Phylogeny (adj., Phylogenetic). The *evolutionary* lineage of *organisms*; their *evolutionary* history.

Physiology. The science of organic function, of the processes of *organisms* that constitute their life.

Piriform Lobe. A lobe of the *archipallium* associated with the analysis of *olfactory input*.

Pithecanthropus. A generic name previously given to certain Javan hominid fossils now usually classified as *Homo erectus*.

Pitocin. A *hormone* produced by the *posterior pituitary gland*, responsible for muscular contractions of the *uterus*.

Pituitary Body. A compound *gland* lying beneath the base of the brain close to the *hypothalamus*; it has been described as the "master" gland of the *endocrine system*.

Placenta. An *organ* peculiar to *mammals* consisting of *embryonic* tissues evolved to absorb nourishment from the wall of the *uterus* and there to discharge waste products; it is connected to the *fetus* by the umbilical cord and produces *hormones* that keep the *uterus*, and indeed the mother, adapted to the state of pregnancy.

Placentalia. A *subclass* of *mammals* with worldwide distribution, in which a *placenta* is formed from the *fetal allantois* for the nourishment of the *fetus*. This most widespread *subclass* includes *primates*, *ungulates*, and many other *orders*.

Plantar. Refers to the soles of the hindlimbs; compare PALMAR.

Plantar Ligaments. The *ligaments* of the sole of the foot that maintain its arched form.

Plantigrade. A type of locomotion in which the whole *ventral* surface of the foot comes into contact with the ground. (See DIGITIGRADE.)

Planum Temporale. An area on the upper surface of each *temporal lobe* which lies in the *Sylvian sulcus* and shows differential development on the two sides in many individuals.

Plasticity. Refers here to the *adaptability* of the *phenotype* during *onogeny;* its variable response to differences in *environment* in spite of constancy of the *genotype*—a more obvious characteristic of plants than of animals.

Platyrrhini. An alternative term for the *Ceboidea,* or New World monkeys.

Pleistocene. A geological *period* of the *Quaternary era,* believed to have lasted from about 1.6 million to 10,000 years ago.

Plesiomorphic. Describes characters of the members of one taxonomic group shared with other taxonomic groups as a result of their common ancestry. (Compare APOMORPHIC.)

Pliocene. A geological *period* of the *Tertiary era,* believed to have lasted from about 5 to 1.6 million years ago.

Pollex. The first *digit* of the *manus,* the thumb.

Polyandry. That form of polygamy in which one woman is formally permitted to marry more than one man.

Polygyny. That form of polygamy in which one man has several wives. (See POLYANDRY.)

Pongidae (adj., Pongid). The *family* of *primates* including the orangutan (*Pongo*), a *genus* now confined to Sumatra and Borneo.

Population. A local or breeding group; a group of individuals so situated that any two of them have an equal probability of mating with each other, which generally find their mates within the group, but which are also able to mate with members of neighboring populations.

Positional Behavior. The postures and locomotion of animals in feeding, sleeping, running, climbing, etc.

Posterior. Dorsal, nearer the back of an organism.

Power Arm. That part of a beam which lies between the pivot and the point at which (muscular) forces are exerted.

Precision Grip. A grip, characteristic of the human hand, in which the tip of the thumb can be opposed to the tips of the other fingers to give a precise, yet firm, grip.

Precocial. A reproductive strategy characterized by small litters, slow development, extended gestation, and the birth of well-developed, capable young.

Prefrontal Area. The most *anterior* part of the *cerebral cortex,* so named because it lies forward of the frontal area on the *frontal lobes.*

Prescription. Instruction.

Primates. An *order* of the *class Mammalia,* characterized by *arboreal adaptations* and including humans.

Proconsul. A *genus* of fossil ape from the Miocene of Kenya.

Progenesis. *Paedomorphosis* produced by precocious sexual maturation of an organism still in a morphologically juvenile stage of growth.

Progesterone. A *hormone* that is *secreted* mainly by the *corpus luteum* and that prepares the *endometrium* for *implantation* and brings about many of the changes associated with pregnancy.

Prognathism. With jaws projecting beyond the rest of the face.

Prolactin. A *hormone secreted* by the *anterior pituitary,* responsible for the onset of milk production after *parturition.*

Pronation. Position or rotation of the forearm so that the *manus* is palm down. (Compare SUPINATION.)

Propliopithecus. A *genus* of fossil ape from the *Oligocene epoch* of Egypt.

Proprioceptors. A sensory *organ* of the nervous system that is found in muscles as well as other parts of the body and detects stretch or contraction.

Prosimii (adj., Prosimian). A *suborder* of the *order Primates* containing various Old World *genera*, including the lemurs, lorises, and tarsiers.

Protein. A very complex organic compound containing chains of *amino acid* molecules; proteins occur in infinite variety and are the basis of most living substances.

Protoculture. Behavior spread between members of a social group by observational learning, but without any symbolic content. (Compare CULTURE.)

Prototechnology. Categories of primitive technology found among animals not involving tool-making, but including tool use and modification. (See Table 9.2.)

Proximal. Part of the body nearest to the trunk or to the midline. (Compare DISTAL.)

Psychology. The science of conscious life, of mental and emotional processes.

Pterygoid Muscles. Muscles controlling movement of the lower jaw. (See LATERAL PTERYGOIDS and MEDIAL PTEROGOIDS.)

Ptyalin. A digestive enzyme present in the saliva of some *mammals*, including humans, which brings about the breakdown of starch into sugar.

Pubic Symphysis. The *anterior* area of articulation of the two hip bones—the two pubic bones.

Pubis. A bone that, with the *ilium* and *ischium*, forms the hip bone; it is the most *ventral* of the three.

Pulvinar. The most *posterior* part of the *thalamus*; it is significantly larger in higher *primates* than in lower forms.

Punctuated Evolution. A phylogenetic pattern in which species evolve by rapid spurts separated by periods of stability or equilibrium.

Pyramidal System. Descending *nerve tracts* that form a relatively direct link between the cerebrum and the muscles. (Compare EXTRAPYRAMIDAL SYSTEM.)

Quadrupedalism. Locomotion upon four feet.

Quantum Evolution. A phylogenetic pattern in which species evolve by quantum jumps from one adaptive zone to another, first discussed by Simpson (1943). (Compare PUNCTUATED EVOLUTION.)

Quaternary. The most recent geological era, believed to date from about 1.6 million years ago and containing the *Pleistocene* period, as well as recent time.

r-**Selection.** Selection which favors rapid rates of population increase, commonly operating on species that specialize in colonizing short-lived or unstable environments and who undergo large fluctuations in population size.

Race. A group of *populations* of a *species* that are distinct in at least a few *characteristics* from other races of the same *species*. Very similar in meaning to the terms variety and subspecies.

Radiation. Refers here to *evolutionary* radiation of several *species* from a single *species*, all bearing a proportion of characteristics in common.

Radius (pl. Radii). The *lateral* of the two bones of the forearm.

Ramapithecus. A genus of fossil ape found in Asia and commonly believed to be ancestral to the orangutan. Currently, this is considered part of the genus *Sivapithecus*.

Ramus (adj., Ramal). Literally a branch: here commonly applied to the vertical part of the *mandible* upon which are *inserted* the muscles of mastication.

Receptors. A term applied to *organs* of the nervous system with a sensory function.

Reciprocal Altruism. The trading of altruistic acts by individuals at different times; not true altruism as there is an expectation of repayment in kind—"one good turn deserves another."

Reciprocity. The exchange of goods or services over a period of time based on tradition and expectation, but without contract.

Rectus Abdominis. The segmented muscle of the anterior abdominal wall *originating* on the *pubic* bones and *inserted* in humans upon the lower *sternum* and rib cartilages. It maintains tension between *pelvis* and *thorax*, and flexes the spine and pelvis.

Rectus Femoris. One of four *extensors* of the knee joint. This muscle *originates* from the *anterior inferior iliac spine* and ilium above the *acetabulum*, and *inserts* into the *patella*.

Referent. An object referred to; an external object that is perceived.

Reflexes. An involuntary reaction on the part of an *organism* to a particular stimulus; reflexes may be *innate* or conditioned by learning.

Regulatory Genes. Genes believed to determine the growth rates of different organs of the body by controlling the activity of the structural genes.

Relative Dating. Dating of the age of rocks in relation to others which may overlie or underlie them. (Compare CHRONOMETRIC DATING.)

Reptiles. A class of *vertebrates*, usually *terrestrial*, dominant in the *Mesozoic era*. They evolved from amphibians, and in turn *mammals* evolved from them. Examples include alligators, lizards, and snakes.

Respiration. Refers here to the inhalation of air for the absorption of oxygen into the bloodstream, and its exhalation, together with carbon dioxide, by the lungs.

Reticular System. An indirect and *phylogenetically* old system of ascending *nerve tracts* forming a relay of *neurons* between the sensory *neurons* and the *thalamus* and *cortex*.

Retina (adj., Retinal). Part of the eye containing nerve *receptors* (*rods* and *cones*) sensitive to light. These receptors form a dense layer on the inner surface of the eyeball.

Rhinal Sulcus. A *sulcus* on each *cerebral hemisphere* that separates the *archipallium* and *neopallium*.

Rhinarium. The moistened, hairless, tactile-sensitive skin that surrounds the nostrils of many mammals, seen typically in the dog.

Rhinencephalon. The *olfactory* brain; that part of the brain concerned with the sense of smell (olfaction).

Rhodesian Man. A fossil skull and other bones discovered in a mine at Kabwe (Broken Hill), Zambia in 1921.

Rites de Passage. Rituals connected with important stages in the development of individuals as members of society.

Rodentia. An *order* of *mammals* characterized by teeth evolved for gnawing; it is the largest *order*, with 350 *genera*, including rats, mice, squirrels, and porcupines.

Rods. A sensory cell of the nervous system stimulated by light, especially of very small intensities. (See CONES and RETINA.)

Sacculus. A sensory *organ* that, with the *utricle*, is sensitive to the direction of gravity and changes in that direction; part of the *inner ear*.

Sacrum (adj., Sacral). A single curved bone that is part of the *pelvis*, formed in humans by the fusion of five sacral *vertebrae*.

Saddle Joint. A joint with a saddle-shaped *articular* surface allowing movement in two planes as well as rotation.

Sagittal Crest. A crest that develops along the *sagittal* line on top of the skull in certain *primates* and serves to increase the area of *origin* of the *temporal muscles*.

Saltation. Jumping; a mode of progression, usually with the backbone erect, found among certain *prosimians*.

Sanction. A mechanism whereby society as a whole enforces *behavior* patterns upon individuals, for example, by punishment.

Scapula (pl., Scapulae). A bone of the *pectoral girdle*, the shoulder blade.

Sebaceous Glands. Oil-producing *glands* of the skin.

Secondary Sexual Characteristic. A *characteristic* peculiar to males or females but without a function directly related to reproduction.

Secretion (vb., Secrete). An activity involving the passage of a substance produced within specialized gland cells to the surrounding tissues.

Sectorial. Cutting; referring to teeth that have evolved a cutting edge.

Selection (adj., Selective). As used here, this term refers to *natural selection*. (See also GENE SELECTION, GROUP SELECTION, KIN SELECTION.)

Selection Pressure. The effect of any feature in the environment that results in *natural selection*; e.g., food shortage, predation, competition.

Sematic. Acting as a signal to other animals.

Semicircular Canals. Sensory *organs* of the *inner ear* which detect the direction and acceleration of movement.

Sensory Cortex. See SOMATIC SENSORY CORTEX, VISUAL CORTEX, AUDITORY CORTEX.

Serratus Anterior. In humans, this muscle *originates* from the upper ribs and *inserts* on the vertebral border of the *scapula*; it rotates the *scapula* and pulls it forward laterally, tipping the *glenoid cavity* upward.

Sexual Dimorphism. The characteristic differences between the sexes of a single *species*.

Simian Shelf. A small bony shelf on the inner surface of the bottom of the *symphyseal* region of the *mandible*, serving to strengthen it.

Simian Sulcus. See LUNATE SULCUS.

Sinanthropus. A generic name previously given to certain Chinese hominid fossils now usually classified as *Homo erectus*.

Sivapithecus. Genus of fossil ape, possibly ancestral to the orangutan; found in Asia, Europe, and perhaps Africa.

Sinus. A hollow chamber in the bones of the skull.

Sociobiology. The scientific study of the biological basis of all social behavior.

Soft Palate. The back of the palate not directly supported by bone, which separates the back of the mouth from the nasal cavity.

Soleus. A muscle of the calf of the leg *originating* in humans on the *proximal* parts of the *tibia* and *fibula* and *inserting* by the *achilles tendon* into the *calcaneus*.

Somatic Sensory Cortex. That part of the *cortex* in which are located *neurons* which receive and transmit the *somesthetic* input to the brain.

Somesthetic. Refers to sense *receptors* of the skin (e.g., receptors of touch, temperature, and pain).

Sorex. A *genus* of small animals of the *order Insectivora* found in both the Old and New World, e.g., a shrew.

Specialization. In the context of *evolutionary* studies, a *characteristic* evolved for a particular and limited function—the opposite of a generalized *characteristic*.

Speciation (vb., Speciate). The division of a *biospecies* over a period of time as a result of geographical isolation. Speciation results in the establishment of two or more *biospecies*.

Species. A group of *populations* of organisms between which gene flow can occur, and which is reproductively isolated from other such groups. (See BIOSPECIES, CHRONOSPECIES.)

Sperm: (pl., Spermatozoa). The male sex cell or *gamete*.

Spermatogenesis. The generation of *spermatozoa* that occurs in the *testes*.

Sphenoid Bone. A butterfly-shaped bone situated at the anterior part of the base of the skull, of which it forms the floor. It articulates with all the other cranial bones.

Spinal Cord. The extension of the *central nervous system* within the vertebral column, and consisting of neurons and nerve tracts in the form of a hollow tube.

Spinous Process. Connotes the bony projection on the *dorsal* side of most *vertebrae*.

SQC. Slow quadrupedal climbing; mode of locomotion found among some prosimians, e.g., slow loris.

Status. Social position, or position in relation to related objects.

Sternoclavicular Joint. *Articulation* of *sternum* and *clavicle*.

Sternomastoid. Paired muscles which effect rotation of the head and have their origin in the *sternum* and *clavicle* and are inserted in the *mastoid process* and *occipital bone*.

Sternum. The breastbone; the bone that *articulates* with the ribs on the *ventral* side of the *thorax*.

Strategy. The adaptive options evolved by an evolutionary lineage, e.g., life-history strategy. As used by biologists, the term does not imply conscious choice.

Structural Gene. A gene that controls the synthesis of a protein that builds a part of the body. (Compare REGULATORY GENE.)

Subfamily. A taxonomic rank, for example, the Homininae. (See TAXONOMY.)

Suborder. A taxonomic rank, for example the Anthropoidea. (See TAXONOMY.)

Sulcus. A fissure; one of the grooves or furrows on the surface of the brain, for example, on the *cerebral hemispheres*.

Superfamily. A taxonomic rank, for example the Lemuroidea. (See TAXONOMY.)

Superior. Higher, in relation to another structure. (Compare INFERIOS.)

Supination. Position of the forearm such that the *manus* is palm down. (Compare PRONATION.)

Supraorbital Torus. A rounded transverse thickening of the *frontal* bone across the upper edge of the *orbits*, evolved to carry some of the forces developed by a powerful masticatory apparatus.

Supraspinous Ligaments. *Ligaments* connecting the tips of the *spinous processes* of the *vertebrae*.

Sweat Glands. *Glands*, situated in the skin of *mammals*, that secrete perspiration. (See APOCRINE and ECCRINE.)

Sylvian Sulcus. A deep infolding of the *cortex* of the *cerebral hemispheres* which separates the *temporal lobes* from the *frontal* and *parietal lobes*. (See Figure 8.14.)

Symbol. An object, activity, or concept representing and standing as a substitute for something else.

Symphysis (adj., Symphyseal). A union of two bones (e.g., *pubic symphysis*).

Synapse (pl., Synapses). The point of contact of an *axon* and another *neuron* between which nerve impulses can pass; the connecting point between two nerve cells.

Synthetic Theory. Current theory of evolution by *natural selection*, still generally accepted by biologists, based on Darwin's work, but modified by new knowledge of classical and population genetics, cytology, paleontology, systematics, etc.

Talus. The ankle bone, which *articulates* with the *tibia* and *fibula* to form the ankle joint.

Tapetum Lucidum. Light-reflecting membrane that lies behind the photoreceptors in the retina of the eye and increases sensitivity to light. Found in many nocturnal primates.

Tarsus (adj., Tarsal). The short bones of the *pes*, consisting of seven bones, including the *talus* and *calcaneus*.

Taxonomy. The science of the classification of plants or animals which involves placing them in groups according to their relationships and ranking these into hierarchies. The conventional ranks used in this book are shown in Table 1.1.

Temporal Bone. A bone found on each side of the skull between the *occipital*, *parietal*, *sphenoid*, and *frontal* bones.

Temporal Line. A ridge on the side of the skull, which delineates the edge of the *origin* of the *temporalis* muscle.

Temporal Lobe. A lobe of the *cerebral hemispheres* lying low down on each side of the brain.

Temporal Muscle. See TEMPORALIS.

Temporalis. The largest of the muscles that close the jaws, *originating* on the sides and roof of the skull and *inserting* on the *coronoid process* of the *mandible*.

Terrestrial. Living or moving on the ground.

Tertiary. A geological *era* believed to have lasted from about 63 million to 1.6 million years ago characterized by the *radiation* of the *Mammalia*.

Testis (pl., Testes). The *gonad* of male animals that produces *spermatozoa*, and, in *mammals*, the male sex *hormones*.

Thalamus. A large egg-shaped mass of tissue within the brain that serves as a relay for sensory stimuli to the *cortex*.

Thorax (adj., Thoracic). The chest; the upper part of the trunk between neck and abdomen containing a basketlike structure of ribs and *sternum*. (See Figure 5.11.)

Tibia (pl., Tibiae). A bone of the calf that carries all the weight transmitted down the hindlimbs in higher *primates*.

Tool. An *artifact* that may be considered as an extension of the manipulative *organs*.

Trachea. The windpipe or breathing tube which connects the bottom of the *larynx* with the two bronchi that lead to the lungs.

Trapezius. A muscle of the neck, back, and shoulder which *originates* in the *nuchal* area of the skull, the *cervical vertebrae*, and upper *thoracic vertebrae*, and *inserts* on the *acromion* and lateral part of the spine of the *scapula*.

Triceps Brachii. A muscle that extends the forearm at the elbow joint; it *originates* from the *scapula* and shaft of the *humerus* and *inserts* on the olecranon process of the *ulna* at the elbow.

Trochanter. Bony protuberance on the *proximal* end of the *femur*. (See GREATER TROCHANTER; LESSER TROCHANTER.)

Trochlea. The grooved and rounded surface of the lower end of the *humerus* that *articulates* as a hinge joint with the *ulna.*

Turbinal Bones. The fine bones inside the nose that carry the olfactory *mucoperiosteum.*

Ulna (pl., Ulnae). The *medial* and larger of the two bones of the forearm.

Ungulates. A group of *mammals* including two *orders* of plains-living *species*, the *Perissodactyla* and *Artiodactyla.*

Urethra. Tube leading from the urinary bladder of mammals to the exterior; in female humans it opens within the vulva and in males at the end of the *penis.*

Uterus. The womb; a hollow muscular organ evolved from the oviduct in which the fertilized egg develops into a *fetus.*

Utricle. A *membranous* fluid-filled sac lying in the *inner ear*, which functions with the *sacculus* to detect the direction of gravity and movement.

VCL. Vertical clinging and leaping; a mode of locomotion found among certain prosimians.

Vagina. The lowest part of the ancient *oviduct* in *mammals* which joins the *uterus* to the exterior. Receives the *penis* in *copulation.*

Variation. Naturally occurring differences between individuals of a single *species* that are due to differences in *genotype* and *environment.*

Vascular. Containing blood vessels.

Ventral. Anterior or front of an organism; opposite of *dorsal.*

Versatility. Applied here to the flexibility of *behavior* that results from evolved learning ability, skills, and perception.

Vertebra. Complex structure of bone, a number of which form the backbone or vertebral column in *vertebrates*; beside constituting the core structural components of the body, the vertebrae also protect the *spinal cord.*

Vertebrates. A major *subphylum* of the animal kingdom; containing all animals with back bones, including fish, amphibians, reptiles, birds, and *mammals.*

Vesicular. Sacklike; a small cellular structure like a vessel or sack, often containing fluid.

Vestigial Characteristic. A rudimentary structure in an animal corresponding to a fully formed structure in an earlier or related form, sometimes assumed to have lost its function.

Vibrissa (pl., Vibrissae). A long hair, such as a whisker, with sensory nerves at its base, evolved as a highly sensitive detector of mechanical stimulus.

Visual Cortex. That part of the *cerebral cortex* involved in the reception and analysis of nerve impulses from the eyes.

Viviparity (adj., Viviparous). The ability to give birth to living young, as contrasted with laying eggs.

Volar Pads. The hairless pads with special skin found on the friction surfaces of the *manus* and *pes* in *primates* as well as on the tails of some New World monkeys.

Wernicke's Area. Part of the *cerebral cortex*, on the upper surface of the *temporal lobe* and adjoining the angular gyrus. It plays an important part in the symbolization involved in human speech and is found only in humans.

Word Memory. That part of the memory which records words and the motor patterns for speech and writing.

Zinjanthropus. A *genus* of the family *Hominidae*, currently believed to be indistinct from the *genus Australopithecus.* The name was given to a skull discovered in Tanzania in 1959.

Zygomatic Arch. A bar of bone from the cheek to the ear region, called the cheekbone, accommodating the *origin* of the *masseter* muscle.

Zygomatic Bone. The cheekbone; the bones of the face which form the zygomatic arches, and articulate with the *frontal, sphenoid,* and *temporal bones* of the *cranium* and the maxillary bones of the face.

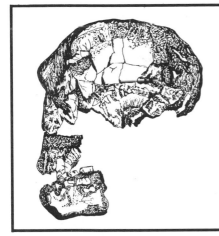

Index